Transcor

Brazilian company specialized at colors

PIGMENTS

DYES

DISPERSIONS

Transcor Ind. de Pigmentos e Corantes Ltda
www.transcor.com.br

DONGRAY® WAX

丰顺东丽精细化工有限公司
Fengshun Dongray Fine Chemical Industry Co.,Ltd.

丰顺东丽精细化工有限公司是丰顺正通高分子材料有限公司的旗下公司，成立于2006年，是科莱丽化工(泰国)有限公司在中国投资的一家外商独资企业。是一家专业集研发、生产、销售聚乙烯蜡产品的技术实力雄厚的现代化企业。

FengShun DongRay Fine Chemical Co., Ltd. is a subsidiary of FengShun ZhengTong Polymer Material Co., Ltd. Founded in 2006, registered capital of 2 million U.S. dollars, is a wholly foreign-owned company invested by Ke Lai Li Chemical (Thailand) Co., Ltd. in China. In August 2008, FengShun DongRay Fine Chemical Co., Ltd. built the office buildings, laboratories, factories, warehouses and etc., which covers an area of 16 thousand square meters in GuangDong Province MeiZhou FengShun Economic Development Zone (Industrial Park Phase II). It is a modern enterprise with strong technical strength, the company researches, invents produces and sells polyethylene wax products.

中国广东省梅州市丰顺县经济开发区工业园二期
Fengshun Economic Development Area Project 2, Meizhou City, Guangdong, 514300 China
电话TEL：0753-6683688　　传真FAX：0753-6659989　　Web: www.dongraywax.com

山东鲁燕色母粒有限公司
Shandong Luyan Color Masterbatch Co., Ltd

　　山东鲁燕色母粒有限公司成立于1994年6月，主要从事色母粒的研究、开发、生产、销售，年生产能力15000吨，并且通过了ISO9001-2000国际质量体系认证和ISO14001环境管理体系认证。本公司主要生产用于各种改性塑料的无载体黑色母粒及各种用途的黑色母粒、白色母粒、彩色母粒、功能母粒。

　　主要应用于PE、PP、ABS、AS、HIPS、PS、POM、PA、PC、PET、PBT、PC/ABS、PC/PET等，同时可以根据客户要求研发新产品。

　　欢迎使用色母粒的用户、代理商前来洽谈或投资合作。

联 系 人：刘玉莲 13605389538
邮　　箱：13605389538@163.com
公司网站：www.sdlycm.com ，www.sdlycm.net
地　　址：山东省泰安市青春创业开发区振兴街。

Shandong Luyan Color Masterbatch Co., Ltd was founded in June 1994, mainly engaged in the research, development, production and marketing of color masterbatch. The annual production capacity of our enterprise has reached 15,000 tons, meanwhile, our enterprise has passed the attestation of the international quality system certification ISO9001-2000and the environmental management system certification ISO14001. We mainly produce carrier-free black masterbatch for a variety of modified plastics and black masterbatch, white masterbatch, color masterbatch, functional masterbatch for various uses.
Our products are mainly used in PE, PP, AS, ABS, PS, HIPS, POM, PA, PC, PET, PBT, PC/ABS, PC/PET, etc. In addition, we have the ability to develop new products based on customer requirements. Sincerely welcome the users, agents of color masterbatch to negotiate business, investment and cooperation.

Contacts: Adela Liu　　Mob:13605389538
Email: 13605389538@163.com　　Company Website: www.sdlycm.com, www.sdlycm.net
Address: Zhenxing road, Development Zone of the Youth entrepreneurship, Taian City, Shandong Province, China.

南京培蒙特科技有限公司
Nanjing Pigment Tech. Co., Ltd.

南京培蒙特系列复合无机颜料（CICP），是由多种无机盐经高温复合而成，是已知的热稳定性、耐光性、耐化学腐蚀性、耐候性最好的颜料。这些颜料被广泛用于户外建构筑物上，如PVC滑道、PVC塑钢窗、PVC甲板、PVC栅栏、金属屋顶和墙面用卷钢；还可用于工程塑料、伪装涂料等各种高性能户外涂料等等。

Nanjing PMT Tech. Co., Ltd. is a high-tech company offers a wide range of CICP, including the yellow, green, blue brown and black families. They have outstanding fastness to light and weathering, and are alkali, acid, and heat resistant. Our CICP are noted for their stability, ease of dispersion, non-bleeding,

Pigment Name	C.I. No.	Appearance	Heat Resistance (℃)	Light fastness	Weather Resistance	Oil Absorption (cc/g)	pH	Mean Particle Size(μm)
PY-53	77788	Yellow	1000	8	5	14	7.1	≤1.1
PG-17	77288	Green	1000	8	5	13	7.1	≤1.2
PG-26	77344	Green	1000	8	5	16	7.5	≤1.2
PG-50	77377	Green	1000	8	5	13	7.5	≤1.1
PBR-33	77503	Brown	1000	8	5	17	7.1	≤1.0
PY-164	77899	Brown	1000	8	5	19	7.2	≤1.0
PBR-24	77310	Buff	1000	8	5	17	7.4	≤1.1
PB-28	77346	Blue	1000	8	5	28	7.4	≤1.0
PB-36	77343	Blue	1000	8	5	22	703	≤1.0
PBK-26	77494	Black	1000	8	5	18	7.5	≤1.0
PBK-28	77428	Black	1000	8	5	17	7.0	≤1.0
PBK-30	77504	Black	1000	8	5	17	7.6	≤1.3

Non-migrating. These properties make our pigments an excellent choice for a variety of applications, from plastics, paints, and inks, to coil coating industries.

PRODUCTS INFORMATIONS:

PMT Complex Inorganic Pigments (CICP) are synthetic minerals that are defined as the most heat stable, lightfast, chemical and weather resistant pigments known to exist. These pigments are designed for use in a variety of exterior building applications, such as PVC Siding, PVC Window profiles, PVC decking and fencing, Coil Coatings for metal roofs and sidewalls. These pigments are also highly recommended for use in Engineering Plastics, Camouflage and a variety of other exterior coatings where high performance is required.

http://www.chinapigment.net   Email: fxd331@163.com
Tel : +86-25-84199537   Fax: +86-25-56615698   Cell : 13901598982   13770843838
Add: Dongping Industrial Park, Lishui, Nanjing 211211 China

FangChen testing machine

Zibo FangChen specializes producing masterbatch testing equipment, the products are: lab spinning machine, BCF spinning machine, BCF three color machine, lab FDY/POY spinning machine, lab POY spinning machine, lab bicomponent conjugate spinning machine, lab spinning and non-woven machine, lab non-woven machine, lab melt-blown machine, lab SMS machine, lab filter performance tester, single-screw extruder, DTY test machine, film blowing test machine, dust collector, grinder and masterbatch etc., provide scientific research equipment for more than 40 universities, exported to The U.S., Indonesia, Vietnam, Mexico, The Republic of South Africa, The Czech Republic, Thailand, Russia, India etc.

Film blowing test machine

lab filter performance tester

lab spinning machine

lab bicomponent conjugate spinning machine

lab SMS machine

lab non-woven spinning machine

lab spinning and nonwoven machine

BCF spinning machine

DTY test machine

lab POY-FDY spinning machine

lab POY spinning machine

Melt blown nonwoven test machine

山东省淄博市临淄方辰母料厂
Zibo FangChen masterbatch Co., LTD
Address: Linzi District, Zibo City, Shandong China
Http://www.cnfangchen.com   Email:zbfangchen@sina.com
Tel:+86-533-7082276   Fax:+86-533-7084005

扫一扫 加微信

中国驰名商标

好事成双乐在共赢

專業的藍（酞菁藍）
綠（酞菁綠）顏料製造商

A professional pigment manufacturer of
blue (phthalocyanine blue)
green (phthalocyanine green)

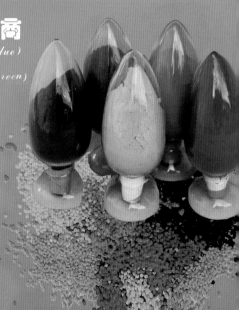

江苏双乐化工颜料有限公司
地址：江苏省兴化市张郭镇人民路2号
邮编：225722
电话：0523-83998555
传真：0523-83760948
HTTP: WWW.Shuangle.com

东莞市锡华检测仪器有限公司
DONGGUAN XIHUA TESTING MACHINE CO.,LTD.

东莞市锡华检测仪器有限公司是一家专业设计制造各类精密品管检测仪器的高科技合资企业,公司始建于1995年,致力于产品开发,市场推广,网络营销和售后服务,极大的满足了广大客户的需求。公司不仅拥有先进的生产设备及雄厚的技术力量,更先后引进了北京,上海等科研单位的高新技术成果,研制出较具前瞻性的检测仪器,符合ISO、GB、BS、ASTM、DIN、JIS、CEN、EN等测试标准。凭借一流的品质,实惠的价格及优良的售后服务,深得用户信赖。

公司产品广泛适用于皮革业、鞋业、电子电器业、电线电缆业、纺织业、橡塑料业、汽车业、自动、电动车、摩托车业、婴儿车业、运动器材业、五金电镀业、包装容器业等。

主营项目:非金属材料试验机、双辊塑炼机、开炼机、混炼机、平板硫化机、压片机、密炼机、双螺杆挤出机、流延机、吹膜机、过滤性测试机、老化试验箱、热压成型机、拉力试验机等众多材料物性试验设备。

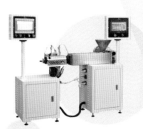

联系人:李本松
地　址:广东省东莞市厚街镇厚街村石角路133号
电　话:0769-85037377　13751313737　传　真:0769-85084177
网　址:http://www.dgxihua.com　邮　编:523960
邮　箱:E-mail:xihua@dgxihua.com　lbs0622@163.com

探索行业至高科技　开创领域精彩未来
YTCM Dual Rotor Continuous Mixing Machine

YTCM Dual rotor continuous mixing machine, with strong distributive mixing and dispersive mixing ability, can satisfy different mixing process requirement; Split barrel, easy to clean material and convenient to maintenance equipment; Two-stage type layout, single screw can be used with a variety of granule cutting mode, technology of wide adaptability; With-full automatic weight-loss metering feeding system, and interlock control system, the mixing operation can be always at the best state. The man-made interference to the production can be reduced, to make sure the automation and high effective and safe production; The inner wall of charging barrel adopts W6Cr5Mo4V2 busing, the rotor adopts the high quality alloy steel, and the surface overlays hard alloy, which solves the problem of abrasion and ensure the stability in the long run.

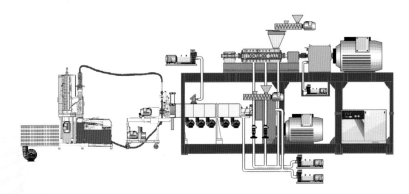

	Model	YTCM50	YTCM75	YTCM100	YTCM13
FIRST STAGE CONTINUOUS MIXER	Normal diameter (mm)	51	72	101	130
	Max rpm (r/min)	600	600	600	600
	L/D	8	10	10	10
	Motor power (kw)	45	75	160	250
SECOND STAGE SINGLE-SCREW	Model	SJ70	SJ100	SJ150	SJ200
	Normal diameter (mm)	70	100	150	200
	Max rpm (r/min)	90	90	90	90
	L/D	11	11	11	11
	Motor power (kw)	22	45	90	90
	Output (Kg/h)	30-200	50-350	400-1200	600-1500

Nanjing Yongteng Chemical Equipment Co.,Ltd.
ADD: No.8 Shitao Road, Shiqiu Town, Lishui District, Nanjing, Jiangsu China
TEL: 86-25-52337410、52414133、52873100、52337141　Fax: 86-25- 52336140

世界因我更精彩　The World Becomes More Wonderful By Us

塑胶专用条状铝颜料
Strip Aluminum Pigments for Special Plastic

旭阳产品	**Sunrise Products**
铝银浆	*Aluminium Paste*
铝银粉	*Aluminium Pigment Powder*
条状铝颜料	*Pelletized Aluminum Pigment*
水性铝颜料	*Aqueous Aluminium Pigment*

安徽旭阳，创立于2002年，是中国大型铝银浆、铝银粉、球型铝粉研发、生产和供应商机构，各种铝银浆年产能力达10000吨，铝银粉年产能力达1500吨，球型铝粉年产能力达6000吨。公司旗下有5个生产基地、5个直属大区销售办事处，1个国际销售中心和1个省级金属浆料工程技术研究中心，并已获得ISO9001:2008、ISO14001:2004及OHSAS18001：2007等管理体系认证。

Sunrise was established in 2002, which is a state- level high- tech enterprise in China with R&D, production and sales of aluminium pigment, fine spherical aluminium powder and bronze pigment, with annual capacity of 10,000 tons aluminium pastes, 6,000 tons fine spherical aluminium powders and 1,500 tons bronze pigments by 5 factories loacted in Anhui and Shandong Province.Sunrise owns five Domestic sales branches and one International sales center in Headquarter,and one State- level Functional and Technical Service Platform of Metal Slurry and Province - level Engineering Techonology Research Center, and ISO9001: 2008 / ISO14001:2004 / OHSAS18001: 2007 etc were awarded.

安徽旭阳铝颜料有限公司（安徽省合肥市双凤开发区金沪路18号）
ANHUI SUNRISE ALUMINIUM PIGMENTS CO.,LTD.　(Shuangfeng Development Zone, Hefei, Anhui, China)
T.（86 551）66399111　（86 551）65778259（Int'l Marketing Dept.）　F.（86 551）66399118
E. xuyang@xuyang.cc　**www.xuyang.cc**

Coloring of Plastics
Fundamental-Application-Masterbatch

Xinhua Chen
Hui Qiao
Yimin Xu
Dalu Yun
Junjie Yuan

  Sponsor : Hongda Group

图书在版编目（CIP）数据

塑料着色——基础·应用·色母粒＝Coloring of Plastics: Fundamental-Application-Masterbatch：英文/陈信华等编著．—北京：化学工业出版社，2016.5

ISBN 978-7-122-26506-7

Ⅰ.①塑⋯　Ⅱ.①陈⋯　Ⅲ.①塑料着色-英文　Ⅳ.①TQ320.67

中国版本图书馆CIP数据核字（2016）第049286号

责任编辑：赵卫娟　　　　　　　　　　装帧设计：张　辉
责任校对：王素芹

出版发行：化学工业出版社（北京市东城区青年湖南街13号　邮政编码100011）
印　　刷：北京永鑫印刷有限责任公司
装　　订：三河市宇新装订厂
710mm×1000mm　1/16　印张26¼　字数525千字　2016年6月北京第1版第1次印刷

购书咨询：010-64518888（传真：010-64519686）　售后服务：010-64518899
网　　址：http://www.cip.com.cn
凡购买本书，如有缺损质量问题，本社销售中心负责调换。

定　　价：98.00元　　　　　　　　　　　　　　　版权所有　违者必究
京化广临字2016——8号

本书编译人员

陈信华	乔 辉	徐一敏	云大陆	袁俊杰	丁 筠	张 雯
赵金榜	宋秀山	谢新辉	宋奇忆	王贤丰	刘晓梅	姚志卿
赵瑞良	吴克宇	陈 悦	张小明	季 青	刘维松	徐伟鑫
黎 力	杨金兴	李艳东	陈 侃	王树建	赖金琼	袁乐乐
吕伟亚	徐剑波	张薇波	徐海红			

Foreword

Masterbatch industry in China, along with development of the rests of plastics industry, has experienced a historical growth: from non-exiting to its nascent appearance, from a feeble status in the past to a strong identity today. The author has accumulated invaluable experience in plastic coloring from 40 years of theoretical research and experimental activities. Thanks for publication of this book, every reader can share and benefit from such experience. As referred by the author, plastics coloring is a systematic project. It has a huge and complex industrial connection, which requires employees to have knowledge regarding color, materials, mechanical engineering, legal concerns, regulatory requirements and so on. This book offers a systematic introduction of the masterbatch industry and related knowledge. It was written to be professional, informative and comprehensive to its readers. It is an instrumental reference book in plastic coloring.

I first got to know the wonderful world of masterbatch in 1982 and witnessed the development of Chinese masterbatch industry. Such industry in China started from the seventies of last century. After nearly 40 years of development, the annual output has reached more than 1 million tons. Now it has become an indispensable strategic partner of plastics and chemical fiber industry, and ubiquitous worldwide. Taking advantage of publishing English version of this book, I wish to bridge Chinese masterbatch industry and those around the world. Together, we can communicate, develop and make the world to radiate her best ever beauty !

Hui Qiao
February 2016

Preface

Plastics has merged into our daily lives everywhere. We enjoy benefits of these products all the time. A list of examples includes children's toys, instrument components, computer housings, automotive parts, toothbrush, tooth-cylinder, hygiene products, aircraft parts, aerospace applications and so on. It is a miracle that the plastics products has advanced to such level of industrial success in just 100 years, and continues to grow at an incredible rate.

Plastic coloring industry chain is composed of 3 key sectors: pigment and dye manufacturing, masterbatch and modified plastics manufacturing, and plastics processing. Masterbatch and modified plastics manufacturing industry is a connecting link of upper stream and downstream industries. Therefore it plays an important role to the whole plastic coloring industry chain. The historical experience shows that masterbatch and plastic processors do not have good understanding of complexity of colorants. On the other hand, pigment and dye manufacturers have limited understanding of plastics and plastics processing. Therefore, a sufficient information exchange between industrial chains is very important. It can significantly enhance mutual understanding between upper and down-stream industries to achieve higher standards of product quality.

For colorants users, the value of colorants depends on color appearance, color retention and its processing performance. If a colorant has high-value of color performance but poor processing performance, the value of application will be lower. In the same way, if the colorant has high-value of color and processing performances but unsuitable color retention for the applications, the usage values are reduced as well. Therefore, a significant portion of this book is on introduction of structure and properties of colorants. In such way, colorant users could have a good choices to ensure coloring quality, considering the pigment properties, application formulation, application equipment, technology and environmental conditions. As a result, an optimal coloring cost can be achieved. If colorant producers and users want to choose the stable performance of the colorants, this book explains well the technical indicators and test methods of plastic colorants with unified testing methods to improve the quality of products. However, only quality testing itself is not enough. If we want to achieve good coloring effect in various types of plastics, typical plastics processing conditions needed to be simulated to accomplish application testing.

Overall, plastics coloring is a systematic project, involving colorants, raw materials, processing equipment, processing technology, and the final product quality assurance. Although the comprehensive level of plastics coloring industrial chain has been improving, there are still many unanswered questions, problems and errors. Due to special needs of customers, some issues are contradictory and others are too difficult to be resolved. Every part in this industrial chain needs to communicate and develop together to meet customers' needs. This book has a systematic introduction of the plastics coloring and the relevant knowledge, trying to be more professional, informative and comprehensive. It can be used as a good reference book in plastic coloring.

Invaluable help from many warm-hearted experts and colleagues in plastics coloring industry has poured in the development of this book. The translation of English version has been mainly accomplished by a team of Professor Hui Qiao from *Beijing University of Chemical Technology (China Dyestuff Industry Association Color Masterbatch Speciality Committee)*. They are Hui Qiao, Yun Ding, Wen Zhang, Weisong Liu, Li Li, Weixin Xu, Yandong Li, Jinxing Yang, Kan Chen, Shujian Wang, Jinqiong Lai etc. The translation of chapter VII was completed mainly by Jinbang Zhao, Xiushan Song. Hereby, I want to express my heartfelt thanks to all of them.

Xinhua Chen
January 2016

Table of Contents

⊃ Chapter 1 Introduction of Plastics and Color / 1

1.1 Plastics 1
    1.1.1 Properties of Plastics 1
    1.1.2 Applications of Plastics 3
    1.1.3 Classification and Variety of Plastics 3
    1.1.4 Modified Plastics 14
1.2 Coloring of Plastics 14
    1.2.1 The Meaning of Plastic Coloring 14
    1.2.2 Plastic Colorants 16
    1.2.3 Types and Properties of Plastic Colorant 18

⊃ Chapter 2 Basic Requirements of Color Preparations / 22

2.1 Basic Requirements of Color Preparations 22
    2.1.1 Requirements of Colorant Based on Types of Plastic......................... 22
    2.1.2 Requirements of Colorant Based on Plastic Molding Process 24
    2.1.3 Requirements of Colorant Based on the Application of Plastic 26
    2.1.4 Basic Requirements of the Plastic Colorant 27
2.2 Color Performance 29
    2.2.1 Tinting Strength 29
    2.2.2 Saturation, Lightness, Hiding Power... 29
    2.2.3 Dichroism... 30
2.3 Thermal Resistance 31
    2.3.1 Definition 32
    2.3.2 Concentration.. 32
    2.3.3 Chemical Structure, Crystal Modification, Particle Size 33
    2.3.4 Resins and Additives 35
    2.3.5 Applications 36

2.4 Dispersion 37
2.4.1 Impact on Dispersion of Pigment Surface Properties 38
2.4.2 Impact on Dispersion of Particle Size and Distribution 38
2.5 Migration 39
2.5.1 The Main Reason for Migration 39
2.5.2 Impact of Chemical Structure and Concentration on Migration 40
2.5.3 Application of Migration Indicators 41
2.5.4 New Application Requirements for Migration in Consumer Goods Regulations 41
2.6 Light Fastness/ Weather Resistance 42
2.6.1 Chemical Structure and Particle Size of Pigment 43
2.6.2 Applications of Light Fastness (Weather Resistance) Indexes 44
2.6.3 Difference Between Light Fastness (Weather Resistance) Index and Practical Applications 47
2.7 Shrinkage/Warpage 47
2.7.1 Effect Based on Crystallization 48
2.7.2 Effect Based on Chemical Structure, Crystal Modification, Particle Size and Concentration of Pigment 50
2.7.3 Application of Shrinkage/Warpage Index 53
2.8 Chemical Stability 53
2.8.1 Acid and Alkali Resistance 53
2.8.2 Solvent Resistance 54
2.8.3 Oxidation Resistance 54
2.9 Security 55
2.9.1 Acute Toxicity of Chemicals 55
2.9.2 Organic Pigments Impurities 55
2.9.3 Security of Double Chloride Benzidine Pigment 56
2.9.4 Heavy Metal in Inorganic Pigments 59

Chapter 3 Main Types and Properties of Inorganic Pigments and Specialty Effect Pigments / 61

3.1 Development History of Inorganic Pigments 61
3.2 Classification and Composition of Inorganic Pigments 63
3.2.1 Classification of Inorganic Pigments 63
3.2.2 Composition of Inorganic Pigments 63

- 3.3 Plastic Coloring Properties of Inorganic Pigment 65
  - 3.3.1 Requirements of Inorganic Pigment in Properties 65
  - 3.3.2 Impact of Inorganic Pigment Particle Characters on Coloring Properties 66
  - 3.3.3 Safety of Inorganic Pigments in Plastic Coloring 69
- 3.4 Main Use of Inorganic Pigment in Plastics 70
  - 3.4.1 Coloring for Plastics 70
  - 3.4.2 Special Functions 71
- 3.5 The Main Varieties and Properties of Inorganic Pigments and Effect Pigments 76
  - 3.5.1 Achromatic Pigments 76
  - 3.5.2 The Main Varieties and Properties of Color Inorganic Pigment 87
  - 3.5.3 The Main Varieties and Properties of Effect Pigments (Inorganic) 113
  - 3.5.4 Main Varieties and Properties of Effect Pigments (Organic) 124

Chapter 4 Main Types and Properties of Organic Pigments / 132

- 4.1 History of Organic Pigments 132
- 4.2 Classification of Organic Pigments 133
- 4.3 Importance of Organic Pigments in Plastics Coloring 136
  - 4.3.1 Differences Between Organic and Inorganic Pigments 136
  - 4.3.2 Difference of Organic Pigments and Solvent Dyes 139
- 4.4 Relationship between Color Performance and Pigment Chemical Structure, Crystal Structure, Application Media 141
- 4.5 Azo Pigment 142
  - 4.5.1 Monoazo Pigment 142
  - 4.5.2 Disazo Pigments 176
  - 4.5.3 Disazo Condensation Pigments 185
- 4.6 Phthalocyanine Pigments 201
  - 4.6.1 Blue phthalocyanine pigments 202
  - 4.6.2 Green Phthalocyanine Pigments 206
- 4.7 Heterocyclic and Polycyclic Ketonic Pigments 209
  - 4.7.1 Dioxazine Pigments 210
  - 4.7.2 Quinacridone Pigments 213
  - 4.7.3 Perylene and Pyrene Ketone 220
  - 4.7.4 Anthraquinone and Anthraquinone Ketone Pigments 227
  - 4.7.5 Isoindolinone and Isoindoline Pigments 231
  - 4.7.6 Diketopyrrolo-pyrrolo Pigment 237

    4.7.7 Quinophthalone Pigments...244
    4.7.8 Metal Complex-based Pigments...246
    4.7.9 Thiazide Pigments 248
    4.7.10 Pteridine Pigments...249

Chapter 5 Main Varieties and Properties of Solvent Dyestuffs / 252

5.1 Development History of Solvent Dyestuffs............................252
5.2 Types, Properties and Varieties of Solvent Dyestuffs............253
5.3 Application Characteristics of Solvent Dyestuffs in Coloring of Plastics..256
    5.3.1 Solubility..257
    5.3.2 Sublimation..258
    5.3.3 Melting Point..259
5.4 Market and Produce of Solvent Dyestuffs.............................259
5.5 Main Varieties and Properties of Solvent Dyestuffs 260
    5.5.1 Anthraquinone Solvent Dyestuffs 261
    5.5.2 Heterocyclic Solvent Dyestuffs 275
    5.5.3 Methine Solvent Dyestuffs 287
    5.5.4 Azo Solvent Dyestuffs 290
    5.5.5 Azomethine Solvent Dyestuffs...292
    5.5.6 Phthalocyanine Solvent Dyestuffs..294

Chapter 6 Plastic Colorants Inspection Method and Standard / 296

6.1 Inspection and Standard 296
    6.1.1 Importance of Inspection..296
    6.1.2 Adopt International Standards, Improve Enterprise Competitiveness 297
    6.1.3 Quality Control Test and Application Performance Test................. 298
    6.1.4 Standards 299
6.2 Quality Tests and Standards of Color Performance..............302
    6.2.1 Composition of The Basic Mixtures of PVC 303
    6.2.2 Test of Color Performance...304
    6.2.3 Evaluation of Color Performance...305
6.3 Evaluation of Color Stability to Heat During Processing of Coloring Materials in Plastics 307

- 6.3.1 Evaluation by Injection Moulding (HG/4767.2—2014, EN BS 12877-2) ... 307
- 6.3.2 Evaluation by Oven Test (HG/T 4767.3—2014, EN BS 12877-3) ... 308
- 6.3.3 Evaluation by Two-roll Milling ... 309

6.4 Standards and Methods of Assessment of Dispersibility in Plastics ... 310
- 6.4.1 Evaluation by Two-roll Milling ... 311
- 6.4.2 Evaluation by Filter Pressure Value Test (HG/T 4768.5—2014, EN BS 13900-5) ... 312
- 6.4.3 Evaluation by Film Test (EN BS 13900-6) ... 314

6.5 The Test Methods and Standard of Migration (HG/T 4769.4—2014, EN BS 14469-4) ... 315

6.6 The Test Methods and Standards of Light Fastness and Weather Resistance ... 316
- 6.6.1 Sunlight Fastness and Weather Resistance Test ... 317
- 6.6.2 Light Fastness to Artificial Light and Weather Resistance Tests ... 320

6.7 Evaluation of Chemical Stability ... 326
- 6.7.1 Test for Acid Resistance and Alkali Resistance ... 326
- 6.7.2 Solvent Resistance Test ... 327

6.8 Evaluation of Warpage and Deformability ... 328
- 6.8.1 Evaluation of Degeneration or Shrinkage ... 328
- 6.8.2 Evaluation of Warpage ... 329

Chapter 7 Pigment Dispersion in Plastics / 331

7.1 The Purpose and Significance of Pigment Dispersion in Plastics .. 331
- 7.1.1 Improving Pigment Coloring ... 331
- 7.1.2 Meeting the Requirement of The Plastic Processing For The Pigment Dispersion ... 332

7.2 Dispersion of Pigments in Plastic ... 334
- 7.2.1 The Types and Properties of Particles Before Pigment Dispersion ... 334
- 7.2.2 Dispersion of Pigment in Plastic ... 336

7.3 Wetting ... 338
- 7.3.1 Wetting Phenomena and its Evaluation Contact Angle and Young's Equation ... 338
- 7.3.2 Pigment Wetting ... 339
- 7.3.3 The Judgment Basis of Pigment Surface Properties——Wetting Angle ... 347

7.4 Pigment Dispersion ... 348

7.5　The Dispersion Stability of Pigments..................................350
　　7.5.1　The stabilization Mechanism of Double Electronic layer...........351
　　7.5.2　The stabilization Mechanism of Steric Hindrance..................354
7.6　Color Masterbatch and Pigment Preparation 358
　　7.6.1　The Definition of Masterbatch 358
　　7.6.2　Pigment Dispersion in the Color Masterbatch of Polyolefin.......359
　　7.6.3　The Jewel on the Crown of the Color Masterbatch Technology—
　　　　　Chemical Fiber Dyeing..367
　　7.6.4　Pre-Dispersed Pigment Preparations for Plastics 374

Chapter 8　Practical Technology and Quality Control of Color Matching for Plastics Coloring / 376

8.1　Basic Theory and Principles of Color Matching 377
　　8.1.1　Basic Theory...377
　　8.1.2　Basic Principles of Color Matching.............................378
　　8.1.3　Color Matching for Plastics Coloring is a Complex Process......379
8.2　Basic Knowledge That Color Matching Staff Should Have 380
　　8.2.1　Comprehensive Understanding the Performance of Colorant........380
　　8.2.2　Comprehensive Understanding the Performance of Plastics........381
　　8.2.3　Fully Grasp the Process Conditions of Plastic Molding Process..381
　　8.2.4　Fully Understand the Type of Plastic Processing Aids and Master
　　　　　the Content...381
　　8.2.5　Comprehensive Understanding of the Safety Regulations of
　　　　　Plastic Products at Home and Abroad 382
8.3　Basic Steps of Color Matching 382
　　8.3.1　Preparatory Work of Color Matching 382
　　8.3.2　Specific Steps of Color Matching...............................383
8.4　Quality Control 389
　　8.4.1　Chromatic Aberration...390
　　8.4.2　Quality Control of Formula Design..............................394
　　8.4.3　Quality Control of Production 394
　　8.4.4　Establish Quality Control System 396
　　8.4.5　Problems and Solutions 397
8.5　Practical Technology of Color Matching............................399
　　8.5.1　How to Formulate Special White Colorant for Plastic Coloring...399
　　8.5.2　How to Formulate Special Black Colorant for Plastic Coloring...400
　　8.5.3　How to Formulate Gray for Plastic Coloring....................400

8.5.4 How to Formulate Colorant for Outdoor Plastic Products 401
8.5.5 How to Formulate Colorant for Transparent Products 402
8.5.6 How to Formulate Pearlescent Colorant for Plastic Coloring 403
8.5.7 How to Formulate Gold, Silver Colorant for Plastic Coloring 404
8.5.8 How to Formulate Fluorescent Colorant for Plastic Coloring 405

⮕ Reference / 406

Chapter 1
Introduction of Plastics and Color

It has been a century since the first kind of plastic was invented by an American Belgians named Baekeland, currently plastic products exist in every area in our daily life, and keep growing with an incredible rate. There are no other materials have ever experienced such a miracle in the history of human industrial development. In addition, the invention and use of plastics have been recorded in history as an important innovation in the twentieth century, which has influenced the human's development. The products produced by pure resin can't draw consumer's attention, for the colorless, transparent or natural white. Therefore, it is an inescapable and sacred mission for the plastic processing practitioners to get the plastic products colorful.

1.1 Plastics

1.1.1 Properties of Plastics

As a new material, plastic can development rapidly in just 100 years, mainly rely on its excellent performance.

(1) Mass production

The raw material to produce plastic is oil, which can be largely produced, so the cost to produce plastic is low. We can greatly reduce the production costs of plastic products with its plasticity, even though the oil is not so cheap.

Plasticity is the ability to form a new shape by cooling in the die after softening by heat. That can make the products with complex shapes, more easy to be produced, and the processing efficiency is much higher than metal. A large amount of inventions and innovations of plastic molding process has been emerged with the plastic development in the last 100 years, especially the injection molding, which can produce products with complex shapes easily. With the development of the plastic industry, wood, steel, cotton, paper and a series of other traditional materials was replaced by plastic, for it is easy to color, better processing, lower energy consumption and lower cost. Thus, it accelerates the development rate of plastic industry.

(2) Low density and High Strength

The density of plastic is about $1g/cm^3$, only 1/5 of aluminum, 1/10 of steel. Many polyolefin plastics is less than $1g/cm^3$, such as polyethylene and polypropylene, which could floating on the water. The density of plastic foams is only about $0.1\ g/cm^3$. The density of few polymer is heavier, for example, PVC is $1.4\ g/cm^3$, PTFE is $2.2\ g/cm^3$, but they are still lighter than metal and ceramic.

Although the density of plastics is low, the strength is high. The density of nylon is 1/10 of the steel, but the fracture strength is only less than half of the steel wire. Although plastic is not as hard as metal, compared with metal and ceramic, it has light weight, good mechanical properties, high mechanical strength and abrasion resistance, so it can be made into high strength products with light weight. At present, the plastic parts are used for the auto parts in large quantities, about 50% of the volume of automotive materials, so that the weight of the car reduced 1/3. The weight of cars is light and the fuel consumption is reduced.

(3) Corrosion resistance

Corrosion resistance is an important environmental issue to solve. Every year in our country, the direct economic losses caused by corrosion is at least RMB 20 billion. Most of the plastics have strong corrosion resistance and does not react with acid and alkali, so that they often be used as chemical infusion pipelines. For instance, synthetic fiber can be used for the preparation of industrial filter cloth and plastic containers can be used to store and transport high corrosive liquid.

Plastic neither rust in moist air, nor rot in the humid environment or eroded by microbial. Therefore, plastics are largely used to produce building doors, windows and so on.

(4) Excellent electric and heat insulation

The molecular chain of plastic is the combination of atoms by covalent bonds, neither ionization nor transmit electrons in the structure and the electrical resistivity up to $10^{14} \sim 10^{16} \Omega$. So plastics have excellent electrical insulation, can be used as electric switches and household appliances, insulation sheath of wire and cable, also widely used in electronic, electrical, radar, television, broadcasting, communications, computers and other electronics and instrumentation industry.

Plastics have low heat conductivity, good heat preservation and insulation, so it is widely used in the field of thermal insulation. With the agricultural film built greenhouses, even the northeasterner can eat fresh vegetables in winter. The surface temperature of missiles, rockets, space shuttles and some other aircraft can reach to 1300℃, and their protective coating surface is some plastic material with low thermal conductivity, such as phenolic - epoxy and organic silicone resins.

It is important to know the shortcomings as to know the advantages, such as the poor heat resistance, deformed in high temperature, burning and aging under the condition of light, oxygen, heat, water and atmospheric environment. Moreover, the surface hardness is low and is subject to damage. Due to the insulation, plastics are static electricity and dust absorption.

The main shortcoming of plastics is non-perishable, which is extremely harmful to the environment with its invariability buried in the earth for hundreds of years. As the pace of life accelerate, in order to comply with the demand for convenient and healthy social life, disposable foam lunch boxes, plastic bags and other products are flowering in our daily life. Although these convenient and cheep packaging materials have brought a lot of convenience to people's lives, they are potential hazard to ecological environment if they are discarded casually which resulted "white pollution".

1.1.2 Applications of Plastics

Owing to the excellent features, plastics can be processed to various products as flexible as silk, as firm as iron, and as transparent as glass. At present, plastic products are closely related to daily lives. It is no exaggeration to say: no plastic, no life. The main application of plastics are listed in Table 1.1.

Table1.1 Main applications of plastics

Structural materials	Household appliances, automobile parts, mechanical parts
Insulating materials	Cables, insulating boards, electrical parts
Building materials	Plastic building materials, drain pipes, gas pipes
Packing materials	Film bags, packaging films, foamed plastic, plastic containers
Daily supplies	Office equipments, furnitures, toys
Transportation	Transportation facilities, vehicle parts
Textile fibers	Clothing fibers, carpets, lawns

1.1.3 Classification and Variety of Plastics

1.1.3.1 Classification

So far more than three hundred kinds of plastics have been produced around the world. The classification methods are diverse and complex which even crossed by each other. The conventional methods are classified by physicochemical properties, application properties, and processing methods.

(1) Classified by physicochemical properties

According to physicochemical properties, plastics can be divided into two categories: thermoplastic and thermosetting plastic.

Thermoplastic has resin molecular structure with linear or branched. It is softened by heating, hardened by cooling. And this process is reversible and repeatable, which only involves physical change. Therefore, the scraps and wastes produced in the plastic processing can be recycled to reuse by pulverizing into particles. Thermoplastic includes polyethylene, polypropylene, polyvinyl chloride, polystyrene, ABS, polyamide, polyoxymethylene, polycarbonate, PMMA, and so on.

The thermosetting plastics have plasticity and the structure is linear before heated. The linear polymer molecular chains can combined by chemical bonds to form a cross-linking network when being heated, to make a three-dimensional network polymer, which neither be melted nor dissolved, and the structure cannot be changed again. There are both physical and chemical changes in that process. The wastes and leftover materials in the processing can not be reused, for the properties mentioned above. Such as the phenolic resins, amino plastics, epoxy plastics, silicone plastic etc.

The structures of thermoplastic and thermosetting plastic are shown in Figure 1.1.

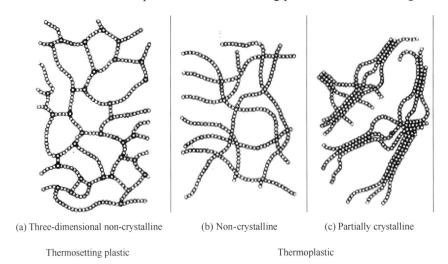

(a) Three-dimensional non-crystalline　　(b) Non-crystalline　　(c) Partially crystalline

Thermosetting plastic　　　　　　　　　Thermoplastic

Figure 1.1　The structures of thermoplastic and thermosetting plastic

(2) Classification by the applications

General plastics　General plastics are generally refer to the materials that high yield, wide applications, good formability and low prices. The production of general plastics occupy more than 3/4 of the world market, it is the main part of the plastics industry. The members of the family are polyethylene(PE), polypropylene(PP), polystyrene(PS), acrylonitrile-butadienestyrene(ABS), polyvinylchloride(PVC), polymethylmethacrylate(PMMA), epoxy resin(EP), phenolic resin(PF), polyurethane(PU) and unsaturated polyester.

Engineering Plastics The engineering plastics are normally characterized as materials that can be used as engineering plastic components, which have excellent strength, dimensional stability, and can maintain good properties under the high or low temperature. Engineering plastics include polyamide (PA), poly-butylene terephthalate (PBT), polyethylene terephthalate (PET), polycarbonate (PC), polyformaldehyde (POM) and polyphenylene oxide (PPO). The production of engineering plastics is small and the price is expensive.

Special plastics Special plastic is a kind of plastics with special function (i.e. heat resistance, self-lubrication, etc.) and used in special fields. Such as poly-phenylenesulfide (PPS), polysulfone (PSU), polyether sulfone (PES), polytetrafluoroethylene (PTFE) and polyether ether ketone (PEEK).

The abrasion resistance, corrosion resistance, heat resistance, self-lubrication and dimensional stability for both engineering plastics and special plastics are superior to general plastic. They are some ways similar to metallic materials, thus have a widely use in machinery manufacturing, light industry, electronics, daily use, aerospace, missile, atomic energy and other engineering departments, they gradually replace metal in producing some mechanical parts.

(3) Classification by processing method

Plastics can be subdivided based on the processing in extrusion molding plastic, injection molding plastic, blow molding plastic, molded plastic, rolling plastic, etc.

(4) Classification by products

Plastics can be divided into the filmplastic, pipeplastic, wireplastic, cableplastic, building material plastic,foamplastic, etc.

1.1.3.2 The main varieties of plastics

(1) Thermoplastics

The molecular structure of the thermoplastic resin is in general connected by one or two or more "basic unit" repeatedly according to a certain arrangement through chemical bonds, just as the chain of pearls. Be a linear or branched chain type structure, which called linear polymer. For example, the basic unit of the polyethylene molecules is C_2H_4, and each polyethylene molecule was connected by numerous basic units:

$$-(C_2H_4-C_2H_4-C_2H_4-C_2H_4)_n-$$

① Polyvinyl chloride (PVC)

Formula: $(CH_2CHCl)_n$. PVC based on the polymerization of vinyl chloride monomers, is one of the main global plastic varieties, production capacity is ranking the second place, only less than that of polyethylene, roughly 16% of the synthetic resins in the world.

PVC resin is white or pale yellow powder. Due to the high electronegative chlorine atoms, there is a strong attraction which blocked the relative movement between the molecules, so the PVC is quite hardness and has excellent chemical corrosion resistance, but brittle and lack of flexibility. PVC has good flame-retardant and self-extinguishing properties, for the big content of chlorine (more than 55%). The molecular chain polarity of PVC make it has a higher processing temperature. According to the requirements of the products, different amount of plasticizer being added to decrease the intermolecular attraction and improve its flexibility. PVC can release HCl gas and discoloration under the influence of heat, oxygen and light, result in the decrease of material performance, therefore antioxidants and UV stabilizer are need. The disadvantage of PVC is that the monomer and plasticizers are toxic, HCl can be released when burning, so PVC is difficult to degrade and pollute environment.

The rigid plastic of PVC (plasticized PVC) can be used to produce containers (can not withstand pressure), pipe, plate, wire and cable, etc. Improved the hardness by $CaCO_3$, the plastic can instead of steel or wood to manufacture door, window, floor, ceiling and wire conduit, etc.

Plasticized PVC can produce the packaging of the beverage, pharmaceutical and cosmetics, thin film products, toys, sandals and artificial leather, etc.

② Polyolefin (PO)

Polyolefin is a kind of polymer that build only with carbon and hydrogen. The most common used polyolefins are polyethylene and polypropylene, which has the largest production and using.

Polyethylene (PE)   PE is an ethylene homo-polymer or the copolymer of ethylene with α-olefins, and it is a non-polar polymer with its molecules containing only C—C and C—H bonds. PE is a wax crystalline thermoplastic resin with odorless, tasteless and non-toxic. Besides, it has excellent low temperature resistance ($-70 \sim -140 ℃$), chemical stability, and resistance to most acids (excluding oxidizing), alkalis, salts. At room temperature, it has low hygroscopicity, excellent electrical insulation, medium mechanical properties and insoluble in common solvents. But the heat resistance, aging resistance and environmental stress cracking resistance are poor. The production of polyethylene accounts for more than 30% of the total output of plastics, so PE is the largest kind of common plastics.

PE is often differentiated by density. The different structure of polyethylene is shown in Figure 1.2. The low density polyethylene has long-chain branch, the length of which is even more than main chain, and its molecule is dendritic with poor regularity, low crystallinity and density. The high density polyethylene without long-chain branch, and its molecule is linear with high crystallinity and density. The molecular state of linear low density polyethylene is between LDPE and HDPE, so as the density.

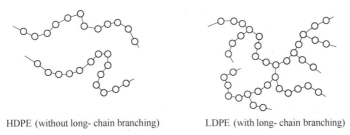

HDPE (without long-chain branching)　　　LDPE (with long-chain branching)

Figure 1.2　The molecular structure of PE

***Low density polyethylene* (LDPE)**　LDPE is an ethylene homo-polymer, which is also called high-pressure polyethylene because of its high pressure polymerization. It is slightly heavier than polypropylene ($0.903 \sim 0.904 \text{g/cm}^3$). LDPE has the general properties of PE, in addition, its transparency, softness, cohesiveness and electrical property (high-frequency insulation) are better than HDPE. Meanwhile, it is especially excellent in processing, and this is the reason that it is widely used and irreplaceable. LDPE has low melting point, good heat sealing, but has slightly inferior mechanical properties, heat resistance and aging resistance.

Owing to its high transparency, glossiness and softness, LDPE is suitable for demanding packaging films, such as textiles, clothing and food packaging. Because of its excellent transparency, it is processed to various agricultural films to improve crop yields; LDPE is processed to adhesive layer of extrusion coating or composite film, which is widely used in packaging field, for the low melting point and excellent cohesiveness, heat sealing and film forming property. Owing to the polymerization without catalyst residual metal component, high insulation and high temperature resistance cross-linked, LDPE is used for cable insulating materials, which the voltage is up to 340 kV; Owing to the excellent fluidity, LDPE is injection molded into various complex shape daily necessities (bottle caps, plastic flowers, sealed containers, kitchen supplies), toys, stationeries, and so on.

***High density polyethylene* (HDPE)**　HDPE is an ethylene homo-polymer or the copolymer of ethylene with small amount (<2%) of α-olefins, and has a molecular weight of $4 \sim 30$ million. HDPE is also called low-pressure polyethylene which is attributed to its atmospheric or low pressure polymerization. It is a white powder or ivory wax particle thermoplastic resin with non-toxic, tasteless and odorless. Besides, its properties are more ideal for its oil resistance, solvent resistance, vapor barrier and not ease to have brittle fracture at low temperature ($-100 \sim -140°C$). But its transparency, electrical insulation, processability, heat sealing and bond strength are inferior to LDPE. Compared with common plastics, HDPE is superior than polypropylene on toughness, low temperature resistance and aging resistance. Moreover, the processability and operating temperature are better than PVC.

The main applications of HDPE are: hollow blow molding containers, cosmetics and oil packaging containers; Gas pipes, pressure delivery pipes for water, oil and mining liquid;Injection containers, pallets, etc.; Seepage, pollution prevention of the geo-membrane.

Linear low density polyethylene (LLDPE)   LLDPE is a copolymer of ethylene with 5%~20% α-olefins, which is also called "third generation polyethylene" following LDPE and HDPE. According to the structure and density, the properties of LLDPE lies between LDPE and HDPE. However, due to the narrow molecular weight distribution, great proportion of medium or high molecular weight, large extent of molecule folding and winding, high intermolecular forces, and low shear sensitivity, LLDPE has its own uniqueness in physical and mechanical properties and processability. Therefore, LLDPE has high melt viscosity (100%~150% higher than LDPE), high extrusion power and energy consumption (20%~120% higher than LDPE), high back pressure, which leads to its small productions. And the critical shear rate of LLDPE is much less than LDPE, thus, which prone to sharkskin melt fracture and other phenomena. In order to resolve the problems mentioned above, many circumstances external lubricant such as the fluorocarbon elastomer can be added, meanwhile blend with LDPE at 7∶3 or 6∶4 can also improve the mechanical properties and productions.

In injection molding, LLDPE has properties of short stress relaxation time, low internal residual stress, little deformation, small shrinkage and buck deformation, which are not easily occurred tensile strain hardening phenomenon like LDPE.

LLDPE is the most rapidly developing resin in polyethylene resin. The main applications are: producing ultrathin films that mainly used for ice bags, frozen food packaging films and garbage bags. High impact strength thin-walled products and rotational molding products which is excellent in low temperature performance, resistance to environmental stress cracking and shrinkage property can be prepared by LLDPE with narrow molecular weight distribution and high melt index. LLDPE has acquired the communication cable sheath and insulation material markets owing to its good toughness, wide operating temperature and excellent resistance to environmental stress cracking.

Polypropylene (PP)   Formula: $(C_3H_6)_n$, PP is polymerized by propylene monomer at a certain temperature and pressure. Its relative density (only $0.9 \sim 0.91 g/cm^3$) is the minimum in general plastics. PP is non-toxic, tasteless, good heat resistance, and can be long-term used at about 110℃.

Currently we usually use isotactic PP that all the methyl are on the same side of the main chain, which has a high crystallinity. Atactic polypropylene is a kind of amorphous polymers, it has little value in single use but can be used as the carrier of filled masterbatch and toughening modifier.

PP is processed to films, fibers, hollow containers and injection products by extrusion, injection molding, blow molding, and so forth. Since PP molecular chain contains tertiary carbon atom and the hydrogen is vulnerable by oxygen, thus, it has poor aging resistance. The applications of PP shown in Table 1.2.

Table 1.2 PP molding process and applications

Molding method	Molding products	Melt index/(g/10 min)
Extrusion	Pipes, plates, sheets, rods	0.15~0.4
	Drafting zone	1~5
	Monofilament, flat fibers	3~6
	Blown films	8~12
Injection molding	Injection products	1~26
Blow molding	Hollow containers	0.4~1.5
Biaxial tension	Films	1~2
Spinning	Fibers	10~20

③ Styrene resins

Styrene resin is a kind of thermoplastic resins that homo-polymerized by styrene or copolymerized by styrene and other monomers. Styrene products include the high transparent SAN to high hiding ABS. Currently, the production of styrene resins is the fourth, fall behind polyethylene, polyvinyl chloride and polypropylene.

Polystyrene (PS)   Formula: $(C_8H_8)_n$, Polystyrene (PS) is a kind of linear polymers, which is polymerized by styrene monomer, and it is also known as general purpose polystyrene (GPPS). Due to the high transparency, it is often called crystalline polystyrene. As well as the benzene ring steric hindrance in molecule, it is a random polymer.

The relative density of PS is between $1.04 g/cm^3$ and $1.09 g/cm^3$. Owing to the excellent dimensional stability, low shrinkage and hygroscopicity, PS can maintain dimensional stability and strength in humid environment without bacteria growth. Moreover, the transparency is 88%~92%, and the refractive index is 1.59~1.60. PS is a good refrigeration insulating material, because its heat distortion temperature is about 70~98℃, and the thermal conductivity does not change with temperature. But it will be decomposed when the temperature exceeds 300℃. PS has good dielectric and insulation properties. Due to its high surface resistance, volume resistance and hydrophobicity, PS is easy to occur static electricity.

Owing to its good transparency, PS is widely used in the optical industry. It can be manufactured to optical glass, optical instrument, and transparent or bright color products such as lampshade, lighting equipment and so forth. In addition, PS is processed to electrical components and instruments which are used in the high-frequency environment.

In order to improve the brittleness of PS, modified polystyrene is prepared by grafting polystyrene on the rubber (butadiene rubber or styrene-butadiene rubber), which is called high-impact polystyrene (HIPS) with both rigidity and toughness. Its impact strength is of 3~4 times higher than ABS. In recent years, HIPS has accounted for 60% of PS productions with the rapid development, and its production in many countries has exceeded PS and ABS.

Moreover, HIPS is easy to extrusion, injection molding and secondary processing. HIPS can be manufactured to packaging containers such as thin glass, extruded dairy containers. It can also be used as household appliance, stationery, washing machine, freezer, refrigerated truck parts and shell of home appliances, such as radios, television and so on.

Acrylonitrile-Butadiene-Styrene copolymers (ABS) Formula: $(C_8H_8C_4H_6 \cdot C_3H_3N)_n$. ABS resin is a copolymer of acrylonitrile, butadiene and styrene with good cooperativity. Moreover, it has properties of chemical resistance, surface hardness and thermal stability of acrylonitrile, toughness of butadiene, rigidity and processability of styrene. Therefore, ABS may be prepared to different properties by regulating proportions of three components. The monomer content of current ABS resin production is that acrylonitrile is 20%~30%, butadiene is 6%~30%, and styrene is 45%~75%.

ABS is an amorphous polymer material. Owing to the excellent tenacity, hard texture, and rigidity, it is a good comprehensive engineering plastic. The melt temperature is 217~237℃, and the thermal decomposition temperature is over 250℃. The impact strength decreases slightly with temperature declines, ABS can long-term use in −40~100℃. Due to the double bond contained in butadiene, it has poor weather resistance, easy aging, discoloration, and cracking, which results in mechanical properties decrease. ABS is high hygroscopicity, in general, the luster will be bright by surface drying. Its melt viscosity is almost the same as HDPE, and it has excellent processability.

The main applications: ABS is widely used in home appliances, automotive and mechanical industries such as liner television, telephone and instrument shell, refrigerator liner, auto parts, pump impeller, handles, battery slot, and so forth. ABS can be processed to metal substitutes and adornments by surface metallization.

Poly-methyl methacrylate (PMMA)   PMMA is plastic material that with amorphous structure, and familiarly known by its trade names, such as Lucite or Plexiglas. It is light, tough and transparent, good luster performance, and transmittance for light and ultraviolet is 92% and 75%, respectively, while the glass is only 85% and 10%. PMMA has good electrical insulation property, excellent dimensional stability, prominent weather resistance and chemical stability. Specifically, the mechanical strength is more than 10 times than that of ordinary silicate glass.

The main applications: instrument parts, aircraft and automobile glazings, optical lenses, transparent models, lighting fixtures, signal equipment, dashboard, blood storage containers, DVDs, light scattering devices, beverage cups, stationeries, etc.

④ Polyamide, polycarbonate, polyethylene terephthalate and poly-oxy-methylene.

This kind of plastics can be used as structural materials, also known as engineering plastics, which are high-performance materials, and can be used in harsh chemical and physical environment, in addition can withstand the mechanical stress in a wide temperature range. These materials have excellent mechanical properties, good dimensional stability, outstanding high and low temperature resistance, and have been the most rapidly growing plastics in the world's plastic industry..

Polyamide (PA, commonly known as Nylon)   Polyamides are produced by the reaction of di-amine with dibasic acid, which with repeating amide groups on the main molecular chain.

There are numerous species of polyamide, like nylon 6, 66, 610, 1010, etc. For polymer molecules, the more carbon atoms they have, the softer they are. Nylon 66 yarns can be used for scrubbing brushes and shoe brushes, while nylon 1010 yarns can be toothbrushes. Wherein the nylon 6 and nylon 66 can be widely used in our life for the cheap price and excellent processing property.

Nylon has excellent mechanical property, easily pigmented and non-toxic. In addition, it is flame retardant and self-extinguishing. Nylon is the earliest engineering plastic and its production is 1/3 of the total output of plastics, for the excellent overall performance.

Nylon can be processed by injection, extrusion, casting and rotational molding.

Nylon is mainly used for the production of transmission parts with abrasion resistance and high strength. It has been widely used in machinery, transportation, instrumentation, electrical, electronics, communications, chemical industry, medical equipment and daily necessities. The specific applications, such as gears, pulleys, turbines, bearings, pump impellers, fan blades, seals and oil storage containers and so on.

Polyethylene terephthalate (PET, commonly known as Terylene)   Terylene is an important species of synthetic fibers, which is based on purified terephthalic acid (PTA) or dimethyl terephthalate (DMT) and ethylene glycol (EG) as raw materials by esterification or trans-esterification and poly-condensation to obtained fiber-forming polymer——PET, then obtained by spinning and post-processing. Terylene is widely used, mainly for the manufacture of clothes and industrial products. Terylene has excellent setting performance. After setting, the terylene yarns or fabrics have very flat and fluffy morphology, and can keep enduring after many times washing.

PET has good optical property, weather resistance, and amorphous PET has good optical transparency. In addition, PET has excellent frictional abrasion resistance, dimensional stability and electrical insulation. Bottles made from PET are non-toxic, anti-infiltration, and have high strength, good transparency, light weight and high production efficiency, etc. Most of them are used as carbonated beverage and edible oil bottles. PET can also be used for the automotive industry (structural components such as the reflector's boxes, electrical components such as headlights, reflectors, etc.), electrical components (motor shells, electrical couplings, relays, switches, internal devices of microwave ovens, etc.), industrial applications (pump housings, hand instruments, etc.) and so on.

Poly-butylene terephthalate (PBT) Molecular formula: $[(CH_2)_4OOCC_6H_4COO]_n$. The molecular chain structure of PBT is similar to PET, and the majority of properties are the same, the difference is that there are two more methylene groups than PET, so the molecule chain is more submissive and processing property is better. PBT is one of the toughest engineering plastics, which is semi-crystalline material, has very good chemical stability, mechanical strength and electrical insulating property. Typical applications of PBT: household appliances (food processing blades, vacuum cleaners components, electric fans, hair dryer shells, coffee containers, etc.), electrical components (switches, motor shells, fuse boxes, keys of computer keyboards, etc.), automotive industry (radiator latticed windows, body panels. wheel covers, doors and window components, etc.)

Polycarbonate (PC) Molecular formula: $(C_{15}H_{16}O_2 \cdot CH_2O_3)_n$. PC is colorless, transparent. PC has a high impact strength, great coloring property, good dimensional stability, excellent electrical insulation, corrosion resistance and abrasion resistance, but poor self-lubricating, tend to stress cracking, easily hydrolyzed at high temperatures, and the compatibility with other resins is poor. Polycarbonate is an amorphous material, has a good thermal stability, wide molding temperature range and poor mobility. PC has low hygroscopicity, but be sensitive to water, subject to drying, and has low mold shrinkage, prone to melt cracking and stress concentration.

The main application of PC: baby bottles, drinking cups (also known as space cups) and purified water buckets, PC can be repeatedly sterilized, has better light-admitting quality than PMMA, also be suitable for the production of small parts of the instruments, insulating transparent parts and impact-resistant pieces.

Polyoxymethylene (Abbreviated as POM) Molecular formula: $(CH_2O)_n$. POM has better overall performance, has a very low coefficient of friction and good geometric stability, high strength and stiffness, low water absorption, but poor thermal stability, being flammable, easy to aging when exposure in the atmosphere. POM is suitable for production of abrasion resistant parts, transmission parts, and chemical industry, instrumentations, and other parts. Because it has a high temperature resistant property, it can be used for pipe devices (like valves, pump housings).

(2) Thermosetting plastics

The heat resistance of thermoset is better than thermoplastic. The phenolic resin, amino resin and unsaturated polyester all belong to thermosetting plastics. The processing technology commonly is compression molding, at 150~190℃, injection molding can also be used.

Phenol-Formaldehyde (PF) Molecular Formula: $(C_{12}H_{10}O \cdot C_{10}H_{14}O \cdot CH_2O)_n$. PF was the earlist plastics to realize. It is familiarly known by its trade name Bakelite. PF has many excellent properties such as: high mechanical strength, prominent abrasion resistance, dimensional stability, corrosion resistance, excellent electrical insulation properties and so on, which is suitable for the production of electrical and instrumentation's insulated parts.

Amino-Plastics (MF, UF) Molecular Formula: $(C_3H_6N_6 \cdot CH_2O)_n$. Amino resin is illustrated by the reaction between formaldehyde and a compound which has amino or acyl-amino. The most commonly used amino resins in the industry are urea-formaldehyde and melamine formaldehyde is non-toxic, odorless, hard, scratch-resistant, colorless and translucent. It can be made into variety of colorful plastic products. MF was widely used in aviation, electronics and other fields.

Polyurethane (PU) Molecular Formula: $C_3H_8N_2O$. PU is the abbreviation of polyurethane, which contains repeating carbamate groups on the main chain. PU is a new organic polymer materials, known as "the fifth plastic", it is synthesized by diisocyanate or polyisocyanate and dihydroxy or polyhydroxy through the method of addition polymerization.

PU is subdivided according to the proportion of NCO and OH in the formulation into thermoset and thermoplastic, for the repeating carbamate groups on the main chain. A further subdivision marks its molecular structure as linear and body structure. Due to the different crosslink density, PU can present a different property, such as hard, soft or between the two, high strength, abrasion resistance, solvent resistance and other characteristics.

Properties depend on materials, polyurethane is generally divided into polyester PU and polyether PU. Polyurethane material has a very wide range of application in our daily life and can be used in the manufacture of plastics, rubbers, fibers, rigid and flexible foams, adhesives and coatings.

Rigid polyurethane foam is mainly used for building insulation materials (insulation pipeline facilities), household items (bed, sofa cushion and other materials, refrigerators, air conditioners and other insulation layer and the core material such as surfboards), and transportation (automotive, aircraft, rail vehicle seat, roof materials etc.). The high tensile strength, great tear strength, impact resistance, outstanding abrasion resistance, weather resistance, hydrolysis resistance and oil resistance of PU rubber suggest that it would make an excellent use as coating material (such as protection of hoses, gaskets, round belts, rollers, gears, pipes, etc.), insulation, shoe soles and solid tires and so on.

1.1.4 Modified Plastics

It is not usual to use a pure resin in a plastic product. The properties and performance of a plastic are usually improved by various types of additives such as fillers, plasticizers, lubricants, stabilizers, modifiers, and coloring agents etc. Modified plastics refers to the performance of plastics changed, or give it a new function accordance to people's expectations to obtain a new material by the method of physical, chemical, or the combine of the both. Modification occurs in plastics processing, such as filling, blending and enhancing modification, sometimes it occurs in the polymerization process.

Currently modified plastics mainly used in the area of automotive, household appliance, agriculture, construction, electrical and electronics, light industrial, military and so on. The growth rate of domestic demand of modified plastics will remain around 10%～20%, driven by the rapid growth of these industries in the next 5～10 years.

1.2　Coloring of Plastics

1.2.1　The Meaning of Plastic Coloring

Coloring is an indispensable part in plastics industry. The main purpose is to beautify the plastic products. The purpose and effect are shown in Table 1.3.

Table 1.3 The Purpose and Effect of Plastics Coloring

Purpose and effect	Examples
Increase the value of goods	Household appliances, automotive interior and exterior decorations, toys, cosmetics bottles, imitation leather
Distinguish colors	Wire coating, packaging materials
Protection	Packaging film, steel coating, food inside packaging
Improve light fastness and weather resistance	Cable sheathing, pressure pipes
Improve the dielectric properties	Using carbon black to improve conductive
Others	Energy saving materials to promote plant growth

(1) Increase the value of goods

Only when the goods look pretty enough, people willing to know its intrinsic quality seriously. Color is the most important external characteristics of products, which determines the fate of the product in the consumer's mind. The low cost and high added value of products competitiveness bring by color is more surprising.

Because of the rising trend of product homogeneity, personalized marketing has become the dominant. Consumer's pursuit is not only the function, but the personality and fashion they can fell from the product. A product with beautiful color makes people feel a kind of artistic enjoyment, which need to rely on coloring technology.

(2) Identification and marking

It is easy to be distinguished after plastic product being colored, which has practical value. It is very convenience to examine and repair after the communication cable being marked by color, which in line with the munsell ten color standard in chromatography. Colors also can be used in safety signs, for example, highway reflective signs, traffic police's uniform with fluorescent yellow, sanitation workers use bold orange etc.

Logo is a symbol of a company, so does the Logo's color. Red represents passion, vitality and movement, Nike (an international brands in sporting goods field) chose it to match their culture, fusion and penetration on each other and make Nike in sports competition charming.

(3) Protection

Colors have been used in a wide variety of plastic products, such as containers (food, medicine, cosmetics and detergent), plastic film, trough, vehicle inside/outside decorations, forms etc, to play a covert protective role.

(4) Improve light fastness and weather resistance

Carbon black plays a very prominent effect to improve the light fastness and weather resistance of polyolefin plastics. As we all known, carbon black is an inexpensive and efficient light-shield agent. A large number of scientific data prove that: 2.6%

carbon black with certain particle size (27nm or less) and spread evenly in the polymer, could make the service life of polymer above 50 years. In addition, add 1.5 parts of carbon black in 100 parts of LDPE matrix, the breaking elongation of the film was still up to 190% after a year and half of outdoor exposure, according to the literature. However, the elongation of pure LDPE film is almost zero under the same test conditions.

(5) Change the properties of material

Plastic can have conductive performance after conductive carbon black added, which has been widely used in the internal and external shielding materials of electric cables. Moreover, clothing could be antistatic made by composite fibers, which contain conductive carbon black. Composite fibers are used for computers and electronic elements factories extensively as labor protective clothing.

(6) Give plastics some special functions

Promote the growth of crops

The color agricultural film could make the light through selectively. Colorful agricultural film can increase the surface temperature and promote the growth of plants, and plays an optimistic role in weeding, avoiding insects and seedling setc.

Energy conservation and environmental protection

A variety of vehicles and buildings could absorb light and convert it to heat energy when exposing in sunlight, which rised the surface temperature of plastics and lead to aging. Plastics would have excellent infrared reflectivity, and not only don't waste energy but reduce the surface temperature effectively, when colored by metal oxide pigments, which preven theat transferring from surface to the interior. From the source to achieve energy conservation and cooling.So it is widely used in the building materials.

(7) Military purposes

It is significant to apply some types of the metal oxide mixed pigments to military uniform and shelter of arms in the defense and military fields because of their imitation function of chlorophyll.

1.2.2 Plastic Colorants

Colorant is a subject that can modify the color of an object or color an object that is colorless. Colorant presents color by selectively absorbing and reflecting certain lights in the colored light waves. By convention, colorants can be classified as dyes and pigments.

Dye is an organic synthesis chemical that is soluble in most solvents and dyed media. Dyes have good transparency, strong coloring capability and small density. Pigment, unlike dye, does not dissolve in water, oil or resin. It usually disperses in plastics to make products take on various colors. In addition, as it has no affinity with materials to be colored, pigment can only disperse uniformly in plastics by mechanical method to achieve ideal coloring performance.

Despite of the distinct differences between dye and pigment, it is still impossible to give them precise definitions. The classification of pigments in colorant is not clear. Few organic pigments can dissolve in certain polymers. For instance, Pigment Red 254 (DPP Red), which offers bright red, is insoluble in most polymers; while it is soluble in polycarbonate where the temperature is above 330 ℃ (626 ℉) and it can offers fluorescent yellow just like dye. Although pigment and dye have different concepts, they can be interchangeable under specific circumstances. For example, certain anthraquinone vat dyes that are insoluble can be used as pigments after pigmentation. A typical case is that Vat Blue 4 turns into Pigment Blue 60 through pigmentation.

Dyes are organic compounds in structure while pigments can be either organic or inorganic. They differ from each other in properties. Inorganic pigments are metal oxides and metal salts, and inorganic pigments of the same class have a lot in common. Inorganic pigment is insoluble in common solvents and has stronger heat resistance than organic pigment. On the other hand, organic pigment generally has better coloring capability.

To sum up, plastic colorants refer to inorganic pigments, organic pigments and solvent dyes. Table 1.4 demonstrates their properties in plastic coloring.

Table 1.4 Comparison of colorants' properties in plastic coloring

Property index	Organic pigment	Inorganic pigment	Solvent dye
Chromatography range	Wide	Narrow	Wide
Hue	Bright	Not bright	Bright
Tinting strength	Strong	Weak	Very strong
Light and weather resistance	Medium—High	High	Low
Migration resistance	Medium—High	High	Low
Dispersive property	Medium—Bad	Medium—Good	N/A
Heat resistance	Medium—High	High	Medium—High

It can be clearly seen from Table 1.4 that organic pigment has irreplaceable properties, such as bright color, that are superior to other pigments. On the other side, inorganic pigment has narrow chromatography range and weak coloring capability, and is restricted by environment protection because of heavy metal. And solvent dye has low migration resistance, it is not used in polyolefin plastic and can be only applied in amorphous polymer (e.g. PS, ABS).

With the diversified development of plastic products and the fast advancement of process technology, people put more emphasis on the color and application performance, usability and safety of these products. Colorant thus plays an important role. It can withstand various technological conditions in plastic processing and make plastic products present better application performance. In this regard, good colorants not only provide plastics colors, but also endow plastics with excellent process performance and usability. Colorant gathers popularity only when it plays an essential role in production and value adding to terminal products, which shown in Figure 1.3.

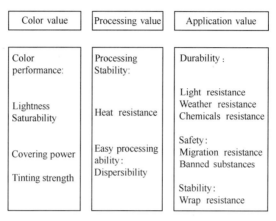

Figure 1.3　Values of Colorants in Plastics

This book will occupy three chapters in elaborating various performances of inorganic pigments, organic pigments and solvent dyes in plastics. All the colorants mentioned in this book are listed in the common dye index which has not indicated colorants' chemical and application performances in plastics. For this reason, the author describes each colorant's performances in detail based on the published data. It is important for colorists to know the chemical performances of colorants so as to give direct assessment on the properties. However, understanding processing and application performances of colorants is very important for colorists to meet the market's demand by freely choosing colorants based on plastic products' molding technology and functions.

1.2.3　Types and Properties of Plastic Colorant

The coloring of plastics aimed at reducing costs. In addition, the different choice of color preparations is depended on the type of plastic, molding method and the requirements of application. The representative color preparations are described as below.

(1) Powdery pigment

Powderypigment is the original state of the color preparations. Due to the large specific surface area of powder and the agglomerate of pigments, it is dusty, automatic metering problematic and incomplete dispersed during the coloring process. It will not destroy the properties of colored resin and the cost is low. Currently, the usage amount of powdery pigment is 12% of the all plastics colorants. Despite the shortcomings, it is easy to process and easy to formulate a variety of colors, especially for small batch production of a variety of colors. Unlike masterbatch coloring it is not limited by the types of colored resin and colors, so widely used in many small and medium sized plastic processing factories. It must be accurate when metering powder pigment, otherwise it will bring a chromaticaberration. Therefore, to minimize the error as much as possible, a higher precision electronic balance should be used.

(2) Sand-like pigment

It is a kind of pigment that could improve the dusty and automatic metering problematic of powdery pigment. The state is sand-like, simple composition. It usually contains dispersing agent and pigment concentration is as high as 30%~70%. The advantages of sandy-like pigments are: excellent distribution, good automatic metering, dust-free, favorable processing adaptability and relatively low cost. Due to the high pigment concentration and small destruction to the properties of colored resin, this color preparation is mainly used in coloring of PVC, EVA, rubber and so on.

(3) Masterbatch

Masterbatch, the granulated form, prepared by attaching the extraordinary amount of pigment to resin carrier evenly. The concentration of organic pigments is usually 20%~40% and inorganic pigments is usually 50%~80% in the masterbatch. The pigments are dispersed completely in the carrier during the extrusion, so it has excellent distribution for coloring plastics. The advantages, such as: dust-free, less pollution, environmentally friendly, easily mixing, precisely metering, high production efficiency, excellent product performance, reduce the quantity of materials during the transformation, extend the shelf life of stored materials,simplify the production process and easy to operate, etc.

The proportion of masterbatch and colored resin is generally about 1:50 in the polymer coloring. Currently it is used a lot in films, electric cables, sheets, pipes and synthetic fibers. It has become the mainstream method for coloring plastics, and the usage amount is above 60%. Nevertheless, masterbatch is relatively expensive of all color preparations for the complex process and high cost. Another disadvantage is there are some degree of incompatibility when polymers blended with each other. Therefore the carrier of colorant and the polymer being colored should be the same type.

(4) Liquid colorants

The concentration of dye and pigment is 20%~60% in liquid colorants. The feature of this colorants is measurability with small syringe, which will reduce the equipment investment compared with the automatic metering mixing machine. This applied to several types of plastics, such as PET. Occupied only 2% of the whole color preparations. The constitution of a liquid colorants is quite similar to masterbatch.

One advantage of a liquid colorants is the very good wetting of each polymer pellet prior to extrusion, which is especially important for pastel or transparent color shades. This is the most important advantage of a liquid colorants in comparison to a mastebatch. Further advantages are the utilization of the full tinting strength because of the complete dispersed colorants, a good metering behavior, dust-free handling, and suitable costs of product. However, a liquid colorants has a heavier damage on a series of properties of colored plastic in comparison to other color preparations. And it is mediocre to change the types of colors. Further disadvantages are poor long-term storage stability, and a high precision requirement of the metering pumps.

Thermosets and elastomers are colored by blending of one of their liquid components with the liquid colorants.

(5) Panchromatic modified color material

Panchromatic modified color material is a colorful particulate that prepared by melting, mixing and extruding after blending the plastic resin, colorant and other modified additives. It can be used in variety of molding methods directly, especially in thermoplastics. Moreover, panchromatic modified color material has excellent distribution and stable quality, so it is widely used in coloring of household appliances and automotive vehicle components, which have strict coloring quality requirements. The cost of panchromatic modified color material is high due to the full amount of colorants used in coloring resin. It is an effective coloring method in the field of modified plastics, engineering plastics, composite resins, high-functional resins and so on.

On basis of Figure 1.4, different methods are suitable for different kinds of colored plastic. The characteristics and applications of color preparations are shown in Table 1.5.

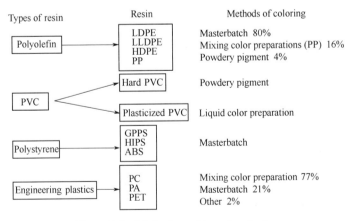

Figure 1.4   Methods used to color plastics

As shown in Figure 1.4, coloring thermoplastic resins, masterbatch is widely used, liquid and powdery colorants are used occasionally by special customers. However, for thermosetting resins only the liquid or powdery colorants is applicable.

Table 1.5 Characteristics and applications of color preparations

Characteristics	Powdery pigment	Sand-like pigment	Liquid colorants	Masterbatch	Panchromatic modified color material
Distribution	△~○	◎	◎	◎	◎
Homogenization	○	△~○	△~○	△~◎	◎
Dust	X	◎	◎	◎	◎
Pollution	X	X~△	○	◎	◎
Metering	△	△	◎	◎	NO
Molding processability	△~○	○	○	○	◎
Effect of properties	○	○	△~○	○	◎
Storage stability	○	△~○	△	○	◎
Cost of storage	○	○	○	○	X
Universality	○	△~○	△~○	△~○	X
Cost of coloring	◎	○	○	X~○	X
Dilution ratio (phr)	0.5~1	1~5	1~1.5	2~10	—
Shape	Powder	Sand-like	Liquid	Particle	Particle
Colored resins	PE PP PS ABS Other	PVC UP PU EP Other	PET PVC Other	PE PP PS ABS PVC Other	PE PP PS ABS PVC Other
Applications	Common molding pipe, film, etc.	Film, sheet, artificial leather, etc.	Common molding, etc.	Electric wire, film, laminated film, monofilament multifilament	Common molding, industrial components, etc.

◎ Excellent, ○ Good, △ General, X Bad.

Chapter 2
Basic Requirements of Color Preparations

2.1 Basic Requirements of Color Preparations

Now, the appearance of products has been an important element to attract consumers. So plastic colorants were asked to have a good color performance to fit the competitive market. In order to develop the commercial value of plastic products, the demands are higher and higher from the pure pursuit of beauty to both the application and security. In this chapter the technical index of the plastic coloring and the affecting factors are described.

2.1.1 Requirements of Colorant Based on Types of Plastic

Plastics can be subdivided based on their properties in thermoplastics and thermosets. The thermoplastic is developing rapidly for the simple molding process and high mechanical strength, such as polyvinyl chloride, polyethylene, polypropylene, polystyrene and so on. Thermoplastics are distributed into polymeric (no by-products generated during polymerization) and condensation (water generated during polymerization) according to the difference of polymerization methods. The classification and types of thermoplastics are shown in Figure 2.1.

Thermosetting plastics will be softened and plasticized into a certain shape after being heated, despite the cross-linking structure. However, it will be cured after adding a small amount of curing agent. In addition, thermosetting plastics, such as phenolic aldehyde resin, amino plastics, and epoxy resin etc., all have perfect heat resistance, less deformation and low cost.

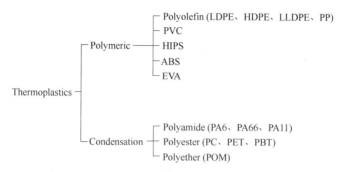

Figure 2.1 Classification and types of thermoplastics

The processing temperature ranges from 120~350℃ for different types of plastics. Among of them, the processing temperature of thermosetting resin is relatively low (but last longer). However, the rigid and plasticized PVC and EVA is 170~200℃ generally. Low density and high density polyethylene, polystyrene, is in the range of 200~260℃. Polypropylene, polyamide, ABS resin, polycarbonate is above 260℃. The processing temperature of some plastics is particularly high, such as fluorocarbons (Fluoropolymers), silicones. Therefore it is necessary to meet processing temperature of different kinds of resins for color preparations, and the processing temperatures of various plastics are shown in Figure 2.2.

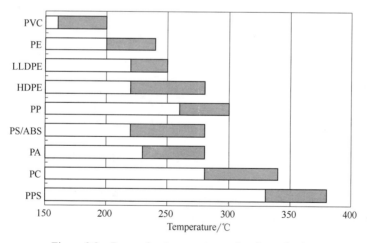

Figure 2.2 Processing temperatures of various plastics

In addition, symmetrical structure or rod-crystalline organic pigments act as a nucleating agent to promote crystallization during the molding of crystalline thermoplastic resin, which leads to high precision plastic products shrinkage ratio increased and dimensional unstable. However, for thermosetting plastics and non-crystalline plastics, all kinds of pigments have no effect on shrinkage ratio.

Migration occurs when the glass transition temperature of polyolefin plastic below room temperature in coloring with some organic pigments, whereas the phenomenon won't happen for engineering plastics whose glass transition temperature higher than room temperature, even use dyes.

The molding processing of variety of plastics for requirements of color preparations are shown in Table 2.1.

Table 2.1 Molding processing of variety of plastics for requirements of color preparations

Plastic	Contraction	Temperature of molding/℃	Requirements of colorant
Polyvinyl chloride	PVC	150~220	Migration caused by plasticizer, the relationship between stabilizer and resistance to weather, heat
Polyvinylidene chloride	PVDC	170~180	Influence of zinc and iron to the aging of PVDC
Polyethylene	PE	120~300	Migration and distribution of colorant, molding shrinkage, under 190~300℃
Polypropylene	PP	170~280	Migration and distribution of colorant, under 170~280℃
Ethylene-vinyl acetate	EVA	160~200	Migration, solvent resistance and distribution of colorants
Polystyrene	PS	190~260	Transparency, impact resistance, under 220~280℃
Acrylonitrile-butadiene-styrene	ABS	230~280	Impact resistance, under 250~300℃
Polyamide	PA	160~240	Under 250~300℃, reducing resistance of colorant
Polycarbonate	PC	350~400	Influence of water, pH and metal to the aging, under 250~300℃
Polyethylene terephthalate	PET	250~280	Water, under 250~280℃
Polyurethane (foaming)	PU	—	Reaction caused by pH, activity of reactant affected by metals, eliminate water
Urea formaldehyde resin	UF	150~180	Under 150~180℃
Unsaturated polyester	UP	—	Influence of catalyst to weather resistance, influence of colorants to cure

2.1.2 Requirements of Colorant Based on Plastic Molding Process

In the last 100 years' development of the plastics industry, innovation has improved the plastic molding processes and equipment greatly, which promotes automation, industrialization, large-scale manufacturing, low cost in the processing of plastic products and makes plastic products permeate into every corner of society quickly.

The purpose of the plastic molding processing is to give products the value of applications based on the inherent properties of the plastic and all the possible conditions. There are extrusion molding, injection molding, calendaring molding and compression molding. Plastic molding processing is to make the plastic melt after heating, then get through die or mold to obtain the desired shape, finally be cooled to the solid state. There are a lot of plastic molding methods, and the processing temperatures are different according to molding methods. The molding process of plastic and corresponding products are shown in Figure 2.3.

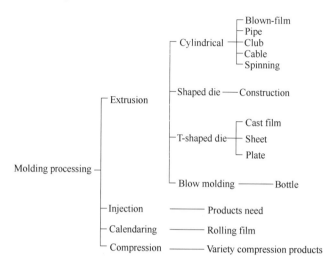

Figure 2.3　Molding process of plastic and corresponding products

For the same type of plastic, the processing technology is different, the temperature is different too. Take LDPE for example, the temperature of extrusion molding is 120~140℃, blow molding is 170~205℃, calendaring molding is up to 285~300℃, casting film is above 300℃. Therefore, the requirements of heat resistance of pigments are different for the same plastic based on the different molding processing. The temperatures of different molding processing of polyolefin resin are shown in Figure 2.4.

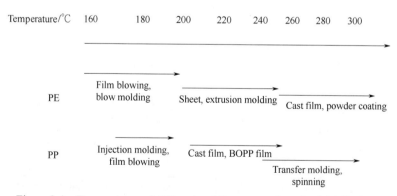

Figure 2.4　Temperatures of different molding processing of polyolefin resin

Plastic molding processing is different, the requirements of colorant dispersion are different too. For example, the requirements of colorant dispersion are relatively higher in the processing of spinning, cast film and communication cable, especially in the processing of superfine denier fiber. Poor dispersion of pigments will block the filter and lead to the fiber broken, which will affect regular production. However, the requirements of colorant dispersion are relatively lower in the compression molding and sheet molding processing.

2.1.3 Requirements of Colorant Based on the Application of Plastic

Plastic is a kind of widely used polymer materials, and can be found everywhere in our daily life. Such as the toiletries, dishes, transports stationeries, mattresses, communication tools, computers and stationery, mattress, and the televisions, washing machines, air conditioners and so on. Plastic is a material, hardness of metal, lightness of wood, transparency of the glass, corrosion resistance of ceramic, elasticity and toughness of rubber. It is therefore more widely used in the fields of aerospace, medical equipment, petrochemical, machinery manufacturing, national defense and construction. With excellent performance, plastics replace the traditional steel, wood, paper and cotton gradually to become indispensable in our life, which will be seen in Table 2.2.

Table 2.2 Application of different kinds of plastic products

Plastic	Contraction	Application
Polyvinyl chloride	PVC	Toy, bar, pipe, plate, oil-conveying pipe, wire insulation, sealing element
Low density polyethylene	LDPE	Packing bag, cable
High density polyethylene	HDPE	Packing materials, construction materials, bucket, toy
Ethylene-vinyl acetate	EVA	Shoe sole, film, sheet, commodity
Chlorinated polyethylene	CPE	Construction materials, pipe, wire insulation, packing materials
Polypropylene	PP	Non-woven, packing bag, filament, commodity, toy
Chlorinated polypropylene	PPC	Commodity, electrical appliance
Polystyrene	PS	Lampshade, plastic shell and cover of instruments, toy
High impact polystyrene	HIPS	Commodity, electrical parts, toy
Acrylonitrile-butadiene-styrene	ABS	Shell of electrical appliances, commodity, high-class toy, sports equipment
Acrylonitrile-styrene	AS(SAN)	Daily transparent ware, transparent household electrical appliances
Acrylic acid-styrene-acrylonitrile	ASA	Outdoor furniture, shell of car side-view mirror
Polyformaldehyde	POM	Good wearability, as mechanical gear and bearing

continued

Plastic	Contraction	Application
Polycarbonate	PC	Transparent parts with high impact, as high strength and impact resistance of component
Polytetrafluoroethylene	PTFE	High frequency electronic instruments, insulation parts of radar
Polyethylene terephthalate	PET	Bearing, chain, gear, tap
Polyamide	PA	Bearing, gear, oil pipe, container, commodity, motor vehicle, chemical, electrical installation
Polymethyl methacrylate	PMMA	Transparent decorative material, lampshade, windshield glass, shell of instrument
Phenol formaldehyde	PF	Silent gear, bearing, helmet, television, parts of communications equipment
Melamine formaldehyde	MF	Food, commodity, component of switch
Polyurethane	PU	Shoe sole, chair cushion, bed mattress, artificial leather

The requirements for the properties of plastic products are different based on different usage conditions. For example, plastic products that are long-term used in outdoor such as artificial turf, construction materials, advertising boxes, circulating boxes, rolling shutters, profiles and plastic auto parts are required colorants to possess the excellent weather resistance and light resistance. Moreover, in these products, color is required to remain stable on exposure full sunlight at least ten years. In addition, colorants should be acid, alkali and solvent resistance for the household cleaning products for the health. The mechanical strength of household appliances should not be reduced after coloring. Colorants are required to have a small impact on the deformation of products in the use of bottle caps and circulating boxes, otherwise it will affect the seal. What is more, the migrations of colorants of all plastic products must not occure during using.

In order to ensure the safety, environmental protection and health, plastic products must obey the law and regulatory of every country and region (especially toys, chemical fiber textile materials, electronic products, food containers, food-contacted materials and automotive materials). Moreover, it is the most important to control the amount of chemicals (colorants) in plastics products. The color preparations should be met the safety requirements of product.

2.1.4 Basic Requirements of the Plastic Colorant

Coloring plastics is a systematic project, as shown in Figure 2.5. Colored resins, color recipe, processing and application all make various demands on the color preparations.

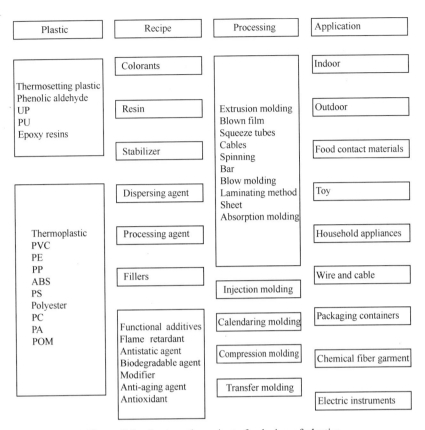

Figure 2.5　Systematic project of coloring of plastics

In this system, it is not enough to color the plastic. Not only do the colorants maintain stable in the molding processing, but also have good application performance. In summary, the colorants should have following properties: color properties, thermal stability, light resistance, weather resistance, distribution, migration fastness, shrinkage and warpage, acid, alkali, solvent, chemical resistance and safety.

The performance of colorants is not only closely related to its chemical structure, but also the crystal structure, size and distribution of pigments particles, concentration, working conditions, the types of colored plastic and additives. Therefore, coloring plastic is really a systematic project.

In this chapter the relationship between requirements of plastic colorants and the chemical structure, physical properties will be described in detail. The purpose is to help the readers to understand the performance indicators of colorants comprehensively. In addition, it is beneficial to choose the appropriate colorants and add the value of plastic products.

2.2 Color Performance

The color characteristics of colorants include relative tinting strength, saturation, lightness, transparency (or hiding power) and dichroic optical phenomena. Plastic colorists often have an insufficient understanding of the importance of the colorant optical properties. It is the optical properties that determine the success or failure of the color matching of plastics.

2.2.1 Tinting Strength

Tinting strength is given by a measure of the color depth of the coloring matters. Tinting strength refers to the grams of colorant when achieving the color standard depth (SD) with containing 5% TiO_2 of polyvinylchloride (PVC) or 1% TiO_2 of polyolefin (PO) per kilogram in the plastics.

The maximum absorption wavelength of colorants determines the color, and the absorptive capacity in the maximum absorption wavelength determines tinting strength.

The main factors influencing tinting strength are chemical and crystal structures. Inorganic pigments such as ultramarine, chromium, cadmium, iron oxide have low tinting strength, however, most of the organic pigments and dyes have high tinting strength.

In addition to the structures, the higher saturation, the higher tinting strength, the higher lightness, the lower tinting strength. It is associated with the components, materials and conditions of coloring matters.

Tinting strength is an important property of the colorant and closely related to the costs. The lower standard depth, the higher tinting strength, whereas the lower tinting strength. Generally, the level of tinting strength is judged by 1/3 standard depth (1/3 SD) of colorants.

Tinting strength and color matching have an extremely important significance. For example, the need of heavy (deep) shade should choose high tinting strength varieties (organic pigments), when preparing with light shade need to select low tinting strength varieties (inorganic pigments). When the lack of a kind of pigment or the price is expensive, other pigments can be used to instead of the same color. Because of the different tinting strength, the amount of pigments will different to achieve the same shade when instead.

2.2.2 Saturation, Lightness, Hiding Power

Color chromaticity coordinate parameters (tone, saturation, lightness) are the reference of the value of colorants (Figure 2.6). When the color phase diagram is from yellow to red, violet to blue anticlockwise, the color is from pale to dark, the saturation is from high to low.

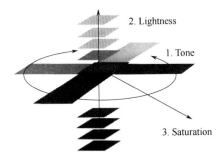

Figure 2.6   Chromaticity coordinate and phase diagram of colorants

In saturation coordinate, the colorants located farther from coordinate origin would have higher saturation and always blended with other colorants to cover the colorants with low saturation near the coordinate origin. So the higher the saturation, the greater the value, and the more widely applications.

For the same chemical structure of colorants, the saturation increases and the lightness decreases with the tinting strength increasing, and due to the various colorants the color phase diagram changes in different.

The hiding power or transparency of the colorants is closely related to the tinting strength, generally, inorganic pigments are high hiding power while dyes are transparent for the solubility in the resin. The hiding power of colorants is an important performance for most plastic coloring products, and it depends not only on hiding power, but also on the effective concentration, material and thickness of the colorants. Therefore, the higher of the colorants hiding power, the more value of applications.

If the colorant is high saturation, high hiding power and high tinting strength, it will have an extremely high commercial value.

2.2.3 Dichroism

Dichroism is a property of the pure tone varies with the concentration of colorants or the thickness of products when transparent colorants for the coloring of plastics. Dichroism is the inherent characteristic of colorants according to the shape variation of transmission curves. Owing to the asymmetry of spectral transmission curve, the dichroism of yellow, orange, red and violet colorants is followed by seriousness. In contrast, the spectral transmittance curve of blue and green colorants tends to symmetry, so the dichroism is minimal or inexistence (Figure 2.7).

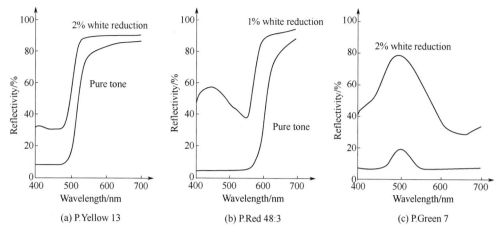

Figure 2.7 Spectral reflection curve of pure tone and dilution

Dichroism will cause difficulties in color matching process. In color matching of transparent plastics, dichroism phenomenon is occurred frequently, when the concentration of colorants changes the color of products may occur different degrees of changes, and will cause changes in tone. These cases will also happen in the color matching process of translucent or even opaque plastics.

Owing to the solubility in solvents, solvent dyestuffs are commonly used in the coloring of transparent hard plastics. Therefore, the dichroism test of solvent dyestuffs may use a simple method. First it should prepare some different concentration solution, then observe their tone changes, and you can know the color changes of dyes in plastics.

2.3 Thermal Resistance

Thermal resistance refers to the plastic colorants have no obvious changes in shade, tinting strength and property within a certain processing temperature and time.

Coloring of plastics has a heating process in most plastic molding which is different from inks and coatings. The colorant often occurs thermal decomposition, color changes in plastic molding and these phenomena also affect light fastness and migration fastness. Therefore, thermal resistance is a very important index in coloring of plastics.

In the early development of the plastic industry processing temperature above 200℃ is rare, but now even 300℃ or higher is common. Various plastic processing temperature ranges are different from each other. Actually requiring thermal resistance of all colorants to reach 300℃ is meaningless. Generally thermal resistance of inorganic pigments is excellent but organic pigments and dyes are at different levels. The most widely used organic pigments that can remain stable at high temperature are phtalocyanine green, phtalocyanine blue, carbazole violet, quinacridone red, isoindolinone yellow and perylene red, however, the thermal resistance of classic azo pigments is much lower. So it is particularly important to select appropriate colorants by thermal resistance index.

2.3.1 Definition

Most manufacturers of colorants follow EU standard method EN BS 12877-2 to check the thermal resistance of each colorant in different types of plastics and provide customers with references. This method defines as follows. The test starts at the processing temperature 200℃ and the standard color plate of colorant at a concentration is injected by injection molding machine. The temperature will increased step by step and at each step there is an increase of 20℃. The residence time is always 5 minutes. By definition the temperature that gives a difference in color of △E=3, is the thermal resistance of the colorant in the type of tested concentration. In Figure 2.8 the thermal resistance test curve of a colorant is shown schematically, owing to △E=3 between injected color plate at 260℃ and standard color plate at 200℃, its thermal resistance is 260℃.

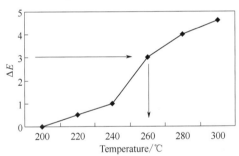

Figure 2.8 Schematic introduction of colorant thermal resistance

2.3.2 Concentration

Currently many manufacturers of colorants provide customers with thermal resistance index refers to 1/3 standard depth of shade, which is quite different from colorant thermal (heat) resistance at all concentrations. It is well known that the thermal (heat) resistance of a colorant declines with decreasing concentration, but there is no general rule that can prove it reach to what extent.

Figure 2.9 is thermal (heat) resistance of various chemical structures of yellow pigments at different concentrations. As can be seen from the figure some pigments (such as yellow metal lake) show no decline of thermal (heat) resistance with decreasing concentration. In Table 2.3 various chemical structures of yellow pigments are listed.

Table 2.3 List of various chemical structures of Yellow pigments

Color index number	Chemical structure	Color index number	Chemical structure
P. Yellow 191	Azo metal lake	P. Yellow 181	Benzimidazolone
P. Yellow 183	Azo metal lake	P. Yellow 162	Azo metal lake
P. Yellow 110	Isoindolinone		

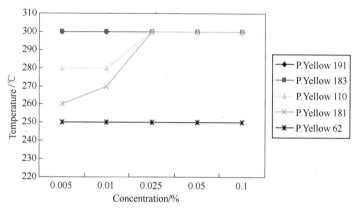

Figure 2.9 Thermal resistance of various chemical structures of yellow pigments at different concentrations

Thermal resistance datas of colorants at different concentrations are very important for the daily work of a plastic colorist. A very low concentration of colorants is not only required for pastel color shades but also when it is necessary to match a special nuance in the shade. Pigments can be used for applications when they show no or just a minor decline of thermal resistance with decreasing concentration. Plastic colorist should find appropriate colorants under the process conditions and provide a basis for the regulation of production process by thermal resistance test. A relatively inexpensive and suitable for production colorant can be chosen by colorist, which will reduce the cost and enable enterprise to stay ahead in the fierce market competition.

Currently quantifying study of thermal (heat) resistance has two series which is either 1%, 0.1%, 0.01% and 0.001% or 1%, 0.1%, 0.05%, 0.025% and 0.001%. The test results supplied as a graph to customers.

2.3.3 Chemical Structure, Crystal Modification, Particle Size

(1) Thermal resistance and chemical structure of colorant

Generally inorganic pigments are metal oxides or metal salts and they are reaction products of high-temperature calcination, for example, up to 700 ℃. The thermal resistance of inorganic pigments is far beyond the temperature of each plastic molding. Thermal resistance of organic pigments and solvent dyes has a great relationship with their chemical structure, for instance, molecular structure of pigments directly determines its color and application, and various molecule substituents of pigments directly affect thermal (heat) stability and decomposition reactions in a certain temperature. For example, the chemical structures of organic pigments are divided into monoazo pigments, azo pigment lakes, disazo condensation pigments, phthalocyanine pigments, quinacridone pigments, dioxazine pigments, isoindolinone pigments, diketopyrrolo-pyrrolo pigments (DPP), indanthrone pigments, and so forth. Pigments of different chemical structure also have different heat resistance. Thermal (heat) resistance of different chemical structures of pigments in HDPE is listed in Table 2.4.

Table 2.4 Thermal resistance of different chemical structures of pigments in HDPE

Color index number	Chemical structure	Thermal resistance Temperature/°C	
		Pure tone 0.1%	White reduction 1:10
P. Yellow 13	Disazo	200[①]	200
P. Yellow 62	Azo lake	250	260
P. Orange 61	Isoindolinone	300	300
P. Red 114	Disazo condensation	300	300
P. Red 254	Diketopyrrolo-pyrrolo (DPP)	300	300
P. Red 122	Quinacridone	300	300
P. Violet 37	Dioxazine	270	260
P. Blue 15:3	Phthalocyanine	300	300

① P. Yellow 13 is dichloro benzidine pigments and decomposes harmful substances with the processing temperature above 200°C, therefore, the operating temperature is less than 200°C.

Changing the chemical structures of pigments is the most important method to improve thermal resistance of organic pigments. Typically adopting the following methods: increasing the molecular weight, introducing halogen atoms, introducing polar substituents to a condensed ring structure, and introducing metal atoms.

(2) Thermal resistance and crystal modification of colorant

Some pigments have multiple crystalline phases, that is to say, the chemical composition of same crystal cells may be arranged in different ways in crystal lattice, and pigments with various crystal modifications have different hues. For example, pigment violet 19 of its β-modification is violet, however, the γ-modification is bluish red. Thermal resistance of pigments are also affected by crystal modification. Pigment blue 15 is unstabilized, nonsolvent resistance and its thermal (heat) resistance is only 200°C, however, the pigment blue 15:3 with a stabilized β-modification is up to 300°C. Thermal resistance of phthalocyanine blue with various crystal modifications in plastics is listed in Table 2.5.

Table 2.5 Thermal resistance of phthalocyanine blue with various crystal modifications in plastics

Color index number	Crystal modification	Thermal resistance temperature (1/3 standard depth of shade)/℃
Phthalocyanine Blue 15	Unstabilized	200
Phthalocyanine Blue 15:1	Stabilized α	300
Phthalocyanine Blue 15:3	Stabilized β	300

(3) Thermal resistance and particle size of colorant

The original particle size of organic pigments also has a great influence on the thermal resistance. In general, pigments with small particle size show large specific surface area, high tinting strength, poor thermal resistance and dispersion. On the contrary, pigments with large particle size show small specific surface area, low tinting strength, excellent thermal resistance and dispersion. The influence of the particle size of P. Red 254 to thermal resistance is shown in Table 2.6.

Table 2.6 Thermal resistance of P. Red 254 with various particle sizes

Project	Cromophtal Red BOC	Cromophtal Red 2030	Cromophtal Red BTR
Specific surface area/(m²/g)	19.9	26.7	93.8
Particle size	Large	Middle	Small
Pigment content 0.01%	280	300	280
Pigment content 0.05%	300	300	280
Pigment content 1%	300	300	280
Pigment content 0.01%、TiO$_2$ 1%	280	300	280
Pigment content 0.05%、TiO$_2$ 1%	300	300	290
Pigment content 1%、TiO$_2$ 1%	300	300	300

2.3.4 Resins and Additives

(1) Thermal resistance and the types and grades of resins

The thermal resistance of perylene Pigment Red 149 is various in different resins, such as up to 310℃ in PC and only 250℃ in ABS (Table 2.7).

Table 2.7 Thermal resistance of perylene Pigment Red 149 in different resins

Resins	Thermal resistance temperature /℃
Polycarbonate (PC)	310
Polyolefin (PO)	300
Polystyrene (PS), polymethyl methacrylate (PMMA), polyethylene terephthalate (PET)	280
Acrylonitrile-butadiene-styrene copolymers (ABS)	250

Owing to the thermal degradation, many resins show discoloration and yellowing by heating which also affect the shade of plastic products, and this discoloration should be taken into account especially when determining the maximum limit of the thermal resistance of the colorant. In order to achieve color stability antioxidants must be added to resins.

(2) Thermal (heat) resistance and additives

Plastics to be colored is different from the standard type of polymer used in the test. In fact, in addition to plastic and colorant, the plastic may also contain filler, plasticizer, dispersing agent, antioxidant, stabilizer and flame-retardant. Chemically all of these components are not inert, each of them may influence the thermal resistance of the colorant and plastic more or less. For example, titanium dioxide is often used to adjust the hue or increase hiding power, thermal resistance of some pigments will change when added titanium dioxide. The thermal resistance of pigment violet 37 with sharp decrease is shown in Figure 2.10.

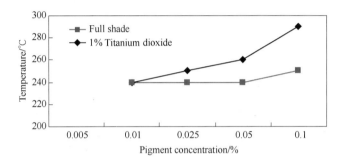

Figure 2.10   Thermal resistance of Pigment Violet 37

2.3.5 Applications

(1) Thermal resistance related to heating time

Any process of colorant discoloration by heating is actually a function of temperature multiplied by the residence time. In the standard test method (EN BS 12877-2), the residence time is fixed at 5 minutes, a time that is much longer than normal plastic processing time. Therefore, experience shows that colorants can be processed at a slightly higher temperature than the lists of thermal resistance, provided that the residence time is very short. During the injection molding process, each injection machine requires a minimum size and volume of screw, but for the molding of very tiny plastic products, it is inevitable to use the injection machine that is really too big in comparison to the size of products. For technical and economic reasons the size of plastic processing machine can't be reduced indefinitely, so the residence time of the plastic melt in the injection molding machine may be above 5 minutes with the result that discoloration of the colorant start at a lower temperature than the lists of thermal resistance. Generally injection molding materials are used repeatedly and particularly deserve attention because each heating will cause an increase in thermal damage of colorants in injection molding.

Another reason for discoloration is the use of a hot runner in injection molding process. The residence time in the hot runner should be added to the residence time in the injection molding machine. The total residence time is generally short enough to avoid any discoloration. Therefore this parameter should be considered carefully. Unfavorable designs in the construction of a hot runner such as small dimension of the hot runner nozzle or other parts of the hot runner are the main reason for discoloration. And the resulting frictional heat can't be calculated or controlled and is normally high enough to cause thermal damage of the organic colorants and plastic materials.

(2) Colorants heating in color masterbatch process

Thermal damage of colorants is possible during the production of masterbatch. When using a twin-screw extruder for production of masterbatch, in order to improve the dispersion increasing aspect ratio and mixing section should be chosen, and it will increase the pigments residence time in the screw and the risk of thermal decomposition. Therefore, it should take preventive measures for thermal sensitive pigments to avoid thermal damage.

Also in the high-speed twin-screw extrusion process, the temperature of materials increased significantly by high shear forces to achieve the partial or complete dissolution temperature of pigments, causing the color change or even complete decoloration, but also the serious consequences of pigment migration and blooming.

2.4 Dispersion

Dispersion, in coloring process of plastics, refers to the ability that pigments are equally distributed in plastics. However, the dispersion here means the ability of reducing the aggregate dimension to the ideal size after the pigments are wetted. Dyes are defined as the colorants that can be completely dissolved in plastics at the processing temperature. Therefore, solvent dyestuffs have no concept of dispersion in the coloring of plastics in principle. Contrary to dyes, pigments are highly dispersed in the form of particles in the coloring of plastics, so there always be original crystalline state. And that is why the crystalline particle states of pigments have a great relationship with the dispersion.

The dispersion of pigments not only affects the tinting strength and shade (Table 2.8 and Table 2.9), but also has a direct impact on the optical properties of plastic products.

Table 2.8 Relationship between particle size and tinting strength for ultramarine pigments
(By default, the tinting strength of original unprocessed ultramarine is 100%)

A variety of particle size distribution					Tinting strength/%
20~10μm	10~5μm	5~2.5μm	2.5~1.5μm	<1.5μm	
26	62	12	0	0	35
0	8	77	12	3	110
0	3	32	52	13	145
0	3	1	3	93	190

Table 2.9 Dispersibility of pigments effect on shade

Color	Excellent pigments dispersion (small particle size)	Poor pigments dispersion (large particle size)
White	Blue light	Yellow light
Yellow	Green light	Red light
Red	Yellow light	Blue light
Blue	Green light	Red light
Green	Yellow light	Blue light
Black	Blue light	Yellow light

The poor dispersion and uneven coloring may cause streaks and specks that not only affect the appearance but also seriously impact on mechanical properties of products. More importantly, the dispersion of pigments affects application value in plastic processing, especially in the chemical fiber spinning (fiber) and ultrathin film. Compared to the ink and coating processing, the pigments are subjected to a much smaller shear force in the melt extrusion processing. And the requirements for dispersion in ultrathin film and fiber spinning are far higher than that in ink and coating. Therefore the dispersion of pigments is a particularly important indicator in plastic applications.

2.4.1 Impact on Dispersion of Pigment Surface Properties

The dispersion of pigments is related to surface properties and the surface properties of organic pigments are associated with molecular packing and arrangement, and different crystal structures show different surface properties.

According to the similar dissolve mutually theory, if the surface is non-polar, the pigments are easy to disperse in the non-polar plastics, whereas if the surface is polar, the pigments are very easy to disperse in the waterborne coatings and high polarity jet inks.

2.4.2 Impact on Dispersion of Particle Size and Distribution

If the organic pigments have a same structure, the dispersion also has a great relationship with the original particle size. With the original particle size reducing, the transparency increases while the dispersion decreases. More specifically, the original particle size of pigments impact on dispersion is that small particles of pigments are filled in the space of the larger particles and cause the arrangement of aggregates more closely. Therefore, the wetting agents (polymers) can't penetrate and the pigment particles can't be sufficiently wetted and covered, then the shear stress can't reach the surface of pigments in the dispersion process. As a result, the aggregates abundantly remains in the final products.

The dispersion of pigments is related to the particle distribution, and the pigments are easy to disperse during spinning coloring process with a narrow size distribution of particles.

2.5 Migration

Migration means that the colorants move from the inside to the surface of the plastics or migrate from one plastic to another through the interface. There are four forms in coloring of plastic.

Bleeding When the colored plastics are plying-up with the white or light shade plastics, the pigments migrate from the colored products to the others.

Plate out Polluting the molds and rollers in plastic molding process.

Blooming The surface of the colored plastic products will be hazy and whitish as time goes on, and the migratory colorants can be erased.

Bronzing The surface of plastic products shows an obvious metallic luster of the colorants.

For plastics, the migration of colorants will greatly affect the applications and can pollute other products seriously. If the migration is serious, it may cause huge economic losses with plenty of products recalled.

2.5.1 The Main Reason for Migration

(1) The supersaturation of pigments in the coloring of plastics

In the case of heating for plastic molding process, a mixture of plastics and the additives become true solution, and the solubility of colorants in the system increase to form a supersaturated state, then the migration will occur after the cooling process of molding.

(2) The molecular motion of pigments in the coloring of plastics

Thermoplastics can be divided into two types: crystalline and amorphous. During the solidification of amorphous plastics, there is no crystal nucleus or grain growth process, only the freedom macromolecular chains are "frozen" (Figure 2.11).

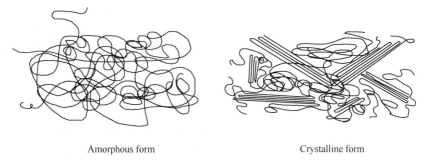

Amorphous form   Crystalline form

Figure 2.11   Different types of thermoplastics

Owing to the rigidity of molecular chains and the tightness between molecules, the network structure of amorphous plastics prevents the colorants movement at room temperature, even if the solvent dyestuffs would not migrate. However, because of the loose molecular structure, the crystalline plastics are prone to migration.

2.5.2 Impact of Chemical Structure and Concentration on Migration

(1) Migration and molecular weight

The migration is related to the chemical structure of colorants, and it would be improved with increasing the molecular weight of the pigments. For example, because the molecular weight of disazo condensation pigments macromolecules is double that of monoazo lake pigments, the migration is obviously improved.

(2) Migration and molecular polarity

The molecular structure of the pigments should avoid introducing hydrophilic substituents, but in can introduce the insoluble carbon amide groups to improve the migration. For example, compared with Pigment Red 170, the migration resistance of Pigment Red 187 is improved significantly. Pigment Red 48∶4 (sulfonic acid: alkaline earth metal =48∶4) form the insoluble polar salts after the lake and the migration is obviously improved.

(3) Migration and concentration

In the coloring of polyolefin plastics, the severity of the occurrence of bleeding and blooming is proportional to the concentration of the colorants. This is because in the heating process of pigments, the colorants of high concentration are partially dissolved in plastics and formed a supersaturated state during cooling, thus it is easy to crystallize on the surface of plastics and spread to other mediums which is in contact with it.

2.5.3 Application of Migration Indicators

(1) Migration and additives

The adhesion between pigment molecules and resins varies with the addition of the additives (e.g., plasticizers and other additives). When the glass transition temperature of PVC (rigid) up to 80 ℃, the migration does not occur with adding dyes. But after adding polar plasticizers, the molecular distances increase and the structures become more loose, thus the interaction between polymer chains reduces. Then the rate of pigment migration increases, and with the increasing dosage of plasticizers, the migration would be more serious. Therefore, the selection of pigments should be serious in the coloring of plasticized PVC.

(2) Migration, density and molecular weight of polyolefin

For example, the migration of optical brightener OB-1 would be more serious in LDPE than HDPE, and it would be more less in polypropylene that has higher molecular weight.

(3) Migration and wettability

Staining refers to the colorant pollution for processing equipment (rollers and molds), which is a kind of migration. It is different from blooming because contaminants can be erased. Staining in PVC processing is directly related to wettability and the pigments which are completely wetted by PVC will not migrate to the surface along with exudates.

(4) Migration and processing temperature

The possibility of pigment migration will increase with the operative temperature increasing in process. Therefore, when the operative temperature is close to the thermal resistance temperature of pigments or the concentration is up to saturation, the possibility of migration should be noted.

(5) Migration and glass transition temperature

The migration of colorants is closely related to the glass transition temperature of plastics, and in the coloring of plastics the pigments are prone to migration when the glass transition temperature below normal temperature, such as polyethylene, polypropylene. When pigments applied to plastics whose glass transition temperature above normal temperature such as polystyrene, polyester, the migration would occur only above the glass transition temperature of plastics. Various glass transition temperatures of plastics are shown in Table 5.4.

2.5.4 New Application Requirements for Migration in Consumer Goods Regulations

In recent years, the indicator of migration is not only used in color migration of plastics, but also appears on consumer goods regulations. Facts have proved that any substance migrating from a plastic article may be harmful to health. In order to avoid any hazard to consumers, many countries published series of rules and regulations for all plastic materials and containers that come in contact with food and beverages during processing and transportation. Such as American Code Federal Regulations 21CFR, in "Requirements for the colorants using in the polymers which come in contact with food" it is specified: the colorants can't migrate into the food.

The migration test method is published in the German food BGVV requirements by German Federal Bureau of Public Health, and it contains a definition of the test liquids to be used to simulate the different kinds of food. The test liquids include that distilled water, acetic acid (2% by weight), ethanol (10% by volume), coconut oil, coconut fat or peanut oil.

Children often put toys in the mouth and perspire during play, so the fastness to saliva and sweat is the main criterion for the migration test of the toys for children. In Germany the test is defined in DIN 53160, and also including the test liquids.

2.6 Light Fastness/Weather Resistance

The light fastness is defined by the ability to retain the initial color values on exposure to daylight for the pigment-polymer system. The changes of pigment in sunlight are mainly due to the coloring product is destroyed by ultraviolet rays and visible rays.

Weather resistance is defined as the color change of the pigment-polymer system with atmospheric impacts such as natural temperature and wetting caused by rain and dew under daylight.

The humidity and atmospheric composition also have effect on the pigment changes in the atmosphere except the sunlight. In addition, rapidly changing moisture and different temperatures obviously accelerate pigment changes and destruction. Generally stable humidity and temperature can slow the rate of destruction. Some gases in the atmosphere and the light will lead to pigment changes, such as oxygen and ozone. Oxygen is the root cause of product oxidation. Ozone destructed the chemical structure of the pigment which results in discoloration. The water-soluble acid compound which is contained in automobile exhaust and factory waste gases also can cause serious erosion of the color.

Weather resistance of pigment contains light fastness, however, light fastness does not cover weather resistance. Some kinds of pigments show good light fastness, but poor weather resistance, due to the climatic factors which have effects on pigment weathering resistance have humidity, atmospheric composition and time besides the sunlight and currently the humidity is the most important atmospheric parameter.

Light fastness index system is evaluated with 8 levels which 8th level is the best and 1st level is the worst.

Weather resistance index system is evaluated 8 levels which 8th level is the best and 1st level is the worst.

The importance of light fastness (weather resistance) depends on its application, sometimes is extremely important, such as outdoor architectural panels, billboards and taillights. The products are expected to keep stable at least a decade which are exposed to the sunlight.

2.6.1 Chemical Structure and Particle Size of Pigment

(1) Relationship between light fastness (weather resistance) and pigment chemical structure

Under the irradiation of light the color will change to a certain exetnt. Most inorganic pigments have very excellent light fastness (weather resistance), but only a few species will be dim because of its crystalline form or chemical composition changes.

Compared with inorganic pigments, the light fastness (weather resistance) of organic pigments and solvent dyes has a strong dependence on their chemical structures. According to the chromophoric principle it shows different colors in the pigment due to the absorption of different wavelengths of electromagnetic waves and their internal electronic transitions which derive from electronic transitions of π-π^* and n-π^* absorbing visible in the pigment molecules. The irradiation of light will lead to pigment molecule conformational changes of organic pigments and other factors which can affect the desaturation and even fade into gray or white. Under irradiation of light the fading process of pigments belongs to gas-solid heterogeneous reaction and the reaction rate is mainly related to the chemical structure.

The pigment with different chemical structure was shown in Table 2.10, and the light fastness in PE was shown in Figure 2.12.

Table 2.10　List of chemical structure of different pigment

Pigments index number	Pigment structure	Pigments index number	Pigment structure
P.Blue 15:1	Phthalocyanin	P.Red 48:3	Azo lake(strontium salt)
P.Green 7	Phthalocyanin	P.Red 179	Perylene
P.Red 53:1	Azo lake (calcium salt)	P.Yellow 138	Quinophthalone

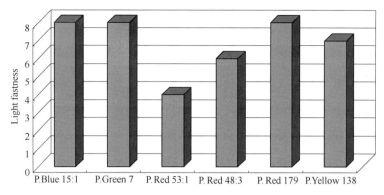

Figure 2.12　Relationship between chemical structure and light fastness

(2) Dependence of light fastness (weather resistance) on the pigment particle size

The fading process of organic pigments under illumination is considered that excited oxygen attack ground state pigment molecules, thereby it will occur photo-oxidation (degradation process). It is a heterogeneous reaction whose reaction rate is related to the specific surface area. Along with the area of the pigment contacted with oxygen increasing the fading process speed up. Small particle size pigment which means a larger surface area leads to poor light resistance.

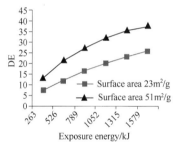

Figure 2.13　Color change of different particle sizes of Yellow 139 after exposure

After illumination, the fading speed of pigment particles which have larger particle diameters is inversely proportional to the square of the particle diameter, and the fading speed of pigment particles which have smaller particle diameters is inversely proportional to particle diameter. The color changes of different particle sizes P.Yellow 139 after exposure was shown in Figure 2.13.

2.6.2　Applications of Light Fastness (Weather Resistance) Indexes

(1) Dependence of light fastness (weather resistance) on the pigment concentration and illumination time

Light fastness (weather resistance) improres along with the pigment addition increasing, and with the volume of pigment increasing the amount of the surface layer increases so that light resistance is better than the small amount of pigment in the same degree of illumination. Therefore, when pigment volume concentration reaches a critical value light resistance reaches the limit.

Light fastness (weather resistance) has strong dependence on the exposure time. Figure 2.14 shows weather resistance of Red 48∶2 in the polyethylene (coloring concentration of 0.05% and 0.2%) at different time.

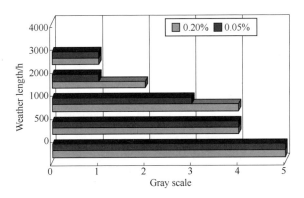

Figure 2.14 Dependence of weather resistance on pigment concentration and illumination time

(2) Dependence of light fastness (weather resistance) on the additive

In color matching of plastic it often needs to add various additives to improve its performance, such as adding titanium dioxide to improve product covering power. Generally adding titanium dioxide into pigment will lead to light resistance decrease in different degrees, and along with the addition increasing light resistance decreases. It is due to two reasons, on one hand titanium dioxide pigment reflect light which make actual light intensity increases, on the other hand the photooxidation degradation will be accelerated with oxide titanium dioxide. Such as 0.05% and 0.2% Red 48∶2, light fastness of pure tone does changes basically, but after adding different amounts of titanium dioxide the difference of light fastness is shown in Figure 2.15.

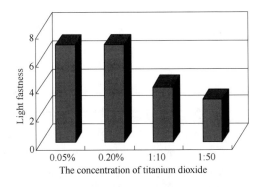

Figure 2.15 Light fastness of Red 48∶2 depends on the concentration of TiO_2

(3) Dependence of light fastness (weather resistance) on the different types of plastic resin

Polymer aging occurs in the sunlight and the most harmful wavelengths for various plastics was shown in Table 2.11.

Table 2.11 the Most Harmful Wavelength for Plastics

Plastic designation	The most harmful wavelength/nm	Plastic designation	The most harmful wavelength/nm
PE	254	PA	254
PP	375	PC	254
PS	254	PET	280~360
PVC	245	CA	254
PMMA	254		

Aging reaction is the same as thermal oxidation autoxidation reaction of polymer under light irradiation along with cleavage and cross-linking of polymer, and it will lead to deterioration of mechanical properties, meanwhile, the color change of polymer will become more serious.. Light fastness of benzimidazolone Yellow 180 in different polymers is shown in Table 2.12. In order to prevent the polymer cracking under UV irradiation, adding UV stabilizers is a viable approach.

Table 2.12 Light fastness of Yellow 180 in different polymer

Type of polymer	Light fastness(full shade/ white reduction)[①]	Type of polymer	Light fastness(full shade/ white reduction)[①]
PA6	7~8 / 6	PO	6~7 / 6~7
PC	5~6 / 5~6	PS/ABS	6~7 / 6~7
PET	6~7 / 7~8	PVC	6~7 / 6~7
PMMA	5~6 / 3~4		

① full shade Yellow 180 0.1%; white reduction: Yellow 180 0.1%, titanium dioxide 1%.

(4) Dependence of light fastness (weather resistance) on the different varieties of the same resin grade

Chemical nature of monomers, polymerization method, antioxygen and light stabilizer for polymer make different grades of plastic have various light fastness (weather resistance).

Through analyzing the same industrial grade polymer, it proves that light induced degradation is caused by breakage of double bond or polymer chain, and it is the primary cause that the defects can absorb energy and result in light-induced damage. Although polymer producers have done a lot of effort, some defects in industrial-grade polymer are still inevitable. Synthetic methods vary widely between various manufacturers. Between the various grades of polymer which is provided by different manufacturers the light resistance of pigment is different, and color matching workers should pay attention to it.

2.6.3 Difference Between Light Fastness (Weather Resistance) Index and Practical Applications

The test station for weathering, for example in Florida, Arizona and Bandol (South of France) are well known. A series of experiments in each location proved that the results are not strictly comparable because of the differences in the climatic conditions of each experimental station. In Florida it is very humid and hot, in Arizona it is very dry and very hot, whereas in Bandol, at moderate humidity and temperature and the impact of atmospheric components due to nearby industrial exhaust plays a major part. But the tendency of fastness is the same everywhere. And every year the climate will occur big change. In the hot summer sun is very strong, so the plastic products bear more exposure than that in the cloudy and rainy day.

Tests done under natural climatic conditions require a long period of time. No customer, however, is willing to wait so long. Consequently methods were developed to determine the light fastness in the laboratory under artificial and accelerated conditions. The basis of any accelerated test is an adequate simulation of the different climatic conditions in the test apparatus and in addition a sufficient correlation between both test principles.

Light fastness (weather resistance) technical specifications of plastics will be in the interpretation of experimental results through frequent discussion between suppliers and customers. On one hand, it is due to demands of customers could not be always fully satisfied. On the other hand, light fastness (weather resistance) technical specifications exist large deviations between theory and practice. However, the color changes and exposure time is not always a linear relationship. Sometimes color changes at the beginning of exposure and stops after a period of time. On the contrary, there is no change at the beginning of exposure, but color changes occur after a period of time. So pigment suppliers should provide comprehensive details about light fastness (weather resistance). Plastic color matching workers should compare the data supplied by the manufacturers with the samples supplied by customers, and utilize light stabilizers to improve the light fastness (weather resistance) which can be up to customers' requirements. At last, the valid data of the ultimate light fastness (weather resistance) can be obtained through combination of test conditions and test system. It is related to authenticity of data supplied by pigment manufacturers which is very important basis of color matching.

2.7 Shrinkage/Warpage

Plastic colorants not only beautify the products also act as nucleating agents and cause the known problems regarding shrinkage and warpage in crystalline plastics, such as PE-HD. The dimensional stability of high precision plastic products can't be guaranteed, so small shrinkage can cause big problems in all those cases where plastic articles have to be assembled, such as screw caps commodities etc.

Because of the heating process in the plastic processing, there exist the shrinkage in plastic coloring and plastics expands with increasing temperature and shrinks again during coloring. During the molding process, the molecular orientation in the flow direction of polymer melt is usually greater than in the cross-flow direction. As a result, rotation- symmetric plastic articles become oval and non rotation-symmetric plastic articles may show a devastating shrinkage. Any additives, including colorants, will affect the plastic deformation, especially in modified reinforced plastics. Normally, the dimensional stability of plastic parts must be kept, no matter what kind of colorant used. Considering the diversity of colorants, the above requirement can't be fulfilled for all colorant, and some kinds of organic pigment play an important role in plastic shrinkage.

There are many factors to consider for plastic molding shrinkage, such as properties of the plastic products (shape, thickness, insert), melt temperature, molding pressure and molding time, mold temperature, the form and size of mold gate, cooling time and so on, and colorant just one of them.

2.7.1 Effect Based on Crystallization

Thermoplastics can be subdivided in crystalline and amorphous. The molecular chains of crystalline plastics line in order, stable, close together. During the solidification of crystalline plastics, crystallization nucleation can initiate crystal growth to form a certain posture. The commonly used polyethylene, polypropylene and polyamide belong to crystalline plastics. The molecular chain of amorphous plastics orders chaotically. And during the solidification, the freeze of macromolecular chain replaces the generation process of the nucleation to grain. PS, PVC and ABS belong to amorphous plastic. Mechanical properties of the non-crystalline plastic in all directions are the same. Crystallization phenomenon of polymer is shown in Figure 2.16.

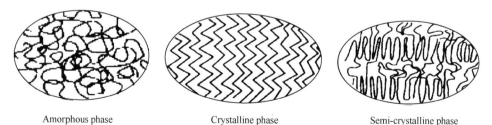

Amorphous phase　　　　　　　Crystalline phase　　　　　　　Semi-crystalline phase

Figure 2.16　Crystallization Phenomenon of Polymer

The apparent characteristics indicate that the general crystalline plastics are opaque or translucent, and amorphous plastic is transparent. But there are exceptions, such as poly-4-methyl-1 is crystalline plastic, with high transparency, and ABS is amorphous plastic with opaque. Properties of two types of resins are showed in Table 2.13 and molding shrinkage of different kinds of plastic are showed in Table 2.14.

Table 2.13 Properties of crystalline and non-crystalline plastics

Properties	Crystalline	Non-crystalline	Properties	Crystalline	Non-crystalline
Proportion	High	Low	Abrasion resistance	Excellent	Low
Tensile strength	High	Low	Creep resistance	Excellent	Low
Tensile modulus	High	Low	Hardness	Hard	Low
Ductility or elongation	Low	High	Transparency	Low	High
Impact resistance	Low	High	Reinforcement with fiberglass	High	Low
The highest temperature	High	Low	Dimensional stability	Poor	Excellent
Brittleness	Brittle	—	Warpage	Easy	—
Shrinkage	High	Low	Coloring	Difficult	Easy
Mobility	Excellent	Low	Heat resistant	High	Low
Chemical resistance	High	Low	Fold mobility	Excellent	Poor

Table 2.14 Molding shrinkage of plastics

Resin	Reinforced material	Molding shrinkage/%
HDPE		1.5~5.0
LDPE		2.0~5.0
PP		1.5~2.5
PP	Fiberglass	0.4~0.8
PA6		0.6~1.4
PA6	Fiberglass	0.3~1.4
PS	Fiberglass	0.2~0.6
PC		0.2~0.6
PC		0.2~0.6
PVC		0.1~0.5

As shown in Figure 2.14, the resin with large crystallization and spherical crystal has a small molding shrinkage. On the contrary, small crystallization and non-spherical crystal has a large molding shrinkage. With injection molding, the molding shrinkage of crystalline resins is larger than non-crystalline plastic and the thermosetting plastic is the least, that is to say, plastic shrinkage/warpage caused by colorants occurs only in crystalline polyolefin plastic, especially in the large injection molding parts produced by HDPE. As shown in Figure 2.17.

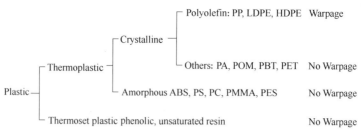

Figure 2.17 The influence of plastic crystallization to warpage

2.7.2 Effect Based on Chemical Structure, Crystal Modification, Particle Size and Concentration of Pigment

(1) Impact of chemical structure of pigment on shrinkage

Organic pigments act as a nucleating agent to promote crystallization in the plastic molding process. Shrinkage of plastic coloring with different structure of pigments is shown in Table 2.15. Shrinkage of pigment molding can be characterized by horizontal (TD) vertical (MD) shrinkage but also aspect ratio (MD/TD). Generally, the closer to 1 the ratio is, the smaller the deformation.

Table 2.15 Impact of Pigment Structure on Plastic Shrinkage

	Name of product	Pigment category	Color index number	HDPE (2208J) Shrinkage /%		HDPE (2208J) Deformation	PP (MH-4) Shrinkage /%		PP (MH-4) Deformation
				MD[1]	TD[2]	D[3]	MD[1]	TD[2]	D[3]
	Blank			2.15	1.99		1.58	1.58	
Inorganic	Titanium dioxide CR-50	Titanium	C.I.PB6	2.18	2.22	−1.8	1.66	1.7	−2.4
	Chrome 2240	Chrome	C.I.PY37	2.22	2.24	−0.9	1.69	1.72	−1.8
	Iron oxide Red 120ED	Iron oxide red	C.I.PR101	2.2	2.1	4.5	1.58	1.67	−5.7
Organic	Carbon Black 4S	Carbon black	C.I.PB7	2.16	2.15	0.5	0.6	1.65	−3.1
	Cromophal Yellow GR	Condensation disazo	C.I.PY95	2.21	2.15	2.7	1.67	1.78	−6.6
	Imgazin Yellow 2GLT	Isoindolinone	C.I.PY109	2.62	1.39	46.9	1.64	1.88	−14.6
	Imgazin Yellow 3RLT	Isoindolinone	C.I.PY110	2.28	1.64	58.1	1.65	1.9	−15.2
	Paliotol Yellow 0961	Quinophthalone	C.I.PY138	1.9	2.02	−6.3	1.6	1.62	−1.3
	Cinquasia Red Y	Quinacridone	C.I.PV19	2.36	2.04	13.6	1.99	2	−0.5

continued

	Name of product	Pigment category	Color index number	HDPE (2208J)			PP (MH-4)		
				Shrinkage /%		Deformation	Shrinkage /%		Deformation
				MD①	TD②	D③	MD①	TD②	D③
Organic	Cromophal bright Red R	Condensation disazo	C.I.PR166	2.85	1.23	53.3	1.6	2	−25
	Heliogenk Blue 6911	Phthalocyanine	C.I.PB15:1	2.66	1.46	45.1	1.59	1.98	−24.5
	Lionol green 2YS	Phthalocyanine	C.I.PG7	2.48	1.68	32.2	1.67	1.97	−18

① MD: Shrinkage of the direction of molding.
② TD: Shrinkage of the direction of vertical molding.
③ D= (MD−TD)/MD*100%.

It can be clearly seen from the table that phthalocyanine Pigment Blue 15:1 and 15:3, Pigment Green 7 isoindoliunone Yellow 110, condensation azo Red 144 and 166 can increase the shrinkage rate. These pigments are common with the feature of symmetry molecular structure. While the inorganic pigments have little influence on shrinkage.

(2) Impact of crystal structure and particle size on shrinkage

The influence of shrinkage caused by pigment crystal and size in plastic molding can be observed by electron microscopy. Table 2.16 demonstrate the results.

Table 2.16 Relationship between Pigment Crystal, Crystal Size, Crystallization and Molding Shrinkage of PE

Pigments	Molding shrinkage deformation	Pigment crystals		Products of PE	
		form	Size	Crystallization	Sphaerocrystal
Natural colored material	Low			Large	Yes
Cadmium	Low	Globular	Small	Large	Yes
Isoindoline	Large	Rod	Large or medium	Small	No
Phthalocyanine	Large	Rod	Minimum	Small	No

In general, the pigment crystal and anisotropic. With the crystalline state of needle, rod in plastic molding, longitudinal direction arranges easily along the flow direction of resin melt, thus generating large shrinkage. Spherical crystal arranges with no direction, thus generating low shrinkage. Inorganic pigments are general with spherical crystal. The transmission electron micrograph of inorganic Pigment Yellow 53 and organic phthalocyanine Pigment Blue 15:3 is shown in Figure 2.18 and Figure 2.19 respectively. As can be seen in figure, the crystal structure of Pigment Yellow 53 is spherical. Spherical crystallization arranges with no direction, while the structure of phthalocyanine pigment is rod.

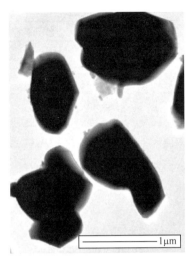

Figure 2.18 Scanning electron microscope of inorganic pigment yellow

Figure 2.19 Scanning electron microscope of organic phthalocyanine pigment blue 15∶3

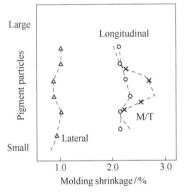

Figure 2.20 Relationship between particle size and shrinkage
△ Transverse shrinkage, ○ Longitudinal shrinkage, × The aspect ratio to shrink (M/T)

In addition to the crystalline structure of pigment, particle size also affects molding shrinkage, as shown in Figure 2.20. For the same isoindoliunone pigment, when the pigment particles are large or small to a certain extent, the molding shrinkage ratio and shrinkage is minimized.

(3) Impact of additive amount of pigment on shrinkage

The additive amount of pigment can also affect shrinkage, the more additive amount is, the larger shrinkage. As shown in Table 2.17.

Table 2.17 Effect of additive amount of phthalocyanine blue on HDPE shrinkage

Additive amount/phr	Longitudinal shrinkage/%	Transverse shrinkage/%	Vertical/horizontal ratio
-	2.26	1.6	1.41
0.01	2.62	1.32	1.98
0.025	2.67	1.18	2.26
0.05	2.71	1.13	2.40
0.10	2.80	1.12	2.50
0.20	2.85	1.10	2.59
0.50	2.92	1.07	2.74

2.7.3 Application of Shrinkage/Warpage Index

(1) Relationship between shrinkage/warpage and plastic processing temperature

The temperature of processing can affect the shrinkage. For instance, at a processing temperature of 220℃ Pigment Brown23 and Pigment Red 149 obviously affect the shrinkage of injection products with the raw material of HDPE, but the effect reduces with temperature elevated. However Pigment Yellow 13 increases the effect of shrinkage in low processing temperature when applied in HDPE.

(2) Effect on mechanical properties of plastics

The very fine organic pigments in the polymer melt provide crystallization nuclei that initiate crystal growth. It is not only the crystallization rate that is influenced by the nucleating agent, but also the morphology and thus the mechanical properties of the polymer. In general, the tensile strength improves with increasing degree of crystallinity, whereas the tear strength decreases.

(3) Phthalocyanine blue can cause warpage to polyolefin

If the pigment is coated with a layer of wax, warpage is no longer observed during the coloring, at least as the coat is not damaged by shear during the processing. Coated grade of pigment blue and other pigments are marked commercially for coloring of partially crystalline polymer.

2.8 Chemical Stability

Chemical stability is a property of the whole system containing at least colorants and plastics that keep stable to acids, alkalis, solvent and chemical subjects. Because of so many different types and grades of plastics and a very broad range of colorants, the importance of chemical stability is not always obvious at first but becomes evident when a plastic products have to fulfill the need daily life and industry.

Because of so many different requirements, it is impossible for the supplier of colorants to run chemical stability tests of every possible application. In conformity to international standards, the supplier of colorants only checks the stability of pigments to acids (HCl, H_2SO_4, HNO_3) and alkalis ($NaOH$, $NaCO_3$), these test results cover a broad scope of applications. In practical applications, especially in the field of packaging, run chemical stability tests of every possible combination is inevitable.

2.8.1 Acid and Alkali Resistance

(1) Many kinds of goods are packed in colored plastic containers, such as medicine, food, perfume, household detergents, chemicals, and so on. Many need-packaged goods

showing weak acid or weak alkalis properties, no forecast is possible because of the variety of possible interactions. Chemical reactions between the packaging materials (plastics, colorants, additives) and the filling good are one possibility of interaction. Another is migration, for solvents as component of the filling good may dissolve the colorant or additives of the packaging material.

(2) Battery container is basically the packing of battery acid. They can have different colors, mainly black, gray and yellow, the pigments must have excellent acid resistance.

(3) Many plastic containers used for storage and transport of chemical raw materials, mainly are blue, the pigment must be acid and alkali resistance.

(4) Wire and cable insulation often used aluminum hydroxide and magnesium hydroxide flame retardant to improve fire resistance. In high-temperature molding process, the pigment must be alkali resistant, few inorganic pigments and rarely organic pigments can be used.

(5) Pigments used for PVA in wet or dry spinning must be resistant to 20℃, 20% sulfuric acid or 100℃, 5% sulfuric acid and 5% NaOH.

2.8.2 Solvent Resistance

(1) In order to reduce weight of car and save energy, the use of plastic materials in automotive industry increased year by year. Many automotive components are made of plastic, these parts at high temperatures must be resistant to fuel, oil, gasoline and the refrigerant. Because of these harsh requirements, most of the parts are colored with carbon black.

(2) EVA plastic used for shoes must be cleaned with solvent before bonding, so the pigments must be solvent resistance, otherwise pigment will precipitate from the solvent.

2.8.3 Oxidation Resistance

(1) Many household disinfectant detergents for cleaning fabric contain peroxide.

(2) Colored textiles should not bleach during washing or dry cleaning. The requirements for colorants are resistant to bleaching substance, mainly peroxide, and resistance to solvent used during dry cleaning.

(3) The test for thermal stability of a colorant is linked to the test for chemical stability. Take polyamide (PA) for example, several colorants can't be used in PA although they withstand much higher temperature when applied in other types of polymer. The reason for the instability of colorants in PA is the reducible force of PA melt. In this case the limitation of use is a question not only of heat stability, but also of

a combination of thermal and chemical stability of the colorant at the temperature of a PA melt.

2.9 Security

In order to meet the requirements of product safety, environmental and health, plastic materials and the products must be in conformity to the regional regulatory requirements all over the world. One of the most important and particular concern is the requirements of chemical substance control, especially for colorants used in plastic materials. There are a wide range of specific consumer products involved mainly include: toys, textile materials (such as some polymer chemical fiber) and accessories (such as zippers and buttons, etc.), electrical and electronic products, food containers and food contact materials or products, automotive materials.

2.9.1 Acute Toxicity of Chemicals

The most commonly used to measure acute toxicity value is LD_{50}. LD_{50} indicates the amount of substance expressed in milligrams perkilogram body weight that has a lethal effect on 50% of the test animals (mainly rats), which also called the median lethal dose. The unit is mg/kg body weight. The smaller the LD_{50} value, the stronger the toxic poisons. On the contrary, the greater the LD_{50} value, the weaker the toxicity of the poison. The European Union defined three acute toxicity substances categories (rat oral):

$LD_{50} \leqslant 25$ mg / kg: very toxic

LD_{50} 25~200 mg / kg: toxic

LD_{50} 200~2000 mg / kg: harmful

There are numerous monographs reviews of toxicological of the colorants (about 194 kinds of pigment), in no case an acute toxicity below 2000 mg/kg, majority are greater than 5000mg / kg. Considering the salt (NaCl) has a LD_{50} of 3000 mg/kg (oral), which is equivalent to the average weight of people to eat 350g pigment, it is impossible, so we can get conclusion that colorants in general are low acute toxicity. Pigments are generally discharged through the gastrointestinal, but not excreted in urine.

2.9.2 Organic Pigments Impurities

Organic pigments have been widely used as plastic colorant in toys, food packaging materials etc. So besides toxicological and ecological properties of pure pigment, organic pigments have some trace amounts of impurities in the production and it must be considered that may affect the use of the consumer products above-mentioned.

Possible trace amounts of impurities are as follows.

(1) Heavy metal compounds. Some heavy metal salts (barium) are lake organic pigments (C.I. Pigment Red 48∶1), are not recommended for food packaging materials and toys.

(2) Aromatic amine. Aromatic amines as a component of a synthetic pigment organic pigment only be allowed a very low amount. There are a limit in the use of food-contact packaging materials.

Aromatic primary amines: <500 ppm (mg/kg, total). 4-aminobiphenyl, benzidine, 2-naphthylamine, 2-methyl-4-chloroaniline: <10 ppm (mg/kg, total)

(3) Polychlorinated biphenyls. Polychlorinated biphenyls (PCBs) mainly due to their persistent residual hazards in the environment is greater than to humans. In the EU chemicals such as PCBs and PCTs (polychlorinated triphenylmethyl) containing with more than 50ppm or 50ppm are prohibited to be sold.

Small amounts of polychlorinated biphenyls may be formed in certain side reaction during the synthesis of two types of red or yellow series organic pigments with dichloro and four benzidine chloride as diazo component. Use two chlorinated benzene or benzene trichloride as a solvent in the synthesis of phthalocyanine blue and green pigments may form polychlorinated biphenyls.

(4) Dioxins. Pigment Violet 23 is condensation of chloranil with N-ethylcarbitol azole, and chloranil will inevitably form a small amount of dioxin in the synthesis process.

2.9.3 Security of Double Chloride Benzidine Pigment

The yellow orange organic pigment synthesised of benzidine chloride with 3, 3 double chloride benzidine (DCB) is an important species of azo pigments because of the bright color, high tinting strength and low price for which it has become an important plastic colorant, such as Pigment Yellow 13,14,17,81,83, Pigment Orange 13, etc.

It has confused people that whether the use of DCB series pigment variety in synthetic non-woven textile materials, accessory such as clothing zipper and button, and packing of food is conformed to the safety request of ecological and environmental protection laws and regulations both domestically and abroad.

(1) Double chloride benzidine may cause cancer. International Affairs of Research in Cancer (IARC) classified compounds into three categories according to their cancerogenic toxicities.

1- Human carcinogen

2A- Probable human carcinogen

2B- Possible human carcinogen

Double chloride benzidine belongs to class 2B carcinogen. The mechanism is that it reacts with nucleic acid (DNA), to cause DNA mutagenesis. Polychlorinated benzidine has a high bioaccumulation, low biological degradation and a strong destruction on human body's endocrine system. So, there are strict requirements internationally. The products content over 10mg/kg of PCB are banned in the European Union; The United States is prohibited to produce, process, sell and use the product that polychlorinated biphenyls (PCBS) content in more than 25mg/kg. International textile ecology research and testing association (Oeko-Tex) formulated the "Oeko Tex-100 General Standard and Special Technical Conditions" provides 24 kinds of aromatic amine should not exceed 20 mg/kg (Table 2.18). And DCB column in it.

Table 2.18 24 Kinds of Aromatic Amine from Oeko-Tex Standard 100

Number	Name	CAS	Number	Name	CAS
1	4-Amino aniline	92-67-1	13	3, 3-Dimethyl-4, 4-diamino diphenylmethane	838-88-0
2	Benzidine	92-87-5	14	2-Methoxy-5-methyl aniline	120-71-8
3	2-Methyl-4-chloroaniline	95-69-2	15	4,4-Methylene-double-(o-Chloroaniline)	101-14-4
4	2-Naphthylamine	91-59-8	16	4, 4-Diamino diphenyl ether	101-80-4
5	4-Amino-2, 3-dimethylazobenzene	97-56-3	17	4, 4-Diaminodiphenyl ether	139-65-1
6	2-Amino-4-nitro toluidine	99-55-8	18	o-Phenylenechiamine	95-53-4
7	p-Chloroaniline	106-47-8	19	2, 4-Diaminotoluene	95-80-7
8	2, 4-Diaminoanisole	615-05-4	20	2, 4, 5-Trimethylaniline	137-17-7
9	4, 4-Diamino diphenylmethane	101-77-9	21	o-Aminoanisole	90-04-0
10	3, 3-Dichlorobenzidine	91-94-1	22	2, 4-Dimethylaniline	95-68-1
11	3, 3-Dimethoxy benzidine	119-90-4	23	2, 6-Dimethylaniline	87-62-7
12	3, 3-Dimethylbenzidine	119-93-7	24	p-Aminoazobenzene	1960-9-3

1. First kind: Carcinogenic to humans (4kinds NO.1-4).
2. Second kind: Carcinogenicity in animals, probably carcinogenic to humans (20kinds NO.5-20).

(2) The safety of organic pigment synthetized of double PCB.

The organic pigment synthetized of double PCB is insoluble in water, sweat, blood, stomach acid and other mediums. Nevertheless, azo pigment was still in the terms of azo dyes, defined by the German Federal Government Agency, and was confirmed by the fourth amendment in July 15 1995 that confined organic pigments which could reduce and decompose to 20 carcinogenic arylamines concluding DCB. However, it is doubtful whether the organic pigment synthesis of DCB is harmful on human body. Chemists has studied on organic pigments' toxicology widely. While the consequence are as follows.

• Fed the experiment animals with C.I. Pigment Yellow 12 and C.I. Pigment Yellow 83 for a long time. Carcinogenic symptom was not found.

- Put the C.I. Pigment Yellow13, C.I. Pigment Yellow17 in the blood and urine of experiment animals. DCB was not detected with the extreme sensitive analyze method. It proved that the pigments were not catalytic decomposed by the enzyme in the blood and urine.
- Experimental animals were lived in the atmosphere with 230mg/m^3 C.I. Pigment Yellow 17 for 4 hours every day. DCB was not detected in the blood and urine samples 14 days later.
- The further research on C.I. Pigment Yellow12, C.I. Pigment Yellow17, C.I. Pigment Yellow83, C.I. Pigment Yellow114 and C.I. Pigment Yellow13 indicated that no potential dangers were existed.

The above research indicated that the pigment synthetized by the intermediate-DCB would not decompose to DCB inside the organism.

In July 23 1996, the German Federal Government issued the Fifth Amendment that explain the definition on organic pigment synthesis by DCB definitely. One of the rules said: since Apr 1 1998, the use of azo pigment which could break and release carcinogenic arylamine under the conditions of the legal analysis was banned. However, the azo pigment would not break under the conditions of the legal analysis, so it would not be confined. For the azo pigment synthetized of DCB, DCB always could not be detected.

The use of DCB is safe if the reaction with the raw materials-DCB and products – organic pigment was complete followed by the completely removing and the detection to control the arylamine within a certain content.

(3) DCB series of organic pigments would decompose to DCB if molding temperature is higher than 200℃.

DCB series of organic pigments will decompose to DCB at high temperatures. Such Pigment 83, although it will change color only when the temperature is higher than 260℃, DCB will be decomposed above 200℃ because of the chemical structure of the pigment. DCB decomposition by different varieties of paint used in polypropylene spinning is shown in Table 2.19.

Table 2.19 DCB decomposition by different varieties of paint used in polypropylene spinning

C.I.	Spinning temperature/℃	Heating time/min	Pyrolysis product concentration/10^{-6}		
			DCB	Single accidentally nitrogen	Aromatic primary amine
C.I.PY13	260	10	8.37	250	
C.I.PY14	260	10	12.3	420	
C.I.PY17	260	10	30.2	319	

continued

C.I.	Spinning temperature/℃	Heating time/min	Pyrolysis product concentration/10^{-6}		
			DCB	Single accidentally nitrogen	Aromatic primary amine
C.I.PY83	200	10	Not detected	0.46	
	220	10	Not detected	0.82	
	240	10	0.11	2.56	58.5
	260	7	0.34	36.3	171
	280	10	182		

1. Manufacturing masterbatch at 180℃, 10 minutes polypropylene spinning at 200~270℃.
2. Take 50g fiber with 700mL toluene quenching 20 hours, after the enrichment in liquid chromatography and nuclear magnetic resonance (NMR) test.

So the supplier of colorants abroad would indicate in the sample. Double benzidine chloride pigments that used in polymer will be decomposed when the processing temperature is over 200℃. Even if the luster of these goods is not be changed in the process of temperature rising, but for the potential thermal degradation (See safety parameter table), it still can't be used if the processing temperature is over 200℃.

DCB series of organic pigment is relatively safe when used at the processing temperature is below 200℃, Such as Red 38, which can conform to the requirements of the FDA when it is used in rubber coloring, therefore it has been used in sausage packaging, because the forming temperature of the sausages materials PVDC is 180℃.

2.9.4 Heavy Metal in Inorganic Pigments

In addition to titanium dioxide, carbon black and ultramarine almost all of the inorganic pigments containing heavy metals. Like other substances, when the concentration of heavy metal exceeds a certain value, it would be considered to be harmful to humans and the environment. The concentration range is depended on the kinds of metal and the form.

2.9.4.1 Chromium

Chromium compound contain trivalent or hexavalent chromium. Hexavalent chromium compounds (chromates) has a strong tendency to be converted to trivalent chromium compound and release oxygen, thus it has a strong oxidation and toxic effects on biological. The toxicity is 1000 times higher than trivalent chromium compound for humans, animals and plants.

(1) Lead chromate pigment. Lead chromate pigments contain lead and hexavalent chromium, both have chronic hazards. Lead chromate is a lead compound with low solubility. The dissolved lead can be found in the hydrochloric acid and gastric acid

which will lead organic to be cumulative in the body. It will disrupt the synthesis of hemoglobin when feeding a high level of lead. As a precaution the EU has already list lead chromate as a category 3 carcinogen (suspected carcinogenic potential).

(2) Chromium oxide green pigments. Chromium oxide green pigment only contain trivalent chromium. Free chromium ion from the chrome oxide green pigment under the condition of natural. There are only a small amount of trivalent chromium released even in a strong acidic conditions (pH 1~2). Chromium oxide (III) only in the case of heating especially under the condition of alkaline can be oxidized to chromium(VI).

2.9.4.2 Cadmium

Cadmium pigments are chemical compounds with low solubility, but a small amount of cadmium dissolved in acid (concentration equivalent acid concentration). Long-term oral intake of cadmium pigments leads to accumulation in the human body. The European Parliament has already listed cadmium sulfide as a category 3 carcinogen, but cadmium pigments are not included.

Chapter 3
Main Types and Properties of Inorganic Pigments and Specialty Effect Pigments

3.1 Development History of Inorganic Pigments

Inorganic pigments have a long history, they have been known since prehistoric times. Around 2000 BC, natural ocher was calcined and mixed with manganese to prepare red, purple and black pigment for ceramic product. And the industrial production of inorganic pigments began in the 18th century. The manufacturing method of iron blue (Prussian blue) was invented by a German named Judith Bach in 1704 and put into production in 1707. Then, cobalt blue was also produced in 1777. Moreover, ultramarine, chromium oxide green, iron oxide pigments and cadmium pigments have been emerged in the 19th century. French Vauquelin exploited chrome yellow pigments successfully in 1809, and put it into industrial production in Germany in 1818. In 1831, France Gime finished industrial production of ultramarine pigments in Lyon, which made ultramarine pigments could meet the needs of market cheaply and plenty. Lithopone was invented by French named Du Haut in 1847 and put into production in 1874 and became the first major white pigment. In 1872, the United States made use of natural gas to bring out industrialization of carbon black. In 1916, the first factory of titanium dioxide was built all over the world, and most of white pigment was replaced by TiO_2 for the excellent performance and TiO_2 become the biggest pigment used in colorant with rapid growth of production.

Since the 20th century, cadmium red, molybdenum red, bismuth vanadate yellow pigments and mixed metallic oxide pigments have entered the marketplace. In addition, the development of specialty effect inorganic pigment (metal effect, pearlescent and interference color pigments) is becoming more and more important. Nowadays the color spectrums of inorganic pigments have been perfect basically. The production of inorganic pigments is made up of around 74% of titanium dioxide, 14.6% iron oxide pigments, 5.0% chrome yellow, 4.6% anticorrosive paints, 1% chromium oxide green, 0.2% cadmium pigments and 0.6% other in the world.

Almost all kind of inorganic pigments have high hiding power, heat resistance, light resistance and weather resistance, and due to the more mature of production process technology, most products of inorganic pigments are in low cost. Most types of inorganic pigment withstood the test of time and are still applied, such as lead chrome yellow, iron blue, ultramarine, zinc oxide, iron oxide and so on, but the white lead, basic copper carbonate have been replaced for the high toxicity and bad performance. In addition, non-environmentally friendly inorganic pigments will fade away gradually with the requirements of international environmental become more and more strict. Therefore, the future development trends of inorganic pigments are as follows.

The chloride process of rutile titanium dioxide will replace the sulfate process gradually because of high quality, great investment and little pollution.

Complex inorganic color pigments will develop rapidly. This kind of pigments has good compatibility with most plastic resins, and present chemically inert, non-bleeding, non-migration, embrace the highest level of light resistance, weather resistance, heat resistance, acid and alkali resistance, excellent performance even if weaken color with white pigments, nontoxicity and environmental-friendly. In addition, this kind of pigment has different varieties and bright color, such as yellow, green, blue, brown and so on. They are used for coloring of building materials and engineering plastics extensively.

The treatment technology of coating each individual pigment particle with colorless inorganic compound or organic compound will develop rapidly, which would change the surface properties of the pigment particles, improve the light (weather) resistance, heat resistance, expand the range of applications and increase use value.

Produce pigments with good processability and replace universal inorganic pigments with special (such as aluminum bar) to make them to be applied directly, save energy consumption and reduce pollution.

3.2 Classification and Composition of Inorganic Pigments

3.2.1 Classification of Inorganic Pigments

There are many ways to classify inorganic pigments. So the classification is carried out according to color, chemical composition and color composition in this book. Inorganic pigments can be divided into three categories: achromatic, colored, specialty effect pigments, as shown in Table 3.1. Achromatic pigments including white and black pigments, only show different amount of reflected light (total scattering or total absorption), that is to say they only show difference of brightness. Colored pigments are able to absorb certain wavelengths of light selectively and reflect the other wavelengths of light to show different colors. The surface of specialty effect inorganic pigments can produce different optical effect of reflection to get different effects.

Table 3.1 Classification of inorganic pigment

Classification	Type	Definition
Achromatic pigments	White pigment	Optical effect caused by total scattering (such as TiO_2, ZnS, lithopone, ZnO pigment)
	Black pigment	Optical effect caused by total absorption (such as carbon black)
Colored pigments	Colored pigment	Optical effect caused by absorbing certain wavelengths of light selectively and reflecting the other wavelengths of light (such as iron oxide red, cadmium yellow, ultramarine blue, chrome yellow, lead chromate molybdate, titanium yellow, titanium brown, cobaltic blue, cobaltic green, bismuth yellow pigment)
Effect pigments	Metal effect pigment	Specular reflection caused by the flat or parallel particles of metallic pigment (such as aluminum powder, aluminum platelet, copper powder)
	Pearlescent pigment	Specular reflection caused by the multiple reflection and transmittance of the sun light on parallel pigment platelets (such as mica coated with TiO_2)
	Interference color pigment	Optical effect caused by interference phenomenon of colored shimmer pigment (such as mica coated with iron oxide)
	Flop- pigments	Light refraction caused by coating of low refractive index and high refractive index medium alternately, color depends on angle of vision (chameleon)

3.2.2 Composition of Inorganic Pigments

Compared with organic pigments and dyes, the palette of inorganic pigments for coloring plastics is relatively small. There are only a few basic chemical formulas, but they are available in numerous variations, so the palette seems larger than it really is. For example, complex inorganic color pigment is a basic type of many variations. The exact color shade is modified within a certain range by adding other metal oxides or changing ratio of the major metal oxides. In a word, the inorganic pigment is usually made of metal oxides, sulfides, metal salts and carbon black. In addition, the composition of white and black inorganic pigment is shown in Table 3.2, colored inorganic pigment is shown in Table 3.3 and specialty effect inorganic pigment is shown in Table 3.4.

Table 3.2 Composition of main white and black inorganic pigment

Chemical class	White pigments	Black pigments
Oxide pigments	TiO_2 ZnO	Iron oxide black (Fe_3O_4), manganese ferrite black spinel [$(Fe, Mn)_2O_4$], Iron cobalt black spinel [$(FeCo)Fe_2O_4$]
Sulfide pigments	ZnS Lithopone ($ZnS \cdot BaSO_4$)	
Carbon and carbonate pigment	White lead [$Pb(OH)_2 \cdot 2PbCO_3$]	Carbon black

Table 3.3 Composition of main colored inorganic pigment

Chemical class	Yellow	Orange	Red	Violet	Blue	Green	Brown
Chrome pigments	Chrome yellow $PbCrO_4, PbSO_4$, $PbCrO_4 \cdot PbO$		Lead chromate molybdate $PbCrO_4$, $PbMoO_4$, $PbSO_4$			Chromium oxide green Cr_2O_3	
Iron oxide pigments	Ferrite yellow α, λ FeO(OH), [$(Zn,Fe)Fe_2O_4$]		Iron oxide red Fe_2O_3				Iron manganese oxide [$(Fe \cdot Mn)_2O_3$]
Metallic oxide pigments	Titanium yellow [$(Ti,Ni,Sb)O_2$]				Cobaltic aluminate blue spinel Co,Al_2O_4, $Co(Cr,Al)_2O_4$	Cobalt green Co_2CrO_4 $(Co,Ni,Zn)_2 TiO_4$	Chrome antimony titanium buff rutile $(Ti,Cr,Sb)O_2$
Cadmium pigments	Cadmium yellow CdS,ZnS	Cadmium orange CdS,ZnS	Cadmium red CdS,CdSe				
Ultramarine pigments				Ultramarine violet $Na_5 Al_4 Si_6 O_{23} S_4$	Ultramarine blue $Na_6 Al_6 Si_6 O_{24} S_4$		
Bismuth vanadium	Bismuth vanadate molybdate $4BiVO_4 \cdot 3Bi_2MoO_6$						

Table 3.4 Composition of main white and black inorganic pigment

Chemical Class	Composition	Color
Metal effect pigment	Al	Silvery shiny
	Cu-Zn alloy	Golden shiny
Pearlescent pigment	Mica coated with TiO_2	Pearlescent silvery white and interference color
	Mica coated with TiO_2 and metal oxide	Gold and bronze
Flop pigments	Aluminum powder coated with iron oxide and SiO_2	Color depends on angle of vision

3.3 Plastic Coloring Properties of Inorganic Pigment

The relative density of inorganic pigment is large which are always in the range of $3.5 \sim 5.0$ g/cm^3, and their tinting strength is poor, heat resistance and light resistance is excellent. Inorganic pigments are micron-sized particulates, the diameter of pigment primary particle is mostly between a few tenths of a micron to several microns, and diameter of the largest particles is less than 100 μm. Inside of these particles, there is a certain arrangement of molecular. Most pigment particles are in the form of crystals, and the coloring of inorganic pigment for plastics is depend on the dispersion of pigment particles. Therefore, the structure, shape, size and distribution of different crystal of pigments are bound to affect their performance and usability.

3.3.1 Requirements of Inorganic Pigment in Properties

Inorganic pigments possess many advantages in application of plastics coloring, such as excellent heat resistance, dispersion, light (weather) resistance. So they are widely used for coloring of different types of plastic, especially for engineering plastic which need high molding temperature and extreme service conditions.

(1) Good dispersion

The relative density of inorganic pigments are large which are always in the range of $3.5 \sim 5.0$ g/cm^3, and because of their large density and small specific surface area,they can easily disperse in plastic matrix. In addition, because the tinting strength of inorganic pigment is relative lower, so the color is very bright and brilliant when the concentration of inorganic pigments is large.

(2) Good hiding power

Hiding power is the ability of pigment to make the colored resin become nontransparent, that is to say, hiding power is the required minimal amount of pigment to cover the black and white check board completely checkered termed hiding power. It changes with the size of pigment particles. In addition, most inorganic pigments possess good hiding power because of large relative density and particle size.

(3) Excellent weather and light resistance

The use value of pigments is affected by weather and light resistance of colored plastics directly. Generally, the influence of sunlight and air atmosphere will lead color of inorganic pigments to darken, but it will not fade. However, for other pigments, the damage of chemical structure of pigment will lead color to fade under the condition of sunlight and air. In general, weather and light resistance of inorganic pigments is far better than common organic pigments.

(4) Excellent heat resistance

Most inorganic pigment possess very excellent heat resistance except chromium pigments, especially those inorganic pigments that produced by high temperature calcination under 700~1000℃. However, to the disadvantages of lead-chrome inorganic pigments in poor heat and light resistance, the surface of chrome yellow pigment is coated to improve the heat resistance of the product in foreign.

(5) Excellent chemical stability

Most inorganic pigment is an inert substance with excellent acid, alkali, salt, corrosive gases and solvent resistance. However, it is hard to ensure that the specific pigment will not react chemically with any substance. For example, chemical structure of ultramarine is not resistant to acid, and the alkali resistance of iron yellow pigment is better than chrome yellow pigments.

3.3.2 Impact of Inorganic Pigment Particle Characters on Coloring Properties

The physical properties of inorganic pigment particle, such as size, crystal structure, surface charge and polarity will of influence the color, hiding power, tinting strength and application of plastic.

(1) Influence of crystal structure

Inorganic pigment has very fine particulate, even if the pigment has the same chemical composition, there will be a different crystal structure due to the different crystal growth environment. The different microstructures of pigment particles have an effect on macroscopic performance directly. The most typical example is titanium dioxide. Although both rutile and anatase are belong to tetragonal, their crystal structure is different, which lead to different crystal properties. Rutile is a spindly pairs of twin crystals, and each of rutile unit cell has two titanium dioxide molecules connected with two edges. In addition, anatase occurs in the form of octahedron, and its oxygen atom is located in the vertex angle of octahedron, each of unit cell has four titanium dioxide molecules connected with eight edges. The unit cell of different titanium dioxide crystal is shown in Figure 3.1, the comparison of rutile and anatase is shown in Table 3.5.

(a) Rutile TiO$_2$ (b) Anatase TiO$_2$

Figure 3.1 Unit cell of different titanium dioxide crystal

Table 3.5 Comparison of rutile and anatase

Crystal form	Rutile (R)	Anatase (A)
Atomic structure	More compact	
Refractive index	2.73	2.55
Hiding power	Higher	
Toning ability	Stronger (10%~30%)	
Weather resistance	Better	
Stability	Better	
Density	3.75~4.15	3.7~3.8
Hue		Bluer
Abrasion ability		Lower

The production of zinc oxide is a similar case in point. Different crystal structures that formed with direct process and indirect process lead to the differences of performance and use. For example the lead chrome yellow pigment with the same chemical composition, the light resistance of monoclinic pigment is better than rhombic pigment. Changes in the crystal structure not only have an influence on the optical properties, particle properties, surface properties, dispersion and stability of the pigment, but also electrical properties, magnetic properties and so on.

(2) Influence of pigment particle size

The most important physical parameters of inorganic pigments are not only the optical parameter, but also the average particle size, distribution of particle size and particle shape. In addition, by optical properties we learn that different size, shape and distribution of the pigment particles will lead the color, hiding power and tinting strength changing under the same chemical composition of pigment. For example, in the production of iron oxide red pigment with wet process, the size of pigment particle change with the oxidation cycle. The iron oxide red pigment appears yellowish red when crystal particle is small, and it appears purplish red when crystal particle is large. It is obvious that they have large differences in tinting strength, hiding power, specific surface area and oil absorption, as shown in Table 3.6.

Table 3.6 Influence of particle size on hue change for iron oxide red

Type of iron oxide red	1	2	3	4	5	6	7	8
Particle size/μm	0.09	0.11	0.12	0.17	0.22	0.3	0.4	0.7
Tonal variation		Yellowish red ----change to bluish red---- purplish red						
Tinting strength		High--Low						
Hiding power		Small--------------------Big------------------Small						
Specific surface area		Big---Small						
Oil absorption		Big---Small						

The changes in the optical properties of the pigment caused by the change of chromium molybdenum red pigment granual size are shown in Figure 3.2. The hue becomes deep and red from light and orange, the brightness and tinting strength become low from high with the increasing of particle size. In a word, the pigment possess high tinting strength and bright colors when the particle size is small, but low tinting strength, dark and deep color when the particle size is too large.

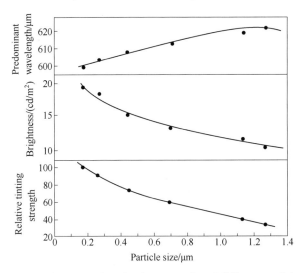

Figure 3.2 Tinting strength and color properties of different particle size of chromium molybdenum red pigment

(3) Influence of pigment particle shape

The surface state of pigment particle is related to the oil absorption. And the oil absorption is related to the specific surface area of pigment particle and the void fraction between the pigment particles. In addition, for inorganic pigments, the oil absorption is also related to pigment particle shape. Generally the oil absorption of needlelike pigment particle is higher than spherical particles because of the bigger specific surface area and void fraction of needlelike particles. The shape of pigment particle is also concerned with the tinting strength. It is clear that the needlelike pigment particles possess greater absorption and scattering as well as higher tinting strength than spherical particles due to larger specific surface area in Figure 3.3.

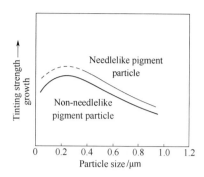

Figure 3.3 Influence of pigment particle shape on tinting strength

3.3.3 Safety of Inorganic Pigments in Plastic Coloring

Most of the inorganic pigments contain heavy metals except titanium dioxide, carbon black and ultramarine. In the past decade, the discussion on heavy metal in our environment had been popular in the world. Therefore, nearly every customer requires explicitly that color preparations must be free of heavy metals. All metals with a density above $3.5 \sim 5 g/cm^3$ are defined as heavy metals in foreign. In addition, all metals with a density above $4.5 g/cm^3$ are defined as heavy metals (including all kinds of metals and precious metals). So, all inorganic pigments except aluminum flakes, carbon black, ultramarine blue and violet contain heavy metals based on the definition.

Actually heavy metal is a natural part of environment, which can be found in rocks and soil widely, even in food because plants absorb them in the soil. Our lives are developed and formed in an environment with natural heavy metals, and the heavy metals have been exist in our tissues. In addition, many heavy metals such as iron, zinc, manganese, molybdenum, chromium and cobalt are the necessary trace elements to maintain life, and humans and animals will die without them. Animal testing had been shown that a lack of chromium will lead to diabetes, arteriosclerosis and growth disorders. Thus it is unscientific to demand extremely that there is no heavy metal in our life.

In fact, like other substances, heavy metal will be harmful to humans and environment when the concentration is over the point. The important inorganic pigments were examined very carefully in foreign with regard to the safety in plastic coloring. Summarizing the results, most inorganic pigments except harmful chromium and cadmium pigments are harmless toxicologically and ecologically. This is because the insoluble inorganic pigments are not bioavailable in the stomach (accidental swallowing) or in the environment, but the toxicological effects of the chromium and cadmium pigments are based on their solubility in gastric acid, and the acid-soluble lead is easy to be absorbed by the body, which causes various symptoms of lead poisoning.

3.4 Main Use of Inorganic Pigment in Plastics

3.4.1 Coloring for Plastics

Application of organic pigments in plastic coloring is enlarged continuously, and inorganic pigments containing heavy metals, such as lead and chromium, are being shrunk and phased out because of the limit of stringent environmental regulations. However, an advantage of inorganic pigments is the irreplaceability in the following fields.

(1) Coloring for engineering plastics

Engineering plastics refer to polymer materials with excellent mechanical properties and dimensional stability that can be used as structure materials, even in trenchant chemical and physical circumstance, and withstand the mechanical stress in a wide range of temperature. The molding temperature is very high in the coloring of engineering plastics, as shown in Table 3.7. There are few inorganic pigments suitable for polyamide coloring due to reducibility. Therefore, inorganic pigments are widely used in coloring of engineering plastics with excellent heat resistance and other properties.

Table 3.7　Processing temperature of some engineering plastics

Resin	Processing temperature/℃	Resin	Processing temperature/℃
PP	180~250	POM	200~260
ABS	220~250	Fluoroplastic	350
PC	270~300	PET	260~280
PA	250~300	PPS	320~360

In the injection molding of plastic, the change of the amount of shrinkage may cause warping or cracking and tolerance (positive or negative tolerance). So, the influence of the color preparations should not be ignored based on the description above. The crystal of organic pigment is oriented easily along the flow direction of polymer melt in the molding for the acicular or rod crystal structure, therefore promote the nucleating effects and large shrinkage after cooling. However, the crystal structure of most inorganic pigment is spherical without orientation. So inorganic pigment has little influence on shrinkage in the molding of plastics, which is another advantage to organic pigment for coloring of plastics.

Now, the carbon black is the second largest consumption of inorganic pigment for plastics after titanium dioxide, and they are the two key types of inorganic pigment for coloring of engineering plastics.

High performance complex inorganic color pigment is short for CICP in foreign. CICP has excellent hiding power and performance, especially is suitable for the engineering plastics that require high performance. The primary hues of iron oxide pigments are red, yellow and black. And orange, brown and green also can be got through coloring match. In addition, iron oxide pigments are widely used in coloring of engineering plastic because of light, weather, alkali and solvent resistance, nontoxicity as well as low cost.

The chromatogram of cadmium pigments is broad, the color shade of light yellow, orange and red, even reddish brown purple is brilliant. Moreover, cadmium pigments can be almost used for coloring of all engineering plastics due to excellent light, weather, heat resistance, high hiding power and tinting strength, and non-migration as well as non-bleeding. However, cadmium pigments are non-environmentally friendly inorganic pigments. Currently, the European Union and the United States have restricted the use of cadmium pigments explicitly, but they are still used in engineering plastics that need high-temperature processing, such as PA, POM and PTFE because of the excellent properties, especially heat resistance.

(2) Coloring for outdoor products largely

Plastic products that long-term used in the outdoor, such as artificial turf, sports equipment, constructional materials, advertising boxes, recycle boxes, rolling shutters, profiled bar and plastic automobile parts, need excellent weather and light resistance. Titanium dioxide, carbon black, CICP and ultramarine are the very good types of inorganic pigments in terms of the weather resistance.

(3) For preparing light-colored types of pigment and adjusting shades

The tinting strength of inorganic pigment is low, but its heat resistance is still very excellent when the concentration of pigment is low. Therefore, inorganic pigment is the first choice to prepare light-colored types of pigment. On the one hand the relative more pigment that added in will decrease as much as possible errors of color, and on the other hand inorganic pigment has very excellent heat and weather resistance even if its concentration is low. So, inorganic pigment is suitable for coloring and shading of plastics.

3.4.2 Special Functions

(1) Conductivity

Plastics are good insulating materials widely used in electrical equipments. The resistivity of plastic is high, which typically in the range of $10^{10} \sim 10^{16} \Omega \cdot m$. Many plastic products require antistatic or conductive properties. In many fields of technology, the electrostatic electricity is a dangerous factor. As we all know, plastics are poor conductors of electricity, of course, including antistatic electricity. Plastic parts must be equipped with antistatic or conductive substances, in order to avoid potentially dangerous. Such as floors, floor coverings, boxes, instruments, sealing materials, containers, pipes and so on.

When the plastic is added a sufficient amount of carbon black, it can be given antistatic properties (resistivity $<10^{10}\Omega\cdot cm$) or the conductivity (resistivity $<10^{4}\Omega\cdot cm$). There are several theories about the conductive mechanism for carbon black in the polymer, the classical one is a chain-type conductive path. This mechanism is based on the contact between the conductive particle chains, so the resistance and the frequency of contact between the particles is a major factor in determining conductive. In the external electric field, the distance between carbon black particles within a few Å (1Å = 0.1nm), then a voltage difference can be generated, so that the π electron of carbon black particles form currents rely on the transfer and move through chains. In order to form passages and appear strong conductive phenomena, there must be a certain amount of carbon black. Therefore, the most important factor for conductive polymer materials is the type and amount of carbon black. In order to improve the conductivity of plastics, a high structural furnace black with small size and low volatile matter content should be chosen.

Conductive plastics have a wide range of applications. Such as, static control products, static eliminator, anti-static conveyor belts, anti-static plastic sheets, anti-static cases, the control boards, anti-static pipes, medical rubber products, carpets, rollers of photocopiers, rollers of printer, packaging film of electronic components, explosion-proof cables. The content of carbon black in static control products is generally at 4%～15%.

The conductive products inclucle conductive foam tubes, charging rollers, shielding materials of cables, conductive films. The content of carbon black in conductive product is 10%～40%.

(2) Anti-aging and UV absorption

Carbon black not only can be a coloring agent, but also can be considered a best UV stabilizer with high quality and inexpensive. Light can aging plastic, especially ultraviolet rays will accelerate the aging process. Consider the use of the environment and the service life of the plastic of the plastics, different approaches are needed to solve the aging problem of plastic products. Generally, adding an ultraviolet absorber and an antioxidant is a good way to improve the usable life.

Carbon black used as a UV absorber is mainly to extend the outdoor service life of plastic products (such as pipes and cable sheathings made by HDPE, LDPE and PVC); also applies in other products (such as mulch used in agriculture manufactured by LDPE). To achieve the above applications, the carbon black with small particle size and a slightly higher concentration must be chosen, as shown in Figure 3.4 and Figure 3.5.

The concentration of 0.5% carbon black with small particle size (20nm) and the concentration of 2% carbon black with a relatively coarse particle size (95nm) have almost the same photoprotection, as shown in Figure 3.6. Taking account into costs and properties, of course, the fine one (20nm) was selected.

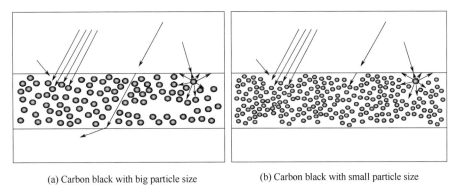

(a) Carbon black with big particle size   (b) Carbon black with small particle size

Figure 3.4 Carbon black with small particle size will improve the aging resistance of polyethylene

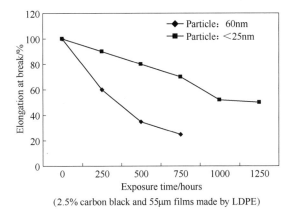

(2.5% carbon black and 55μm films made by LDPE)

Figure 3.5 Aging properties affected by primary particle size

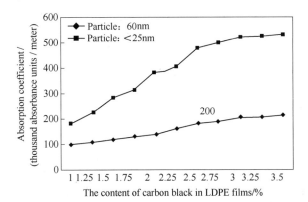

Figure 3.6 UV absorption affected by the adding amount of carbon black

In order to achieve the desired effect, pipe materials made by PE and propylene copolymer have the following standards in the United Kingdom. For security reasons, the particle size of carbon black should be less than 25nm, and the general provisions for the carbon black concentration is in the range of (2.5 ± 0.5)%. In addition, the documents recommended by the German Federal Health Office are defined: the concentration of carbon black in plastic should not exceed 2.5%. After compromising the above two standards, we get the range at (2.25 ± 0.25)%, which is exactly in line with the requirements of producers and consumers. Therefore, the general provision of the current concentration for carbon black is (2.25±0.25)%, used in pressure pipes, cable sheathings and 10kV overhead lines which made by HDPE, the service life is up to 50 years or more.

(3) The function of infrared reflection

The solar spectrum received by earth can be divided into three parts according to its different wavelengths. Each part occupies a different ratio of total energy. The ultraviolet area (UV): 295~400 nm, accounts for 5% of the earth's total energy received by the sun. The visible light area (VIS): 400~720 nm, accounts for about 45% of the earth's total energy received by the sun. The infrared area (IR): 720~2500 nm, accounts for 50% of the earth's total energy received by the sun, as shown in Figure 3.7.

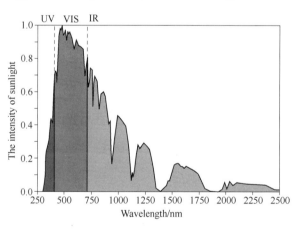

Figure 3.7 Total solar energy received by earth

Everything grows by the sun. The sun passes its own energy to animals and living creatures on earth, so that they can get continuous growth. The sunlight also brings many negative sides to human life while bringing some benefits for people, even devastating effects. The sun's ultraviolet rays can make a lot of organic substances be degraded, and have harmful effects for human skin. Objects receive a lot of thermal radiation which emitted by visible and invisible infrared light, then its surface temperature rises, and cause many problems and inconveniences in human's daily life. The thermal energy received by house roofs and walls, making the indoor temperature rises through a variety of ways (conduction, radiation, convection and other forms), and reducing the comfort of life. In order to make the indoor temperature reduced to an appropriate extent, people use a lot of air conditionings, air conditioners, electric fans and spraying equipment, which requires large amounts of electricity and other energy. For some chemical containers and open air reactors, due to the direct exposure by sunlight in the hot summer, the temperature of the liquid inside the tanks increases as the surface temperature rises, and brings a certain risk. How to reduce and eliminate these negative effects on humans caused by the sun's rays have attracted the attention of many scientists.

The studies found that the composite pigments of inorganic pigments have color and also reflect a portion of the infrared light, reducing heat buildup and having a cooling effect. When using the American Shepherd company's composite pigments called 10C909a, the infrared reflectivity is up to 25% while the average is only 5% for black pigments, as shown in Figure 3.8.

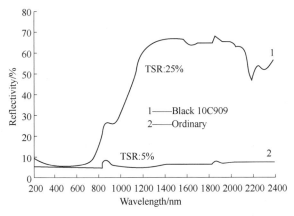

Figure 3.8 Reflectivity between black 10C909a and ordinary carbon black

Germany Heubach have made the surface treatment for inorganic composite pigments and improved the infrared reflectivity, as shown in Table 3.8.

Table 3.8 IR for german heucodur

Names of products	Color index number	TSR/%
Heucodur IR Black 950	PBk.30	10
Heucodur IR Black 945	PBr.29	17
Heucodur IR Black 940	PBr.35	24
Heucodur IR Black 910	PBr.29	21
Heucodur IR Brown 869	PBr.29	22

continued

Names of products	Color index number	TSR/%
Heucodur IR Blue 5-100	PBl.36	24
Heucodur IR Blue 4G	PBl.36	24
Heucodur IR Blue 550	PBl.28	34
Heucodur IR Blue 2R	PBl.28	27
Heucodur IR Green 5G	PG.50	23
Heucodur IR Yellow 3R	PBr.24	58
Heucodur IR Yellow 255	PBr.24	55
Heucodur PLUS IR Yellow 150	PY.53	64
Heucodur IR Yellow 152	PY.53	63

In 2012, the white ALTIRIS®550 and ALTIRIS® 800 infrared reflective pigments were introduced by British Huntsman, which could achieve high solar reflectance in a wider range of colors.

The cooling and energy saving effects of High-performance inorganic composite pigments are more and more attracting people's attention, when the pigments used in plastic building materials. As people improve the requirements for environmental protection and energy savings, the infrared reflectance performance for composite pigments of inorganic pigments will rapidly develop and be used in many areas.

In addition there are varieties of high-performance inorganic composite pigments have the function of chlorophyll imitation, which could be used for army clothing and shelter of weapons, and have great significance in the defense and military fields.

3.5 The Main Varieties and Properties of Inorganic Pigments and Effect Pigments

3.5.1 Achromatic Pigments

3.5.1.1 White Inorganic Pigments

White pigments are mainly used for white pure white coloring, for brightening of colors and for covering shades. Generally, all substances with a refractive index higher than 1.7 are defined as white pigments, the refractive index is lower than 1.7 are defined as fillers, as shown in Table 3.9. Strictly considered, the limiting value of 1.7 is not a constant but depends on the matrix of the pigments, because every matrix has its own specific refractive index. Plastics are no exception. Therefore refractive indices of the white pigments vary from plastic to plastic, and are consequently different for materials, such as polyethylene, polypropylene, polystyrene, PET fibers and so on.

Table 3.9 Refractive Indices of Several White Pigments and Fillers

Pigments	Color index number	Refractive index
Antimony oxide(Sb_2O_3)	Pigment White 11	2.19
Barium sulfate ($BaSO_4$)	Pigment White 21	1.64
White black carbon(SiO_2)	Pigment White 27	1.55
Chalk($CaCO_3$)	Pigment White 18	1.58
Titanium dioxide-anatase (TiO_2)	Pigment White 6	2.55
Titanium dioxide- rutile(TiO_2)	Pigment White 6	2.70
Zinc oxide(ZnO)	Pigment White 4	2.00
Zinc sulfide(ZnS)	Pigment White 7	2.37
Zirconium oxide(ZrO_2)	Pigment White 12	2.40

For white pigments, good optical properties are required, such as high scattering power, a high degree of hiding power, good lightening power, a high degree of lightness and a high degree of whiteness. The scattering power is the most important property, which depends on the refractive index, particle size and distribution and the degree of dispersion. Because of these factors the scattering power is a relative value and not an absolute value. Other indicators, such as hiding power, lightness, undertone and whiteness depend more or less on the scattering power of the white pigments.

(1) **Titanium dioxide** (TiO_2)   C.I. Pigment White 6; C.I. structure number: 77891; CAS registry number: [13463-67-7].

Titanium dioxide (TiO_2) is the most important white pigment currently used in the plastics industry. When it is added to the plastic products, it can effectively scatter the visible light and give products whiteness, brightness and hiding power. Even under the most harsh process conditions, it remained chemically inert and excellent thermal resistance.

Titanium dioxide occurs in nature in the crystal modifications rutile (tetragonal), anatase (tetragonal) and brookite(rhombic). Rutile and anatase are produced industrially in large numbers. Brookite is difficult to produce, and therefore it's not use for the pigments industry.

Titanium dioxide (R-type) has high tinting strength and good hiding power. The tinting strength of titanium dioxide (A-type) was only 70% of that of R- type. The whiteness of A-type is better. The weather resistance of R-type is good, its sample only has a small change on its appearance after ten years. The weather resistance of A-type is poor, and began to crack or peel after only a year, so the R-type should be used for plastic coloring.

There are varies of titanium dioxide with different properties, the main properties of titanium dioxide is tinting strength, brightness and hiding power. The tinting strength of white pigment display whiteness and brightness in plastic coloring, and the hiding power is a capability for white pigment to make plastic being opaque.

Particle size and distribution of titanium dioxide have a great effect on its coloring properties.

The hiding power of titanium dioxide is related to the scattering energy and also related to the ratio of absorption energy and scattering energy. Optical classical theory states that when the pigment particle size is close to the half wavelength of visible light, the scattering energy is biggest. Therefore, in order to improve the scattering ability of titanium dioxide, control the particle size and distribution of titanium dioxide is the most important method. As shown in Table 3.10, the whiteness of colored plastic is related to the color of titanium dioxide, the white color with blue tone gives people a fresh and pleasing feeling, but the one with a yellow tone gives people a feeling of old, therefore the use of yellow tone titanium dioxide in the preparation of white color is less than the blue one.

Table 3.10 Relationship between color tone and particle size of titanium dioxide

Type	Blue light	Green light	Yellow light
Rutile /μm	0.14	0.19	0.21
Anatase /μm	0.16	0.22	0.23

The lighting power of colored products is related to the particle size distribution of titanium dioxide, the latter is narrow, and the former is better. Figure 3.9 shows the particle size distribution of DuPont's titanium dioxide for plastics.

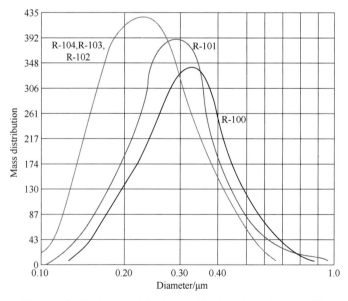

Figure 3.9 Typical particle size distribution of titanium dioxide

Titanium dioxide has the disadvantage of chalking and discoloration as a commodity in market, because the titanium dioxide can cause photochemical reactions, and will be a catalyst for the oxidation of some organics when was exposed in the sunlight and the atmosphere. The aging results can result in loss of mechanical strength and generated a powder layer. This powder layer comprises of loose titanium dioxide powder and the resin dissociation by surface wearing. In order to ensure the fastness properties of titanium dioxide, the surface treatment is necessary. The surface treating agents are divided to inorganic treating agents and organic treating agents based on its basic functions. Using surface treatment agents like aluminum and silicon for coating layers of titanium dioxide, can greatly improve the weather resistance, while its anti-yellowing properties improved too. Using organic surface treatment agents, the resulting surface may be hydrophobic or hydrophilic, but those all are means for improving the dispersion of titanium dioxide in various media, shown in Figure 3.10. The license numbers and properties of American DuPont's TI-PURE titanium dioxide after different surface treatment are shown in Table 3.11.

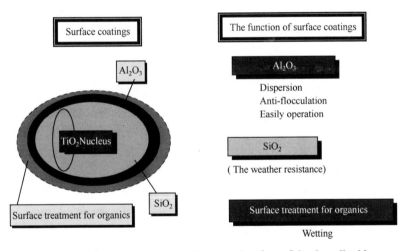

Figure 3.10 Schematic diagram for coated surface of titanium dioxide

Table 3.11 License numbers and properties of DuPont's TI-PURE titanium dioxide properties

Properties	The license number						
	R101/R101	R102	R103	R104	R105	R350	R960
Titanium dioxide (minimum weight fraction) /%	97	96	96	97	92	95	89
Alumina (maximum weight fraction)/%	1.7	3.2	3.2	1.7	3.2	1.7	3.5
Silicon (maximum weight fraction)/%	-	-	-	-	3.5	3.0	-

continued

Properties	The license number						
	R101/R101	R102	R103	R104	R105	R350	R960
Organic treatments	Hydrophobic	Hydrophobic	Hydrophobic	Hydrophilic	Hydrophilic	Hydrophilic	-
CIE L* minimum value	97.9	98.5	97.8	97.5	98.5	98.5	98.5
Tinctorial strength	102/101	109	110	110	105	110	90
Applications	Low volatility (Crack emptiness)	Improve water dispersibility (Anti-coagulation)	Weather resistance engineering plastics	Low volatile (Crack emptiness)	PVC profiles with high weather resistance and high dispersion	Low volatility (Crack emptiness)	Colored PVC profiles with weather resistance

When titanium dioxide being surface treated by aluminum and silicon, the application properties can be improved but also brings dispersed difficulties. Especially the inorganic treatment agents on the surface of titanium dioxide will precipitate as volatiles when extruded at high temperatures, and causing blistering and holes in the products. In order to adapt the high-temperature extrusion process, the anti-hole varieties species should be chosen, such as DuPont titanium dioxide R101, R104 and R350.

Rutile, the best type, is applied to color nearly all kinds of plastics, either for pure white shades or for the lightening of colored shades including colors with a good hiding power. But there is a restriction. The high hardness of the rutile is a drawback in glass fiber reinforced plastics, because the rutile can damage the glass fiber during the coloring process combined with a severe loss of mechanical strength of the final plastic part. Therefore the inferior optical properties, the much softer lithopone or zinc sulfide is the white pigment of choice to color glass fiber reinforced plastics.

Titanium dioxide is inert, insoluble in solvents (except in concentrated sulfuric acid and hydrogen fluorine), and very stable. It is regarded as completely non-toxic and is used in tooth paste and the coat of medical pills.

Acute toxicity: such pigments are considered as non-toxic. oral LD_{50} values for rats> 5000mg/kg, no irritation for contact to skin, but cause little irritation to eyes and respiratory airways due to the mechanical friction.

Chronic toxicity: animals feeding long-term titanium dioxide were observed that no signs of titanium intake. There isn't chronic toxic effects report of manufacturing and using of titanium dioxide environment for many years.

Because the excellent physiological compatibility of titanium dioxide, the United States and the European Union has approved a specific purity titanium dioxide as a coloring agent for food, cosmetics and pharmaceutical products.

The market of titanium dioxide Titanium dioxide is the number one inorganic pigment in worldwide, and has the most outstanding properties for white pigments. The world production of titanium dioxide is estimated at 4.8 million tons in 2013, while the Chinese production is estimated at 1.8 million tons.

The main foreign suppliers DuPont (TI-PURE), Huntsman and US Branch US-based pigment Ltd. etc.

Major domestic suppliers Sichuan Lomon Titanium Industry Co., Ltd., Shandong Dong Jia Group Co., Ltd., Henan Billions Chemicals Co., Ltd., CNNC Hua Yuan Titanium Co., Ltd. and Panzhihua Iron and Steel Co., Ltd. Titanium Company etc.

(2) Zinc oxide (Zinc white, ZnO) C.I. Pigment White 4, C.I. structure number: 77947, CAS registry number: [1314-13-2].

Main properties Zinc oxide is a very fine white powder. The melting point is 1720℃, the refractive index is 1.92~3.00, the density is about 5.67g/cm^3, the Mohs hardness is 4~4.5, the optical properties are inferior to those of titanium dioxide. Zinc oxide is amphoteric, the chemical stability is not very good. It can react with acids and is soluble in alkaline solutions. This is a drawback for the application in the field of packaging materials.

There are two manufacturing processes for Zinc oxide. One is the direct process, a very simple one at low cost. Another is an indirect method, from the beginning of metallic zinc, which is boiled and resulting vapor oxidized to zinc oxide.

Main applications The most important use of zinc oxide is in the rubber industry and molding products. It is generally not used for thermoplastic resin coloring.

Safety data The acute LD_{50} (Lethal Dose, 50%) of zinc oxide > 15000mg/kg, zinc is an essential trace element for humans, animals and plants. Lack of zinc can affect the growth of hair and reproduction. Zinc oxide does not stimulate skins and eyes. It has no potential allergic, and is not be carcinogenic, teratogenic, and reproductive toxicity for the human body. People do not consider zinc oxide as toxic or dangerous, although a few earlier studies have shown some toxic effects. These toxic effects may be the result of impurities, especially of lead. Zinc oxide is often doped with several other metal oxides, mainly cadmium, lead, iron and aluminum oxides. Therefore, it is necessary to use various separation techniques for purifying zinc vapor before the oxidation. Commercial zinc oxide has different degrees of purity, some types still contain a small amount of lead. Before use it should check the quality of zinc oxide to fulfill all consumer requirements regarding impurities of heavy metals.

Zinc oxide is insoluble in water, easily separated from waste water. Zinc ions in water for fish and other aquatic organisms are toxic, so the concentration of zinc ions in waste water is limited.

(3) Zinc sulfide (ZnS) C.I. Pigment White 7; C.I. structure number: 77975; CAS

registry number: [1314-98-3].

Main properties Zinc sulfide is the second important white pigment after titanium dioxide. The refractive index is 2.37, the Mohs hardness is 3. Zns is a kind of soft pigment, and the light fastness is good, but the weather resistance is insufficient. In some plastics, affected by UV radiation and humidity, zinc sulfide is oxidized to a colorless zinc sulfate.

Zinc sulfide is the main component of lithopone (Pigment White 5), and also a major component of some luminescent pigments. There is a luminescent pigment which consists of zinc sulfide doped with silver or copper.

Main applications The optical properties of zinc sulfide are obviously inferior to titanium dioxide, so the application is limited. Zinc sulfide is only used in poly (vinyl chloride) (PVC). In lead-containing systems, zinc sulfide can react with lead forming the black lead sulfide. The structure of zinc sulfide is soft, not easily abraded, the mechanical damage to the fibers could be avoided, especially for glass fiber reinforced plastics coloring. Compared with titanium dioxide that is the main advantage of ZnS, because titanium dioxide in the extrusion process can damage the glass fiber reinforced plastic.

Safety Data Due to the low solubility of zinc sulfide, it is non-toxic to human. Studies have shown that there was no poisoning or chronic damage to health in manufacturing process of the pigment, even exposure to dust during the operation of the very finely ground pigment. US FDA and most European countries allow zinc sulfide in contact with food. Although the human body's metabolism needs small quantities of zinc, soluble zinc is toxic in large amounts, zinc belongs to the essential trace elements.

(4) Barium sulfate ($BaSO_4$)　C.I. Pigment white 21; C.I. structure number: 77120; CAS registry number: [7727-43-7].

Main Properties　For barium sulfate, its density is in the range of $4.3 \sim 4.6 g/cm^3$, the refractive index is 1.64, and the Mohs hardness is 3.5. Barium sulfate is substantially inert toward acids, alkalis and organic solvents. Additionally, the light fastness and weather resistance are very good, the heat stability is high (above 300℃). It is easy to disperse when used in plastics.

There are two types of barium sulfate in the market. One is natural; and the other is a precipitated, also called synthetic barium sulfate.

Main applications　Generally, barium sulfate is not used as white pigment. The main use is as filler or processing aid in coloring products. For example, Barium sulfate can be used in the plastic lampshade because of its transparency. On the one hand it is transparent enough for the light. On the other hand there is a certain amount of light can be scattered. As a result, due to the shade of scattering diffuse light was required.

The synthetic barium sulfate is available in different particle sizes, and suitable for coloring preparations. Especially in the case of a high concentration of organic pigments, barium sulfate improves the flowing properties of the premixed colorants after adding a small amount, and helps to disperse the pigments in plastics. Due to its low Mohs hardness, it is not very abrasion, which is favorable in regard to abrasion of plastics processing machines. Another very special application of barium sulfate is used in toys, which is not very obvious. Barium sulfate can help to locate a toy swallowed accidentally by a child because it is the classic X-ray contrast medium. In addition, barium sulfate is a component of lithopone (Pigment White 5).

Safety data Pure barium sulfate is harmless for toxicology, so its use in plastics in food packaging is permitted in many countries, including the United States (according to FDA) and most European countries.

(5) Lithopone ($ZnS/BaSO_4$) C.I. Pigment white 5;C.I. Structure Number: 77115; CAS Registry Number: [1345-05-7]

Main properties Lithopone is a mixture of barium sulfate and zinc sulfide. The composition formula can be written as $ZnS + BaSO_4$. There are different types for the content of zinc sulfide, 15%, 30%, 40%, and 50%.

Main applications Lithopone is used in many plastics, especially in glass fiber reinforced plastics. Because the soft structure of lithopone can prevent mechanical damage to fibers during the coloring of the reinforced plastic and it is the main advantage in comparison to titanium dioxide.

Lithopone can be formulated a good white pigment, its opacity is relatively strong, which between that of zinc oxide and titanium dioxide. However, due to the superior performance of titanium dioxide, which was widely used in plastic coloring process, so the application of lithopone is restricted a lot.

Safety data The use of zinc sulfide and barium sulfate in plastics for food packaging is permitted in many countries, including the United States (according to FDA) and most European countries. Large amounts of soluble zinc are toxic, but the human body's metabolism requires small quantities. It is harmless in the human because of its poor solubility. Studies have shown that there was no poisoning or chronic damage to health in manufacturing process of the pigment, even exposure to dust during the operation of the very finely ground pigment.

3.5.1.2 Black Inorganic Pigments

Similar to the position of titanium dioxide in the white field, carbon black dominates the market for black pigment. Black pigments are used either for coloring of pure black or for toning of color pigments. Strictly speaking, there are only three kinds of black inorganic pigments: carbon black (Pigment Black 7), iron oxide black (Pigment Black 11) and iron titanium brown (Pigment Black 12). The other black pigments, such as Pigment Black 22, Pigment Black 26 and Pigment Black 30 are variations of iron oxide Black (Pigment Black 11). In this kind of pigments, the iron is partly substituted by one or several other metals such as copper (Cu), manganese (Mn), chromium (Cr), cobalt (Co) and nickel (Ni) (Table 3.12).

Table 3.12 List of black inorganic pigments varieties for coloring of plastics

C.I. index number	Fomula	Product
Pigment Black 7	C	Carbon black
Pigment Black 11	Fe_3O_4	Iron oxide black
Pigment Black 12	Fe_2TiO_4	Iron titanium brown spinel
Pigment Black 22	$Cu(Cr, Fe)_2O_4$	Copper iron chromite black
Pigment Black 26	$(Fe, Mn)_2O_4$	Manganese ferrite black spinel

(1) Carbon black C.I. Pigment Black 7; C.I. structure number: 77266; CAS registry number: [1333-86-4].

Carbon black is a very fine pigment with a high tinting strength. Carbon black is almost composed by pure carbon element (diamond and graphite are other forms of almost pure carbon), and present a nearly spherical colloidal particle shape. In spite of carbon is one of the main components of organic chemistry, carbon black is still classified as an inorganic pigment.

Carbon black is produced by the incomplete combustion and pyrolysis of gaseous or liquid hydrocarbon, the physical appearance is one kind of black and very fine granular or powder. Furnace black, gas black, lamp black, and acetylene black can be obtained by different production of raw materials. Different varieties of carbon black with a wide range of particle size can be obtained by different production process conditions, which surface area is usually in the range of 10~1000 m^2/g. Because of the different production process conditions, the primary particles interact growth for different high structure and low structure carbon black, with very different properties, as shown in Figure 3.11.

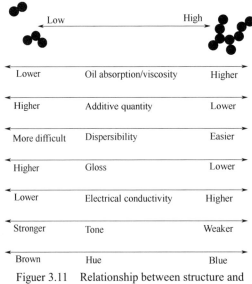

Figuer 3.11 Relationship between structure and properties of carbon black

The particle size of carbon black has a great relationship with its properties, the particle size and he related properties are shown in Table 3.13.

Table 3.13 Relationship between particle size and properties of carbon black

Particle Size	Large	Small
Specific surface area of carbon black	Small	Large
Light aging resisting ability	Low	High
Tinctorial strength	Weak	Strong
Dispersibility	Good	Bad
Filling amount	High	Low
Hygroscopicity	Low	High
Hue	Blue	Red

Main applications   Carbon black is mainly used for plastics and synthetic fiber coloring, Extensively used for black and gray coloring polyethylene, polypropylene, polyvinyl chloride, polystyrene, ABS polymer and polyurethane plastics. In particular, the use in polyolefin plastics is most.

The common require amount of carbon black is 0.5%~2%. For the transparent plastic, 1% of carbon black is sufficient. 1%~2% of carbon black with high tinting strength is required for opaque plastic (e.g., ABS). The carbon black content used in coloring transparent light plastic products is generally 0.02%~0.2%. The grades and properties of Cabot's carbon black are shown in Table 3.14.

Table 3.14 The grades and properties of Cabot's carbon black

Grade	Specific surface area/(m²/g)	Particle size/nm	Tinting strength/%	Volatile/%	Application
Black 120	30	60	50	1.0	Film, sheet
Black Pearl 254					Drinking water pipe, cable
Black Pearl 430/460	80	27	109	1.0	Filament, film, sheet, blow molding
Black Pearl 4560					Filament, short fiber
Black Pearl 800	201	17	150	1.5	Sheet, blow molding
Black Pearl 900	230	15	151	2.0	Blow molding
Black Pearl 4840					Blow molding
Black Pearl 1300	560	13	114	1.5	Blow molding
Black Pearl 6100					Cable
Elftex TP	130	20	1.5	1.5	Drinking water pipe, sheet
Elftex 570	115	22	114	1.5	Filament, film, sheet, blow molding
Vclcan 9A32	140	19	114	1.5	Filament, Cable
Vclcan P	172	20	94	1.5	Drinking water pipe
Vclcan XC72	254	19	114	1.5	Conductive film, sheet

The impact on the environment (security) The LD_{50} value of carbon black is over 5000mg/kg. It did not produce any stimulation of the skin and eyes, when tested on rabbits with carbon black. Long-term research shows that industrial carbon black has no harmful effect. That has been confirmed by several decades of human experience. According to "IARC", "Poison and toxicological planning" (NTP / USA), the European and American chemical legislation, rubber black and pigment black has no potential to cause mutations, deformities and cancer.

There are always has trace impurities, polycyclic aromatic hydrocarbons (PAHS) in commercial carbon black. Polycyclic aromatic hydrocarbons (PAHS) can be extracted with very strict laboratory analysis methods, and it show teratogenic and carcinogenic activities. But the extraction quantities of carbon black in a short period of time is very small. So there is no scientific evidence confirms that there are potentially harm effects such as teratogenic and carcinogenic to the human body for contacting with carbon black normally.

Product market The world production is estimated at 12 million tons of carbon black in 2013, of which China accounted for 6.5 million tons. Due to the adsorption layer structure surface that the rubber formed on the surface of filler particles and the rubber macromolecule adsorbed on the surface of carbon black filler, which producing a reinforcing effect with sliding action. So rubber is the main consumer market of carbon black. And specialty carbon black pigment should be the main useing to coloring the plastics.

The main foreign suppliers Cabot, Orion, Mitsubishi, Columbia, etc.

The main domestic suppliers Qujing Congregation Fine Chemical Co., Ltd. Shanghai Coking Chemical Development Trading Company.

(2) Inorganic composite black pigment Compared with carbon black, the other black pigments are rarely used for coloring of plastics, their main disadvantage is the low tinting strength, and only used for tinted shades or color spell. For the shading only very small amount of a colorant are necesary, and on the other hand, a very even distribution of a small quantity in a larger mixture is difficult to achieve, in this case, a low tinting strength can even be an advantage. In addition, the other black pigments, which are based on metal oxides, are much easier to disperse in a polymer melt than the carbon black. On the other hand, the carbon black is very good UV absorber, which will cause the temperature of the coloring matter increase, accelerate the aging of resin, and cause color change, lead the weather resistance decreased. High-performance inorganic composite black pigment can be used if the black products need a good weather resistance.

The impact on the environment (security) A lot of poison and toxicology tests of composite inorganic black pigments have been done, and no hazardous to human tissue showed. Not be considered toxicity yet.

Physical hazards Black iron oxide containing divalent iron, can be oxidized, and the process is exothermic. It is only initiation at a higher temperature by the limited of kinetics. So the storage temperature of black and brown pigments should not be higher than 80℃. Follow these rules, iron oxide pigments products are safe.

The major properties Generally composite inorganic black pigments have high hiding power, heat resistance could up to 300℃ (except iron oxide, which will generate iron oxide red at a temperatures above 250℃), excellent light resistance and weather resistance, good chemical stability.

The major purposes Suitable for the coloring of polyolefin, ABS, polyamide (PA), polystyrene, phenolic resin, epoxy resin etc.

The main varieties of complex inorganic black pigments are listed in Table 3.15.

Table 3.15 The main varieties of high-performance composite inorganic black pigments

C.I index number	C.I. structure number	CAS registry number	Composition
Pigment Black 11	77499	1317-61-9	Iron oxide black (Fe_3O_4)
Pigment Black 33	77537	68186-94-7	Iron oxide black (Fe_3O_4)
Pigment Black 12	77543	68187-02-0	Iron titanium brown spinel (Fe_2TiO_4)
Pigment Black 22	77429	55353-02-1	Copper chromite black [$Cu(Cr, Fe)_2O_4$]
Pigment Black 26	77494	68186-94-7	Manganese ferrite black spinel [$(Fe, Mn)_2O_4$]
Pigment Black 27	77502	68186-97-0	Iron cobalt chromite black spinel [$(Co, Fe)(Fe, Cr)_2O_4$]
Pigment Black 28	77428	68186-91-4	Copper chromite black spinel [$Cu(Cr, Fe)_2O_4$]
Pigment Black 29	77498	68187-50-8	Iron cobalt black spinel [$(Fe, Co)Fe_2O_4$]
Pigment Black 30	77504	71631-15-7	Chromite iron nickel black spinel [$(Ni, Fe)(Cr, Fe)_2O_4$]

The major foreign suppliers Black iron oxide pigment manufacturers is (Lanxess) COLORTHERM®; Composite inorganic black pigments manufacturers are BASF, Shepherd, FERRO, (ROCKWOOD) SolaplexTM, HEUBACH TICO®, TOMATC®.

The major domestic suppliers Hunan JUFA Technology Co., Ltd. Nanjing Pigments Tech. Co., Ltd.

3.5.2 The Main Varieties and Properties of Color Inorganic Pigment

The transition metal is the main reason for the generation of color in most colored inorganic pigment. Due to the low prices and readily available of many metal oxides, so it is especially important as a coloring pigment. Either be a single component, or be a mixed phase of an oxide-based color pigment. Lead and chromium inorganic pigment, cadmium inorganic pigment, iron oxide inorganic pigment, titanium nickel inorganic pigment, ultramarine blue inorganic pigment, cobalt inorganic pigment, bismuth vanadate yellow inorganic pigment and so on, are the main varieties of colored inorganic pigment.

3.5.2.1 Chromium Inorganic Pigment

Chromium inorganic pigments are lead chrome metal salt, chrome yellow pigment is pure lead chromate or a mixed phase pigment of lead chromate and lead sulfate, the general formula is $Pb(Cr, S)O_4$. Molybdate red is a mixed phase pigment of lead chromate, lead sulfate and lead molybdate, and the general formula is $Pb(Cr, S, Mo)O_4$.

Chrome inorganic pigments are used in coloring of plastics with those pure bright color and high hiding power, but the color could become darker during storage for the unstable crystal and tend to darken after sunlight exposure if without surface treatment, which are the two disadvantages of chrome inorganic pigments that had found. German first expounded that chrome yellow unstable orthorhombic transform to a stable monoclinic is the main reason for lead chrome yellow in particular lemon chrome yellow suddenly becomes dark during storage in 1931. In recent years, scientists gave a long period of in-depth research in the light-fastness problem of chrome yellow pigments, such as the influence of crystal type and particle size.

For the poor heat resistance and light-fastness of lead chrome inorganic pigments, a surface treatment on chrome yellow pigments with aluminum hydroxide, titanium, etc have been studied since 1994, which coated on the surface of inorganic pigment particles with a layer of special film that was uniform thickness and had a continuous and compact structure. So it can be considered not only a physical envelope, but also a chemical bonding. Thus to improve the heat resistance, weather resistance and sulfidation resistance of chrome yellow pigments, especially it can get various performance envelope of lead chrome inorganic pigments due to the different silicon coated dense degree.

The impact on the environment (security) Lead chromate red and yellow pigments contain lead and hexavalent chromium, both with chronic hazards. Lead chromate is a kind of lead compounds with low solubility. Dissolved lead can be found in hydrochloric acid and the acid concentration will lead to the accumulation of lead in organism. The high ingestion amount of lead can disrupt the synthesis of hemoglobin. Therefore, the EU has already classified all lead compounds as Level 1, toxic for reproduction (embryo toxicity). Lead compounds and preparations containing 0.5% lead must be signed with skull icon and crossbones pattern, and writing "Harmful to the fetus" and so on.

Lead chromate pigment used in toys and articles in contact with food coloring are prohibited. The EU "Registration, Evaluation and Authorization of Chemicals" has explicitly classified chrome pigments as carcinogenic, mutagenic, reproductive toxicity substances.

In recent years, a series of statutes had been enacted at home and abroad, such as the US Code of Federal Regulations (CFR) involving the standard for food contact materials (FDA), the US Consumer Product Safety Commission (CPSC), the EU "The Restriction of the use of certain hazardous substances in electrical and electronic equipment." (RoHS), the EU EN71-3 toy standard limit element limited instructions, the EU new Toy Safety Directive requires updating chemistry (EN 2009-48), the EU regulations and technical standards of ecological textiles, and the EU standards AP (89) 1 and Chinese GB-9685 for the food contact items have explicitly stipulate that the chromium content in plastic packaging material would not exceed l000mg/kg. Thus the use of chrome pigments is limited, the amount is greatly reduced.

The main component of chromium oxide green is chromium oxide (Cr_2O_3), a trivalent chromium compound, trivalent chromium oxide is considered harmless. Non-toxic and non-carcinogenic effects were indicated in rats receiving up to 5% trivalent chromium oxide in their food. The LD_{50} value of trivalent chromium oxide exceeds 5000mg/kg body weight, and it does not irritate the skin and mucous membrane. Trivalent chromium is the necessary trace element for human sand animals.

The chromate in lead chrome inorganic pigments though is nonflammable, but due to their oxidizing properties, enabling reduced the ignition of combustible materials. Because of the possibility of fire, we must pay attention to prevent fire to prevent the impact to the environment when using a mixture of such pigment and organic, in particular the iron blue and monoazo pigment mixture.

Product market Lead chrome yellow pigment dominates the three series of chrome pigments. Its consumption had reached a historic peak in the Mid 1970s. But on account of containing acid-soluble lead which harmful to human health (especially children). Since the Late 1970s, with the continuous strengthening of safety and sanitation and environmental regulations, the production and sales started to decline, and will gradually withdraw production especially in the United States, Japan, Western Europe and other developed countries and areas. However, in the world, it has been unsuccessful due to the substitutes of chromium pigments are expensive. So chrome pigments will continue to be applied in a very long period of time, but the usage amount in the worldwide will become increasingly smaller. Currently chrome pigment production is estimated at 20 000 tons, coated products dominate the application on the plastic, Jiangsu Shuangle Chemical Pigment Co., Ltd. is known for the production of coated products in the domestic.

The major foreign manufacturers are as follows.

Red and yellow chrome pigments: BASF Sicomin®, Dominion Colour Corporation (DCC).

Green chrome pigments: Lanxess COLORTHERM®.

The major domestic suppliers: Jiangsu Shuangle Chemical Pigment Co., Ltd. Henan Xinxiang Helen Pigment Co., Ltd. Chongqing Joanna Chemical Co., Ltd.

(1) Yellow chrome pigments

The synthesis research of lead chrome yellow pigment dated from the analysis and research of lead chromate ore founded in Siberia. French chemists first synthesized lead chromate in 1809, and Germany began the industrial production of lead chrome yellow pigments in 1818.

The chemical composition of lead chrome yellow pigment is $PbCrO_4$、$PbSO_4$ and $PbCrO_4 \cdot PbO$. Its color can form a continuous period of yellow chromatography from lemon yellow to orange. It could be divided into tens of thousands of varieties if color is the only standard, but in order to facilitate the production and use, five standard color usually usually be used in pruduction by the factories, those are lemon chrome yellow, light chrome yellow, chrome yellow, deep chrome yellow and orange chrome yellow.

Lead chrome yellow pigment is lead compound with about 53%~64% of lead and 10%~16% of chromium generally.

C.I. Pigment Yellow 34   Chemical constitution: $3.2PbCrO_4 \cdot PbSO_4$; C.I. structure number: 77600, CAS registry number: [1344-37-2].

The major properties   A series of yellow chromatographs from lemon yellow to orange can be acquired from chrome yellow. But the use of chrome yellow in plastic is greatly limited because of the heat-resistant and other factors. In the inorganic pigment particle surface coated chrome can greatly improve the heat resistance, weather resistance and vulcanization resistance, especially different properties of capsule chrome yellow can be obtained by different silicon coated density levels. The properties and application range of BASF Sicomin capsule chrome yellow in plastic are listed in Table 3.16 and Table 3.17.

Table 3.16   The properties of BASF Sicomin capsule chrome yellow in plastic

Grade	Pigment	Titanium dioxide	Heat resistance /℃	Light resistance level	Weather resistance level (3000 h)	Resistance transference level	Warpage
Sicomin Yellow K1630	Full shade	1 %		250	8	5	5	No
	Dilution	1∶4		280	8	5		No

continued

Grade		Pigment	Titanium dioxide	Heat resistance /°C	Light resistance level	Weather resistance level (3000 h)	Resistance transference level	Warpage
Sicomin Yellow K1922	Full shade	1 %		220	7	4	5	No
	Dilution		1:4	220	8	4		No
Sicomin Yellow K1925	Full shade	1 %		240	8	3	5	No
	Dilution		1:4	260	8	3~4		No

Table 3.17 The application range of BASF Sicomin capsule chrome yellow in plastic

LL/LDPE	○	PVC(rigid)	●	PMMA	×
HDPE	○	PS	○	PA	×
PP	○	ABS	×	PC	×
PVC(soft)	●	PBT	×	PUR	●

● Recommended to use, ○ Conditional use, × Not recommended to use.

(2) Red chrome pigments

In 1863, Schultz observed that a special bright orange hue ore would be formed when yellow lead molybdate together with lead chromate. Later some high temperature melting experiments with lead chromate and molybdate aluminum mixed in different proportions have been done, and a valuable red inorganic pigment was invented, patented in 1933. From 1934 to 1935, red chrome molybdenum began to appear in the US markets, and used in the paint industry, plastics industry, printing ink industry.

Pigment Red 104 Commonly known as molybdate red, containing lead, lead chromate and lead sulfate. The formula change between $25PbCrO_4·4PbMoO_4·PbSO_4$ and $7PbCrO_4·PbMoO_4·PbSO_4$, various varieties of orange to red can be obtained from the three ingredients with different molecular ratios.

The major properties The color of molybdate red is bright, with bright red orange. But the use in plastic is greatly limited because of the heat-resistant and other factors. The heat resistance, corrosion resistance and sulfidation resistance of capsule molybdate red products are greatly increased. The properties and application range of BASF Sicomin capsule molybdate red in plastic are listed in Table 3.18 and Table 3.19.

Table 3.18 The properties of BASF Sicomin capsule molybdate red in plastic

Grade		Pigment	Titanium dioxide	Heat resistance (HDPE)/°C	Light resistance level	Weather resistance level (3000 h)	Resistance transference level	Warpage
Sicomin Red K 3023	Full shade	1 %		260	8	3	5	No
	Dilution		1:4	280	8	3~4		No
Sicomin Red K 3030 S	Full shade	1 %		260	8	4	5	No
	Dilution		1:4	280	8	4~5		No

Chapter 3 Main Types and Properties of Inorganic Pigments and Specialty Effect Pigments

continued

Grade		Pigment	Titanium dioxide	Heat resistance (HDPE)/°C	Light resistance level	Weather resistance level (3000 h)	Resistance transference level	Warpage
Sicomin Red 3034 S	Full shade	1%		260	8	4	5	No
	Dilution		1:4	280	8	4-5		No
Sicomin Red 3130 S	Full shade			260	8	4	5	No
	Dilution		1:4	280	8	4~5		No

Table 3.19 The application range of BASF Sicomin capsule molybdate red in plastic

LL/LDPE	○	PVC(rigid)	●	PMMA	×
HDPE	○	PS	○	PA	×
PP	○	ABS	×	PC	×
PVC(soft)	●	PBT	×	PUR	●

● Recommended to use, ○ Conditional use, × Not recommended to use.

(3) Green chrome pigment

In general, chromium oxide green, light olive green have two hues with and dark olive green. Though the hue is not bright enough, hiding comparatively is still good, refractive index is 2.5, and tinting strength is worse than phthalocyanine green, due to the high temperature calcinations. it can crush to a certain fineness, but the particles maintain a certain hardness, Mohs hardness 9.

C.I. Pigment Green 17   Chrome oxide green, chemical constitution: Cr_2O_3; C.I. structure number: 77288, CAS registry number: [68909-79-5].

The major properties   The tinting strength of chrome oxide green in plastics is not high, but the hiding power comparatively is still good. A good light resistance and weather resistance, heat resistant can be achieved to 1000 ℃/5min with no color changing. The grades and properties of Lanxess COLORTHERM® chrome oxide green are listed in Table 3.20 and the application range is listed in Table 3.21.

Table 3.20 The grades and properties of Lanxess COLORTHERM® chrome oxide green

Grades	Cr_2O_3 content/%	Heat resistance /°C	Thermal weight loss (1000℃, 0.5h)/%	pH	Oil absorption/ (g/100g)	Scalpings (45mm,max)/ %	Density /(g/cm^3)	Particle shape
Green GN-M	98.5~99.5	>260	Max0.4	5.7	11	Max0.005	5.2	Globular
Green GN	98.5~99.5	>260	Max0.4	5.7	11	Max0.02	5.2	Globular
Green GX	98.5~99.5	>260	Max0.4	5.7	11	Max0.02	5.2	Globular

Table 3.21 The application range of Lanxess COLORTHERM® chrome oxide green

LL/LDPE	●	PVC(rigid)	●	PMMA	●
HDPE	●	PS	●	PA	●
PP	●	ABS	●	PC	●
PVC(soft)	●	PBT	●	PUR	●

● Recommended to use.

3.5.2.2 Cadmium inorganic pigment

Cadmium pigments are divided into two major categories of red and yellow, the color is mainly determined by the anion of cadmium salt, the color of pigment containing zinc is greenish yellow, and the color of pigment containing selenium is orange and red sauce.

Cadmium pigments are bright color, with high tinting strength, no migration, and can be used for coloring of almost all engineering plastics. But the EU and the United States has made clear restrictions on the use because of the containing of heavy metals cadmium. Due to its excellent properties, especially the heat resistance, it is still used in some special fields, such as polyene amine, polyoxymethylene, polytetrafluoroethylene and other processing of high temperature engineering plastics.

Cadmium pigments should be used for a low water absorption system because the water permeability of medium will reduce the weather resistance of pigment. The presence of titanium white will also reduce the light resistance of pigment.

The impact on the environment (security) Cadmium pigment is a compound with low solubility. However, a small amount of cadmium can be dissolved in acid (The concentration correspond to the concentration of human gastric acid). Long-term ingestion of cadmium pigments would lead accumulation in the human body, particularly in the kidney. Nevertheless, the toxic of cadmium pigments is much lower than the other cadmium compounds (several orders of magnitude). The European Parliament has already listed cadmium sulfide as a category 3 carcinogen.

A series of laws and regulations enacted at home and abroad in recent years had clearly defined that the cadmium content in per kilogram of plastic packaging materials should not exceed 100mg/kg. Thus the use of cadmium pigments is limited, the amount is greatly reduced.

The major foreign suppliers ROCKWOOD.

The major domestic suppliers Hunan Jufa Technology Co., Ltd.

(1) Yellow cadmium pigment

The color of cadmium yellow is bright and saturated. The chemical composition of cadmium yellow basically is cadmium sulfide or solid solution of cadmium sulfide and zinc sulfide. The yellowness of cadmium yellow containing zinc sulfide will become shallow to pale yellow with the increase of the amount of solid solution of zinc sulfide. The industrial production of cadmium yellow such as pale yellow (primrose yellow), bright yellow (lemon yellow), yellow (medium yellow), deep yellow (gold) and orange are listed in Table 3.22.

Table 3.22 List of the typical chemical composition of cadmium yellow

Name		CdS	ZnS
Pigment Yellow 35	Primrose yellow	79.5	20.5
	Lemon yellow	90.9	9.1
	Yellow	93.4	6.6
	Deep gold	98.1	1.9
	Orange	45	($CdCO_3$ 55%)

Cadmium yellow has two stable forms in room temperature. One is the β-CdS, belongs to the cubic type. The other is α-CdS, belongs to the hexagonal type. The former is called low temperature stability type, heat resistance is less than or equal to 500℃. The latter is called high temperature stability type, melting point is 1405℃, heat resistance is greater than or equal to 600℃. At room temperature and a range of 500℃, two crystal types of cadmium yellow can stable co-existence. A single grain size of cubic lattice is less than or equal to l00nm, while the single grain size of hexagonal crystal is100~280nm.

The relative density of cadmium yellow is 4.5~5.9, and the density of pale yellow is smaller than deep yellow, the lipophilic of β-CdS is stronger than α-CdS.

Cadmium Yellow   C.I. Pigment Yellow 35.

Chemical constitution   CdS, ZnS. C.I. structure number: 77205. CAS registry number: [8048-07-05].

The major properties   Pigment Yellow 35 is commonly known as cadmium yellow. The scope of chromatography range from pale yellow, yellow to red yellow. Cadmium yellow has a high tinting strength, bright color. The saturation can be up to 80%~90%, and present translucent when coloring of plastic. The light resistance and weather resistance of cadmium yellow are excellent, but the light stability inferiors to cadmium red in outdoor. The grades and properties of ROCKWOOD cadmium yellow are listed in Table 3.23 and the application range is listed in Table 3.24.

Table 3.23 The grades and properties of ROCKWOOD cadmium yellow

Grade	Color	Light resistance level	Heat resistance(5min)/℃	Acid resistance	Alkali resistance
P7201	Green yellow	7	400	Good	Excellent
P3682	Yellow	7	400	Good	Excellent
P3680	Red yellow	7	400	Good	Excellent
P1101	Red yellow	7	400	Good	Excellent

Table 3.24 The application range of ROCKWOOD cadmium yellow

LL/LDPE	●	PS	●	Rubber	●
HDPE	●	PS-HI	●	PA	●
PP	●	ABS	●	PC	●
PVC(soft)	○	PMMA	●	UP	●
PVC(rigid)	○	CAB			

1. The cadmium pale yellow containing CdS used in polyethylene, should try to shorten the processing time, so that the CdS promote the decomposition of polyethylene plastic and present black.

2. ● Recommended to use, ○ Conditional use.

(2) Orange red cadmium pigment

The color of cadmium red is very saturated and bright, the scope of chromatography can range from yellow red, by red until garnet. The light reflection wavelength of cadmium red increases with the amount of solid solution of CdSe increasing. The higher content of CdSe in cadmium red, the stronger the red light, the darker the color. Typical chemical composition of cadmium red is listed in Table 3.25.

Table 3.25 List of typical chemical composition of cadmium red

Name		CdS	CdSe
Cadmium orange: Pigment Orange 20	Orange red	82	18
Cadmium red: Pigment Red 108	Pale red	69	31
	Dark red	58	42
	Purplish red	50	50

The particle morphology of cadmium red is substantially spherical, and the major crystal structure is hexagonal, also have cubic crystal.

Cadmium Red C.I. Pigment Orange 20, Pigment Orange Red 108. Chemical constitution: CdS, CdSe. C.I. structure number: 77202. CAS registry number: [12656-57-4]

Pigment Orange 20, Pigment Orange Red 108 are commonly known as cadmium orange, cadmium red and cadmium scarlet. The tinting strength and hiding power of cadmium orange and cadmium red are strong. The color is very saturated and bright. Cadmium orange and cadmium red have excellent heat resistance about 600℃, light resistance and weather resistance. The light stability is better than cadmium yellow in outdoor, but it decrease after adding titanium dioxide. Cadmium orange and cadmium red are insoluble in water, organic solvents, oils and alkaline solvent, but its disadvantage is poor acid, slightly soluble in weak acid, dissolved in acid and emit toxic gases H_2Se and H_2S. Equipped with leaded additives may cause black lead sulfide. The grades and properties of ROCKWOOD cadmium orange and cadmium red are listed in Table 3.26 and the application range is listed in Table 3.27.

Table 3.26 The grades and properties of ROCKWOOD cadmium orange and cadmium red

Grade	Color	Light resistance level	Heat resistance(5min)/℃	Acid resistance	Alkali resistance
P5150	Yellow orange	7	400	Good	Excellent
P5155	Orange	7	400	Good	Excellent
P4701	Red orange	7	400	Good	Excellent
P4702	Red orange	7	400	Good	Excellent
P4703	Yellow red	7	400	Good	Excellent
P4704	Red	7	400	Good	Excellent
P4705	Bright red	7	400	Good	Excellent
P4706	Claret	7	400	Good	Excellent
P4707	Crimson	7	400	Good	Excellent
P4708	Purplish red	7	400	Good	Excellent

Table 3.27 The application range of ROCKWOOD cadmium orange and cadmium red

LL/LDPE	●	PS	●	Rubber	●
HDPE	●	PS-HI	●	PA	●
PP	●	ABS	●	PC	●
PVC(soft)	○	PMMA	●	UP	●
PVC(rigid)	○	CAB	●		

● Recommended to use, ○ Conditional use.

3.5.2.3 Iron Oxide Inorganic Pigment

Iron oxide pigment, with wide chromatography, high hiding power, strong tinting strength, is primary composed of red, yellow and black. The increasing importance of iron oxide pigment is based on the non-toxic, excellent light-fastness, weather resistance, acid resistance, alkali resistance and solvent resistance, as well as good cost performance. Iron oxide pigment is widely used in the plastics industry.

The impact on the environment (security)   A large amount of poison and toxicology test showed that iron oxide has no harm to human tissue. Iron oxide pigments produced from pure raw materials can be used for coloring food and pharmaceutical products. Synthetic iron oxide containing no crystalline silica, so even in California, still defined as no toxicity.

Product market   In the worldwide, the production and sales of iron oxide pigments, followed titanium dioxide, is the second inorganic pigments with large production and wide application, approximately one million tons of iron oxide pigment consumption (except the mica iron oxide used as antirust materials, magnetic iron oxide,used as magnetic recording materials), with sales of over 10 billion dollars per year. The production abroad is about 300,000 tons, while 600,000 tons in China. Many foreign companies buy crude iron oxide from China and then sale it after reprocessing.

The major foreign suppliers   (Lanxess) COLORTHERM®, (ROCKWOOD) Solaplex™.

The major domestic suppliers   Shanghai Yipin Pigment Tech. Co., Ltd. Dengqinghuayuan of HuaSheng Group Pigment Tech.Co.,Ltd. Yixing yuxing Tech. Co., Ltd.

(1) Yellow iron oxide pigment

Yellow iron oxide pigment (C.I. Pigment Yellow 42) begin to dehydrate at 150~200℃, and the color gradually become red and be iron red ultimately. With the coating treatment in iron yellow surface, it can be applied in a certain high temperature occasion. The chemical composition of iron oxide yellow pigment (C.I. Pigment Yellow 119) is zinc ferrite with the color of brown, which is produced by the calcination method, and heat-resistant is up to 300℃.

① **C.I. Pigment Yellow 42**   Chemical composition: [α, λFeO(OH)], C.I. Structure number 77492. CAS registration Number. [20344-49-4].

The major properties   Pigment Yellow 42, commonly known as yellow iron oxide, is bright ocher yellow, almost equal to the chrome yellow, with strong hiding power, stable to light effect, light resistance up to 6~8 and excellent weather resistance. Due to the crystallization water, Iron oxide yellow pigments begin to dehydrate when heated to 150~200℃ with poor heat resistance, only suitable for rubber, EVA coloring and so on. What's more, it is relatively cheap and nontoxic. The performance of German Lanxess Bayer Lok Wong (BAYFERROX) 4920, 4960 is showed in table 3.28.

Table 3.28   Performance and model number of German Lanxess Bayer Lok Wong

Model number	Color	Relative density	Particle size/mm	Particle shape	Oil absorption/(g/100g)	Heat resistance /℃
Yellow 4905	Yellowish	4.0	0.1*0.8	Acicular	32	140
Yellow 4910	Yellowish	4.0	0.1*0.8	Acicular	32	140
Yellow 4920	Yellowish	4.0	0.1*0.8	Acicular	32	140
Yellow 4960	Purple-yellow	4.3	0.1*0.8	Acicular	28	140

In order to improve the heat resistance, iron oxide yellow is coated by the mixture of silicon oxide and aluminum oxide, and then add the solution of sodium silicate into the suspension of iron oxide yellow. Then adjust the pH value with acid to neutral. Silicate sodium transformed into silicate, hydrolyzed into silica and precipitated on the surface of iron oxide yellow. The coated iron oxide yellow pigments can resist heat up to 220℃. The performance of German Lanxess Bayer Lok Wong (BAYFERROX) 10, 20 is showed in Table 3.29. Application range is shown in Table 3.30.

Table 3.29   Performance of German Lanxess Bayer Lok Wong (BAYFERROX) 10, 20

Model number	Color index number	Fe_2O_3 content/%	Heat resistance /℃	Oil absorption/(g/100g)	Density /(g/cm^3)	Particle size/mm	Particle shape
Yellow10	P.Y.42	69~71	260	65	3.8	0.1*0.7	Acicular
Yellow 20	P.Y.42	69~71	260	45	3.8	0.1*0.7	Acicular

Table 3.30   Application of German Lanxess Bayer Lok Wong (BAYFERROX) 10, 20

LL/LDPE	●	PS	○	Rubber	●
HDPE	○	PS-HI	○	PA	○
PP	○	ABS	○	PC	○
PVC-P	●	PMMA	○	UP	●
PVC-U	○	PET	○		

● Recommended to use, ○ Conditional use.

Because of the excellent light and weather resistance, coated iron oxide yellow are usually matched with phthalocyanine blue or phthalocyanine green to make artificial lawn, as well as available in light color and coloring.

② **C.I. Pigment Yellow 119** Zinc iron yellow. Chemical composition: [$ZnFe_2O_4$]. Structure number: 77496. CAS registration number. [68187-51-9]

The major properties Pigment Yellow 119 is one kind of yellow pigment mixed iron oxide with the crystal shape of spinel. Pigment Yellow 119 has a good chemical stability, good hiding power and excellent heat resistance of more than 300℃. It can be used in coloring of engineering plastics, available in plastics processing recipes without the pigments of lead and cadmium, as well as in food. What's more, zinc iron yellow can be mixed with iron oxide red to get orange or light brown chromatography. When added into HDPE, it doesn't cause deformation.

The performance of German Lanxess COLORTHERM® iron oxide yellow is shown in Table 3.31. Application range, as shown in Table 3.32.

Table 3.31 Model number and performance of German Lanxess COLORTHERM® iron oxide yellow

Model number	Color index number	Fe_2O_3 content/%	Heat resistance/℃	Oil absorption/(g/100g)	Density /(g/cm^3)	Particle size/(mm)	Particle shape
Yellow 30	P.Y.119	65~67	300	14	5.2	0.15*0.5	slender

Table 3.32 Application of German Lanxess COLORTHERM® iron oxide yellow

LL/LDPE	●	PS	●	Rubber	●
HDPE	●	PS-HI	●	PA	○
PP	○	ABS	●	PC	●
PVC-P	○	PMMA	●	UP	●
PVC-U	○	PET	○		

● Recommended to use, ○ Conditional use.

(2) Red iron oxide pigment The chemical name of red iron oxide is ferric oxide (Fe_2O_3), which is the most stable compound in iron oxide, The red color of iron oxide ranges from reddish yellow to dark red, depended on the purity of iron, the number and size.

C.I. Pigment Yellow 101 Chemical composition: [Fe_2O_3]. C.I. structure number 77491. CAS registration number. [1309-37-1].

The major properties Iron oxide red pigment has high hiding power, strong tinting strength, and has excellent heat resistance while the temperature above 300℃, as well as great light fastness, weather resistance, acid resistance, alkali resistance and solvent resistance. The performance of Germany Lanxess COLORTHERM® red iron oxide pigment are shown in Table 3.33. Application range is shown in Table 3.34.

Table 3.33 Performance of German Lanxess Bayer Lok Wong (BAYFERROX) red iron oxide pigment

Model number	Fe_2O_3 content/%	Heat resistance/℃	Oil absorption/(g/100g)	Density /(g/cm³)	Particle size/mm	Particle shape
Red 110M	94~96	>300	28	5.0	0.09	Acicular
Red 120NM	95~96	>300	28	5.0	0.11	Acicular
Red 120M	95~96	>300	28	5.0	0.12	Acicular
Red 130M	95~96	>300	26	5.0	0.17	Acicular
Red 140M	96~96	>300	24	5.0	0.30	Acicular
Red 160M	96~96	>300	22	5.1	0.40	Acicular
Red 180M	96~96	>300	18	5.1	0.70	Acicular
Red 520M	95	>300	26	5.0	0.20	Acicular

Table 3.34 Application of German Lanxess COLORTHERM® iron oxide red iron oxide pigment

LL/LDPE	●	PS	●	Rubber	●
HDPE	●	PS-HI	●	PA	●
PP	●	ABS	●	PC	●
PVC-P	●	PMMA	●	UP	●
PVC-U	○	PET	●		

● Recommended to use, ○ Conditional use.

(3) Brown iron oxide pigment Iron oxide brown pigments can be regarded as one of the iron oxide red (Pigment Red 101) variants, in which the iron is replaced by manganese with the color ranging from reddish brown to brown.

C.I. Pigment Brown 43 (77536) Chemical composition: $[(Fe·Mn)_2O_3]$. C.I structure number 77536. CAS registration number [12062-81-6].

The major properties Iron oxide brown pigment is inert, with high hiding power, strong tinting strength. It has excellent heat resistance of more than 300℃, as well as great light fastness, weather resistance. The performance of Germany Lanxess COLORTHERM® brown iron oxide pigment 43 is shown in Table 3.35. Application range, as shown in Table 3.36.

Table 3.35 Performance of German Lanxess COLORTHERM® brown iron oxide pigment

Model number	Fe_2O_3 content/%	Heat resistance /℃	Oil absorption /(g/100g)	Density /(g/cm³)	Particle size (maximum)/mm	Particle shape
Brown645T	80~88	>300	28	4.7	0.3	Acicular

Table 3.36 Application of German Lanxess COLORTHERM® iron oxide red iron oxide pigment

LL/LDPE	●	PS	●	Rubber	●
HDPE	●	PS-HI	●	PA	●
PP	●	ABS	○	PC	●
PVC-P	●	PMMA	●	UP	●
PVC-U	○	CAB	●		

● Recommended to use, ○ Conditional use.

3.5.2.4 High-performance composite inorganic pigment

High-performance composite inorganic pigment is one dopant crystal formed by one or several metal ions doped into the lattice of metal oxides. The doped ions can lead to incident light interference, so that some wavelengths are reflected and the rest is absorbed, thus make it to be color pigments. The high-performance color composite inorganic pigment is international called of CICP or MMO. On the chemical structure, the pigment of CICP is regarded as stable solid solution, that is to say, various metal oxides uniformly distributed in the new chemical compounds in the lattice, as it is similar to the solution, but actually a glassy material in the solid state. These compounds have different crystal structures, including rutile, spinel, inverse spinel, hematite and unusual column red stone and pseudo-brookite type structures.

CICP pigment is obtained through high-temperature solid phase reaction, so that it has excellent durability, high hiding power, excellent heat resistance, light resistance, weather resistance, chemical stability, high infrared reflection performance and UV shielding ultraviolet ability. It can improve the weather resistance, light resistance, heat preservation of plastic products, as well as prolonging the service life of products.

CICP pigment has a good compatibility with most of the resin, and widely used in engineering plastics, outdoor architectural components (e.g. PVC slide, PVC plastic windows, PVC deck, PVC fences, metal roof, the crane arm, taxi, telephone booths, postal mailbox, school bus).

The impact on the environment (security) The rutile lattice of high performance composite inorganic pigment absorbs nickel oxide, chromium oxide (III) or manganese oxide as color component. The nickel, chromium, manganese, antimony and other elements in rutile pigment fill the original crystal defect in the titanium dioxide, forming a more complete crystal structure and improved the stability of the crystal lattice. These metallic elements lost their original chemical, physical and physiological properties. So this kind of rutile pigment can't be considered as nickel, chromium, antimony or simple oxide compound. The inertia of high performance composite inorganic pigment is very high, the hot exudation less than 2×10^{-8}. As a result, the gastric acid can't dissolve it, so even into the stomach, it is no harmful to health. It is absolutely safe when personal contact, for this reason, do not classified them as hazardous substances. Described by the manufacturer for compatibility, purity, and security processing, most of these pigments is considered non-toxic, and conformed to the requirements of food contact safety rules.

The major properties CICP pigment has high hiding power, excellent heat resistance, more than 300 ℃, great light and weather resistance as well as chemical stability.

Main applications CICP pigment is applicable in coloring for polyethylene, polypropylene, polystyrene, ABS, polyester, polycarbonate, nylon 6, nylon 66, phenol resin, epoxy resin and other plastic.

Main markets of product It was estimated that the world demand is more than 50000 tons, about 5000 tons produced in China.

The major foreign suppliers BASF, Shepherd, FERRO, ROCKWOOD Solaplex™, HEUBACH TICO®, (TOMATC) TOMATC®.

The major domestic suppliers Nanjing Pigments Tech.Co.,Ltd, HuNan Jufa Pigment Tech.Co.,Ltd.

(1) Yellow and orange composite inorganic pigment

Yellow and orange CICP pigment is made by calcined titanium dioxide and metal oxides in high temperature. The color is caused by that titanium of rutile TiO_2 structure located in the center of the ligand is replaced by the color elements, resulting in a distinctive green or red light, yellow and yellow-brown and orange. The main varieties are shown in Table 3.37.

Table 3.37 The main varieties of Yellow and orange composite inorganic pigment

Variety	Chemical composition	Structure	CAS Registration Number	Color
Pigment Yellow 53	$(Ti,Ni,Sb)O_2$	Rutile	8007 18-9	Greenish yellow
Pigment Yellow 157	$(2NiO \cdot 3BaO \cdot 17TiO_2)$	Rutile	68610-24-2	Yellow
Pigment Yellow 161	$(Ti,Ni,Nb)O_2$	Rutile	68611-43-8	Yellow
Pigment Yellow 162	$(Ti,Cr,Nb)O_2$	Rutile	68611-42-7	Yellow-brown
Pigment Yellow 163	$(Ti,Cr,W)O_2$	Rutile	68186-92-5	Yellow-brown
Pigment Yellow 164	$(Ti,Mn,Sb)O_2$	Rutile	68412-38-4	Yellow-brown
Pigment Yellow 189	$(Ti,Ni,W)O_2$	Rutile	69011-05-08	Yellow
Pigment Yellow 216	$[(Sn,Zn)TiO_3]$	Rutile	85536-73-8	Yellow
Pigment Yellow 227	$(Sn,Zn)_2Nb_2(O,S)_7$	Rutile	1374645-21-2	Bright orange
Pigment Orange 82		Rutile		Orange

① **C.I. Pigment Yellow 53** Chemical composition: $[(Ti, Ni, Sb)O_2]$. Structure No. 77788. CAS registration No. [8007-18-9].

The major properties Pigment Yellow 53 is made by calcination of titanium oxide, nickel, antimony oxide and a small amount of additives in high temperature. The color is greenish yellow, the tinting strength is lower than organic pigment. When coloring for plastic, the hiding power is excellent. The crystal form of Pigment Yellow 53 is very stable, with heat resistance of more than 800℃. Besides, it has great light and weather resistance, reaching to level 8 of light resistance and level 5 of weather resistance. It have good compatibility with most resin. The properties of Pigment Yellow

53 of BASF Sicotan in the plastic is shown in Table 3.38.

Table 3.38 Properties of Pigment Yellow 53 of BASF Sicotan in the plastic

Brand	Pigment	Titanium dioxide	Heat resistance /°C	Light resistance grade	Weather resistance grade (3000 h)	Migration resistance grade	Warpage variant
Sicotan Yellow K1010	Full shade	1 %	300	8	5	5	None
	diluent	1 : 4	300	8	5		None
Sicotan Yellow K1011	Full shade	1 %	300	8	5	5	None
	diluent	1 : 4	300	8	5		None
Sicotan Yellow K2001	Full shade	1 %	300	8	5	5	None
	diluent	1 : 4	300	8	5		None
Sicotan Yellow K2011	Full shade	1 %	300	8	5	5	None
	diluent	1 : 4	300	8	5		None

② **C.I. Pigment Yellow164** Chemical composition: $[(Ti, Mn, Sb)O_2]$. Structure No. 77899. CAS Registration No. [68412-38-4].

The major properties Chemical composition of Pigment Yellow 164 is made of manganese, antimony, titanium and brown, with rutile structure, in which the nickel is replaced by the manganese in Pigment Yellow 53. Pigment Yellow 164 is bright yellow brown, low tinting strength. When coloring for plastic, the hiding power is excellent. The crystal form of Pigment Yellow 164 is very stable, with heat resistance of more than 800℃. Besides, it has great light and weather resistance, reaching to level 8 of light resistance and level 5 of weather resistance. The properties of Pigment Yellow 164 of BASF Sicotan in the plastic is shown in Table 3.39.

Table 3.39 Properties of Pigment Yellow 164 of BASF Sicotan in the plastic

Brand	Pigment	Titanium dioxide	Heat resistance/°C	Light resistance grade	Weather resistance grade (3000 h)	Migration resistance grade	Warpage variant
Sicotan Brown K2611	Full shade	1 %	300	8	5	5	None
	diluent	1 : 4	300	8	5		None
Sicotan Brown K2711	Full shade	1 %	300	8	5	5	None
	Diluent	1 : 4	300	8	5		None

③ **C.I. Pigment Yellow189** Chemical composition: $[(Ti, Fe, Zn)O_2]$. Structure No. 77902. CAS Registration No. [69011-05-8].

The major properties Pigment Yellow 189 is also known as zinc, iron yellow, dark reddish yellow. The hue can be changed by adding aluminum, magnesium, titanium. When coloring for plastic, the hiding power and tinting strength is excellent. It has a good compatibility with most resin, and could coloring them. Because they do not contain heavy metals such as chromium, known as the new generation of green pigment.

④ **C.I. Pigment Yellow216** Chemical composition: $[(Sn,Zn)TiO_3]$. CAS registration No. [85536-73-8].

The major properties Pigment Yellow 216 is a novel mixed metal oxide pigment, the color is bright orange. It fills the gap of high-performance inorganic pigment, the color and performance of the products, conventional inorganic and organic pigments can't match. Pigment Yellow 216 has high hiding power, excellent dispersion, good weather, light and heat resistance, perfect chemical resistance, no migration and warping phenomenon. The pigment is a substitute for lead chrome pigments, which conformed to the environmental requirements and can provide a good cost-effective and high performance. The properties of Pigment Orange 10P320 and 10P340 of American Shepherd in the plastic is shown in Table 3.40.

Table 3.40 Properties of Pigment Orange 10P320 and 10P340 of American Shepherd in the plastic

Brand		Pigment	Titanium dioxide	Heat resistance /°C	Light resistance grade	Weather resistance garde (3000h)	Migration resistance grade	Warpage variant
Orange 10P320	Full shade	1%		320	8	5	5	None
	diluent		1:4	320	8	5	5	None
Orange 10P340	Full shade	1%		320	8	5	5	None
	Diluent		1:4	320	8	5	5	None

⑤ **C.I. Pigment Yellow227** Chemical composition: $[(Sn,Zn)_2Nb_2(O,S)_7]$. C.I. structure No. 777895. CAS registration No. [1374645-21-2].

The major properties Pigment Yellow 227 is a novel composite CICP pigment, which color is yellow. The pigment is a substitute for lead chrome pigments, which is in line with environmental requirements, and a cost-effective, high-performance pigments. The color and performance of the products, conventional inorganic and organic pigments can't match, and it fills the gap of high-performance inorganic pigment. Pigment Yellow 227 has high hiding power, excellent dispersion, good weather, light, heat resistance; perfect chemical resistance, no migration and warping phenomenon. The properties of Pigment Yellow 10P150 of American Shepherd in plastics is shown in Table 3.41.

Table 3.41 Properties of Pigment Yellow 10P150 of American Shepherd in the plastic

Brand	Pigment	Titanium dioxide	Heat resistance/℃	Light resistance grade	Weather resistance garde (3000 h)	Migration resistance grade	Warpage variant
Yellow 10P150	Full shade	1%	320	8	5	5	None
	Diluent	1:4	320	8	5	5	None

⑥ **C.I. Pigment Orange 82** Chemical composition: [(Sn, Zn,Ti) O_2].

The major properties Pigment Yellow 82 is a novel composite inorganic pigment, which color is orange. The properties of pigment Orange 82 of Germany BASF in plastics is shown in Table 3.42.

Table 3.42 Properties of Pigment Orange 82 of BASF in plastics

Brand	Pigment	Titanium dioxide	Heat resistance/℃	Light resistance grade	Weather resistance grade (3000 h)	Migration resistance grade	Warpage variant
Sicopal Orange K2430	Full shade	1%	300	8	5	5	None
	Diluent	1:4	300	8	5		None

(2) Brown composite inorganic pigment

Brown composite inorganic pigment is a mixed phase composite of rutile lattice (TiO_2), colored metal (such as chromium, nickel, manganese) and the colorless antimony, niobium or tungsten. Its color is yellow or brown. The main brand is shown in Table 3.43.

Table 3.43 Varieties of brown titanium and nickel inorganic pigment

Pigment index	Chemical composition	Structure	CAS No.	Color
Pigment Brown 24	(Ti,Cr,Sb)O_2	Rutile	68186-90-3	Taupe phase
Pigment Brown 33	(Zn,Fe) (Fe,Cr)$_2O_4$	Rutile	68186-88-9	Brown
Pigment Brown 29	(Fe,Cr)$_2O_2$	Hematite	12737-27-8	Brown
Pigment Brown 48	Fe_2TiO_5	Spinel	1310-39-0	Dark brown

① **C.I. Pigment Orange 24** Chemical composition: [(Ti, Cr, Sb)O_2]. C.I structure No. 77310. CAS registration No. [68186-90-3].

The major properties Pigment Orange 24 is made by the calcination of titanium dioxide, chromium oxide and antimony oxide in high temperature. Chromium cation can change the color from white to yellowish-brown. The tinting strength of Pigment Orange 24 is equal to ferrite yellow, with excellent hiding power. When coloring for plastic, the hiding power is excellent. The crystal form of Pigment Orange 24 is very stable, with heat resistance more than 800℃.Besides, it has great light and weather resistance, reaching to level 8 of light resistance and level 5 of weather resistance. The properties of

Sicotan yellow nickel and titanium Pigment Brown 24 of BASF in plastics is shown in Table 3.44.

Table 3.44 Properties of Pigment Brown 24 of BASF Sicotan in plastics

Brand	Pigment		Titanium dioxide	Heat resistance /°C	Light resistance grade	Weather resistance grade (3000 h)	Migration resistance grade	Warpage variant
Sicotan Yellow K2001	Full shade	1 %		300	8	5	5	None
	Diluent		1:4	300	8	5		None
Sicotan Yellow K2011	Full shade	1 %		300	8	5	5	None
	Diluent		1:4	300	8	5		None
Sicotan Yellow K2111	Full shade	1 %		300	8	5	5	None
	Diluent		1:4	300	8	5		None
Sicotan Yellow K2112	Full shade	1 %		300	8	5	5	None
	Diluent		1:4	300	8	5		None

② **C.I. Pigment Brown 29**   Chemical composition: $(Fe, Cr)_2O_2$. C.I structure No. 77500. CAS registration No. [12737-27-8].

The major properties   Pigment Orange 29 is made by the calcination of different quantity of chromium oxide and iron oxide in high temperature, with the chemical structure of spinel. Its color is red-brown and the tinting strength is not high. The crystal form of Pigment Brown 29 is very stable, with heat resistance of more than 800 °C. Besides, it has great light and weather resistance, reaching to level 8 of light resistance and level 5 of weather resistance. The properties of German BASF Sicopal Pigment Brown 29 in plastics is shown in Table 3.45.

Table 3.45 Properties of Pigment Brown 29 of German BASF Sicopal in plastics

Brand	Pigment		Titanium dioxide	Heat resistance /°C	Light resistance grade	Weather resistance grade (3000 h)	Migration resistance grade	Warpage variant
Sicopal Brown K2795	Full shade	1 %		300	8	3	5	None
	Diluent		1:4	300	300	3-4		None

③ **C.I. Pigment Brown 48**   Chemical composition: Fe_2TiO_5. C.I structure No. 77543. CAS Registration No. [68187-02-0].

The major properties   The chemical structure of Pigment Brown 48 is spinel. The color is yellow-brown and the tinting strength is not high. The crystal form of Pigment Brown48 is very stable, with heat resistance of more than 800 °C. Besides, it has great light and weather resistance, reaching to level 8 of light resistance and level 5 of weather resistance. Because of no chromium and other heavy metals, it is called a new generation of environmentally friendly paint.

(3) Blue composite inorganic pigments (cobalt blue)

Cobalt blue pigment is basically have the same main ingredient. Blue28 is made by the calcination of cobalt oxide (Co_2O_3) and alumina oxide(Al_2O_3) in high temperature. The variant is that cobalt is partially substituted by alone or together of chromium and zinc. So the products have a great changes on the market. These blue pigments have different tinting strength and hue (red and blue or green and blue). These pigments form spinel lattice. The chemical composition and color of Cobalt blue pigment are shown in Table 3.46.

Table 3.46 Chemical composition and color of cobalt blue pigment

Pigment index	Chemical composition	Structure	CAS No.	Color
Pigment Blue 28	$CoAl_2O_4$	Spinel	1345-16-0	Reddish blue
Pigment Blue 36	$Co(Al,Cr)_2O_4$	Spinel	68187-11-1	Greenish blue

① **C.I. Pigment Blue 28** Chemical composition: $CoAl_2O_4$. C.I structure No. 77346. CAS registration No. [1345-16-0].

The major properties Pigment Blue 28 is commonly known as cobalt blue, bright colors. It has low tinting strength, compared with Pigment Blue 15. Cobalt blue has a good heat resistance, more than 1000℃. Besides, The light resistance of cobalt blue is 8, and the migration resistance is 5. Pigment Blue 28 has excellent chemical resistance and have a good compatibility with most of the resin. But its price is relatively expensive, relatively few applications in plastic, only in special circumstances, such as fluorine plastic. The properties of German BASF Sicopal Pigment Blue 28 in plastics is shown in Table 3.47.

Table 3.47 Properties of German BASF Sicopal Pigment Blue 28 in plastics

Brand	Pigment	Titanium dioxide	Heat resistance /℃	Light resistance grade	Weather resistance grade (3000 h)	Migration resistance grade	Warpage variant
Sicopal Blue K 6210	Full shade	1%	300	8	5	5	None
	Diluent	1:4	300	8	5		None
Sicopal Blue K 6310	Full shade	1%	300	8	5	5	None
	Diluent	1:4	300	8	5		None

② **C.I. Pigment Blue 36** Chemical composition: $Co(Cr, Al)_2O_4$. C.I structure No. 77343. CAS registration No. [68187-11-1].

The major properties Pigment Blue 36 is commonly known as cobalt chrome blue, bright greenish blue. It has low tinting strength, compared with Pigment Blue 15, has a good heat resistance, more than 1000℃. Besides, the light resistance of cobalt blue is 8, and the migration resistance is 5. It has an excellent chemical resistance. The properties of German BASF Sicopal pigment Blue 36 in Plastics is shown in Table 3.48.

Table 3.48 Properties of German BASF Sicopal pigment blue 36 in plastics

Brand	Pigment	Titanium dioxide	Heat resistance /°C	Light resistance grade	Weather resistance garde (3000 h)	Migration resistance grade	Warpage variant
Sicopal Blue K 6710	Full shade	1%	300	8	5	5	None
	Diluent	1:4	300	8	5		None
Sicopal Blue K 7210	Full shade	1%	300	8	5	5	None
	Diluent	1:4	300	8	5		None

(4) Green composite inorganic pigment

The basic chemical formula of cobalt titanium green is CO_2TiO_4, and it is made by the calcination of cobalt oxide (CoO) and hromium oxide (Cr_2O_3) in high temperature. To change the color hue, the pigment may contain one or more of aluminum oxide (Al_2O_3), magnesium oxide (MgO), silicon dioxide (SiO_2), zinc oxide (ZnO) or zirconia (ZrO_2), which belong to Pigment Green 26. The chemical composition and color of cobalt green pigment is shown in Table 3.49.

Table 3.49 Chemical composition and color of cobalt green pigment

Pigment index	Chemical composition	construction	CAS No.	Color
Pigment Green 26	$Co(CrO_2)_2$	Spinel	68187-49-5	Blue-green
Pigment Green 50	CO_2TiO_4	Spinel	68186-85-6	Yellowish green

① **C.I. Pigment Green 50** Chemical composition: Co_2TiO_4. C.I structure No. 77377. CAS registration No. [68186-85-6].

Pigment Green 50 is commonly known as titanium cobalt green, has a unique yellow-green, with bright color. Compared with Pigment Green 7, it has a low tinting strength, but it can resist the heat more than 1000°C. Besides, the light resistance of cobalt blue is 8, and the migration resistance is 5. It has an excellent chemical resistance. In partially crystalline polymers, cobalt green pigment does not cause thermal deformation and it can be used for the polymer to get light green or soft hue. The properties of German BASF Sicopal Pigment Green 50 in plastics is shown in Table 3.50.

Table 3.50 Properties of German BASF Sicopal Pigment Green 50 in plastics

Brand	Pigment	Titanium dioxide	Heat resistance /°C	Light resistance grade	Weather resistance garde (3000 h)	Migration resistance grade	Warpage variant
Sicopal Green K 9610	Full shade	1%	300	8	5	5	None
	Diluent	1:4	300	8	5		None
Sicopal Green K 9710	Full shade	1%	300	8	5	5	None
	Diluent	1:4	300	8	5		None

② **C.I. Pigment Green 26**   Chemical composition: Co $(CrO_2)_2$; C.I. structure No. 77344. CAS registration No. [68187-49-5].

Pigment Green 26, commonly known as cobalt chrome green, has a distinctive bright green. It can resist the heat more than 1000℃. Besides, the light resistance of cobalt blue is 8, and the migration resistance is 5. It has an excellent chemical resistance. The properties of Nanjing Pigments Tech. Co., Ltd Pigment Green 50 in plastics is shown in Table 3.51.

Table 3.51   Properties of Nanjing Pigments Tech. Co., Ltd Pigment Green 50 in plastics

Brand		Heat resistance/℃	Light resistance grade	Weather resistance grade (3000 h)	Migration resistance grade
GT700	Full shade	300	8	5	5
	Diluent	300	8	5	
GT701	Full shade	300	8	5	5
	Diluent	300	8	5	

(5) Purple composite inorganic pigment

Manganese violet is made by heating manganese salt, manganese oxide, phosphate and ammonium phosphate composition and final calcination. In 1900 it has been applied. It is mainly used for color mixing, or covering the yellow color of transparent or white resin. Manganese violet pigment is considered safe to human body, so it can be used in coloring the cosmetics and lipstick.

C.I Pigment Violet 16   Chemical composition: $[(NH_4)Mn(P_2O_7)]$. C.I structure No. 77742. CAS registration No. [10101-66-3].

Pigment Violet 16 is commonly known as manganese violet. Its chemical composition is coke manganese ammonium phosphate, and it has a unique red hue purple, bright colors. Compared with Pigment Violet 23, it has low tinting strength, but it can resist the heat more than 275℃. Besides, it has great light, reaching to level 7 of light resistance and level 5 of migration resistance. Pigment violet has excellent solvent and acid resistance, but poor alkali resistance. The properties of American Shepherd Pigment Violet 16 in plastics is shown in Table 3.52. Application range is shown in Table 3.53.

Table 3.52   Properties of American Shepherd of pigment Violet 16 in plastics

Brand		Pigment	Titanium dioxide	Heat resistance /(℃/5min)	Light resistance grade	Weather resistance grade (3000 h)	Migration resistance grade	Warpage variant
Violet 11T	Full shade	1%		275	7	5	5	None
	Diluent		1:4	275	7	5	5	None

Table 3.53 Application range of American Shepherd pigment Violet 16

PVC	•	PS	•	PA6	•
PUR	•	SB	•	PET/PBT	•
PE	•	ABS/ASA	•	SAN/PMMA	•
PP	•	PC	•		

• Recommended to use.

Manganese violet pigment is non-toxic, safe to use. It can be used for lipstick and other cosmetics, as well as for coloring plastic. Besides, it is no harmful to health coverage, and no impact on the respiratory inhalation.

Physical hazards manganese violet pigment can't be mixed with strong soda, because they will react produce ammonia.

3.5.2.5 Ultramarine Inorganic Pigments

There are only a few places that exist mineral lazurite naturally in the world. A kind of blue pigment can be obtained by crushing lazurite, called ultramarine, which means "across the ocean". There are three colors including reddish blue, purple and pink for chemical synthesis ultramarine commercially. The unique blue pigment is from sodium salt which is formed through capturing and stabilizing free radical of polysulfide in sodium aluminum lattice. Actually ultramarine blue is three-dimensional space lattice of aluminum silicate which can interception sodium and ionized sulfur groups. Ultramarine blue was obtained by calcining at 800℃. Ultramarine purple and ultramarine pink are derived from ultramarine blue through further oxidation and ion exchange which have a very similar structure, as shown in Figure 3.12.

Ultramarine blue is gaudy bright reddish blue. Along with the changes of chemical composition the hue will change from red to green. Purple and pink derivatives possess unsaturated colors.

The heat resistance of ultramarine blue is 400℃, the ultramarine purple and ultramarine pink is 280℃ and 220℃ respectively. Ultramarine blue has excellent light fastness, but acid resistance is poor, strong acid causes pigment decomposed completely and

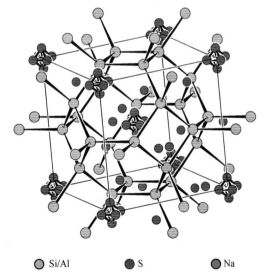

○ Si/Al   ● S   ● Na

Figure 3.12   Structure diagram of ultramarine

fade. Meanwhile, ultramarine purple has excellent light fastness, and have a good chemical stability, except alkali. They do not dissolve in water and organic solvents and have migration resistance and bleeding resistance.

The impact on the environment (security)   Ultramarine inorganic pigments are safe. The LD_{50} >5000mg/kg (rat). There is no harmful chronic pathogenic report in more than a century of production and utilization. And it is widely used in various household to make cloth whitening. Meanwhile, it is also used as sugar brighteners and there is no pathogenic report.

Physical hazards   Ultramarine can't be hybrid storage or placed in a location where there is risk of acid contamination, because the chemical reaction can occur between them and produce hydrogen sulfide. Ultramarine can't combust. It can produce irritating sulfur dioxide gas if involved in the fire.

Main markets of product   Abroad the major manufacturers are Hollday of the United Kingdom and Nubiola of Spain. Major domestic manufacturer is Tianlan pigment (Shandong) Co., Ltd. LAPIS ®.

① **C.I. Pigment Blue29(77007)**   Chemical composition: $Na_6Al_6Si_6O_{24}S_4$; C.I. structure No 77007; CAS registration No. [57455-37-5].

Main properties   Pigment Blue 29 is commonly known as ultramarine blue. Ultramarine has good transparency and unique reddish blue hue. Meanwhile, it has bright colors and good transparency which is completely different from phthalocyanine blue and anthraquinone blue. Ultramarine blue is redder than phthalocyanine blue but less tinting strength. It exhibits good heat resistance up to 400℃, excellent light fastness up to 7~8 and migration resistance up to 5. Ultramarine blue is alkali-resistant and resistant to many chemicals but not acid. The properties of Pigment Blue 29 of Spain Nubiola are shown in Table 3.54, the products which have acid resistance are shown in Table 3.55, the applications are shown in Table 3.56.

Table 3.54　General level of ultramarine grades and performance prepared by Spain NUBIOLA company

Type	Density /(g/cm³)	Oil absorption/ (g/100g)	Light fastness grade	Heat resistance /(℃/5min)	Free sulfur content/%	Sieve residue/%	Acid and alkali resistance	
							Acid resistance grade	Alkali resistance grade
EP-19	2.35	38	8	350	<0.02	<0.05	1	2~3
EP-25	2.35	28	8	350	<0.02	<0.05	1	2~3
GP-58	2.35	39	8	350	<0.02	<0.05	1	2~3
CP-84	2.35	39	8	350	<0.02	<0.05	1	2~3

Table 3.55 Acid-resisting ultramarine grades and performance prepared by Spain NUBIOLA company

Type	Density /(g/cm³)	Oil absorption /(g/100g)	Light fastness grade	Heat resistance /(℃/5min)	Free sulfur content/%	Sieve residue/%	Acid and alkali resistance	
							Acid resistance grade	Alkali resistance grade
RA-40	2.35	46	8	350	<0.03	<0.04	ΔL<0.5	4
Nubiperf	2.35	33	8	350	<0.01	<0.02	ΔL<0.5	4

Table 3.56 Applications of acid-resisting ultramarine prepared by Spain NUBIOLA company

PVC	●	PC	○	PA6	×
PP	●	PMMA	○	PET	○
PE	●	ABS	○	PP spinning	●
PS	○	POM	×	PET spinning	●

● Recommended to use, ○ Conditional use, × Not recommended to use.

② **C.I. Pigment Violet 15** Chemical composition: $Na_5Al_4Si_6O_{23}S_4$. C.I structure No. 77007; CAS registration No. [12169-96-9]

Main properties Pigment Violet 15 is known as ultramarine violet which has less tinting strength than Pigment Violet 23. The heat resistance of ultramarine violet is up to 280℃. Light fastness is up to 7~8 and migration resistance is up to 5. Ultramarine violet is alkali-resistant and resistant to many chemicals but not acid. The properties of Pigment Violet 15 of Spain NUBIOLA are shown in Table 3.57, the applications are shown in Table 3.58.

Table 3.57 Ultramarine Violet 15 grades and performance prepared by Spain NUBIOLA company

Type	Density /(g/cm³)	Oil absorption /(g/100g)	Light fastness grade	Heat resistance /(℃/5min)	Free sulfur content/%	Sieve residue/%	Acid and alkali resistance	
							Acid resistance grade	Alkali resistance grade
V-5	2.35	34	8	300	<0.06	<0.06	1	3~4
V-8	2.35	34	8	300	<0.06	<0.06	1	3~4
V-10	2.35	34	8	300	<0.06	<0.06	1	3~4

Table 3.58 Applications of Ultramarine Violet 15 prepared by Spain NUBIOLA company

PVC	●	PS	●	PA6	●
PUR	●	SB	●	PET/PBT	●
PE	●	ABS/ASA	●	SAN/PMMA	●
PP	●	PC	●		

● Recommended to use.

3.5.2.6 Bismuth Vanadate Yellow Inorganic Pigments

Bismuth vanadate ($BiVO_4$) had appeared in a pharmaceutical patent in 1924, and synthesized in 1964. In the Mid 1970s bismuth vanadate pigments began to be developed. In 1976 Du Pont first issued a patent "bright lemon yellow pigment" which is about a pigment content monoclinic bismuth vanadate. This bright yellow inorganic pigment appeared on the market in 1985.

Although lead chrome yellow pigments have an unshakable position among inorganic yellow pigments, nevertheless, in recent years it is disabled because of the international regulatory restrictions. In addition, the hue of iron oxide yellow is not bright enough and it is too expensive to use organic yellow pigments to obtain superior performance. People do want to develop a new kind of yellow pigment which has bright color, good properties, no lead and non-toxic. Bismuth vanadate/molybdenum bismuth yellow is developed in this context.

Bismuth vanadate/molybdenum bismuth yellow pigment is two-phase pigment which is bright lemon yellow and the chemical formula is $4BiVO_4 \cdot 3BiMoO_4$. This kind of pigment for plastics is stable through binding with silicon envelope or inorganic metal compound (boron, aluminum, zinc oxide), thereby improving the heat resistance, so it is important to avoid excessive shear in the process of plastics coloring.

The impact on the environment (security)　Oral test and inhalation test of animals show that there is no acute toxicity. The LD_{50} of oral dose in rat is more than 5000mg/kg. The skin and eye irritation studies have shown that this type of pigment is feminine. The test of animal inhalation shows that it is toxic due to contain vanadate probably. It is observed that the weight of lung and lung tissue change through studying on rat inhalation for three months and only high concentration will lead to be unrecoverable. Only when exceeding some certain concentrations it can be observed. If it reaches the industrial hygiene standards it would not be so. To further reduce the risk bismuth vanadium pigments should be produced which have high mobility and fine powder-free products. Meanwhile, the sizes of the particles are within the fine dust which will not be sucked.

Major foreign manufacturers: German BASF, Belgium Cappelle.

C.I. Pigment Yellow 184　Chemical composition: $4BiVO_4 \cdot 3Bi_2MoO_6$. C.I. structure No. 771740. CAS Registration No. [14059-33-7/13565-96-3].

Main properties　Pigment Yellow 184 is greenish yellow which have high tinting strength, brightness and hiding power. The tinting strength is 4 times more than nickel titanium yellow, and the hiding power is similar with titanium dioxide. In addition, it can be resistant to various solvents. The hue of bismuth vanadium molybdenum yellow is close to chrome yellow or cadmium yellow, and it is brighter than the nickel-titanium and iron oxide yellow, especially its excellent weather resistance. Its migration resistance is up to 5, acid resistance is up to 4～5 (10% H_2SO_4 or 2% HCl solution), alkali resistance is up to 5 (10% NaOH solution). Due to manufacturing through high temperature calcination it has good heat resistance. It can withstand 200℃ for 30min and its heat resistance is up to 4～5. Pigment Yellow 184 is resistant to various solvents. The properties of Pigment Yellow 184 of German BASF Sicopal series in plastics are shown in Table 3.59. The applications are shown in Table 3.60.

Table 3.59 Properties of Sicopal Yellow 184 prepared by BASF in plastics

Brand		Pigment	Titanium	Heat resistance /℃	Light fastness grade	Weather resistance (3000h)	migration resistance grade	Warpage variant grade
Sicopal Yellow K 1120FG	Full shade	1%		300	8	5	5	None
	Dilute		1:4	300	8	5		None
Sicopal Yellow K 1160FG	Full shade	1%		300	8	5	5	None
	Dilute		1:4	300	8	5		None

Table 3.60 Applications of Sicopal Yellow 184 prepared by BASF

PVC	●	SB	●	SAN/PMMA	●
PUR	○	ABS/ASA	○	PP fiber	●
PE	●	PC	●	PA fiber	○
PP	●	PA6	●	PET fiber	●
PS	●	PET/PBT	●	PAN fiber	●

● Recommended to use, ○ Conditional use.

3.5.3　The Main Varieties and Properties of Effect Pigments (Inorganic)

The main varieties of effect pigments (inorganic) are metallic pigments which mainly produce specular reflection on flat and parallel metallic pigment particles (such as aluminum sheet, copper powder). The specular reflection is produced in highly parallel refractive pigment platelets (e.g. mica titanium dioxide). The interference color pigments (e.g., mica iron oxide) produce optical effects of colored luster pigments because of interference phenomenon, and the optically variable pigments produce birefringent crystal light polarization. The varieties of effect inorganic pigments are shown in Table 3.61.

Table 3.61 Types of (inorganic) effect pigment

Types of pigment	Product variety
Sheet metal body (aluminum, copper, copper and zinc)	Metallic silver, metallic gold
Monolayer mica coated with silica	Silver pearl, all kinds of interference color
Mica coated with silica or iron oxide synchronously	Gold pearl
Mica coated with iron oxide as monolayer	Metal bronze pearl
Mica or metal coated with silica or iron oxide	Chameleon magic color

3.5.3.1 Metallic Pigments

More and more plastics are replacing the metal as structural material, such as automobile parts, household appliances, entertainment devices. Plastics can be used in the preparation of various products of sophisticated design and complex structure which have lighter weight than metal. Because of the decrease of mass it is helpful to save energy when it is used in the car. In order to make plastic parts look like metal parts to meet the sensory needs, and it can achieve the goal through adding metal pigments. Metallic pigments are mainly sheet-like metallic particles which are provided as powder, paste, granular. The typical metallic effect pigments are aluminum and copper and copper alloys.

All of metallic effect pigments will appear pipeline and welding line in the process of injection. The main reason is from less metallic pigments at the weld line and the metallic pigments arrangement of welding departments, shown in Figure 3.13, Figure 3.14. It is suggested to increase the injection rate and pressure, reduce the amount of pigments and select big size of metallic pigment particles and transparent pigments which can effectively reduce the flow marks.

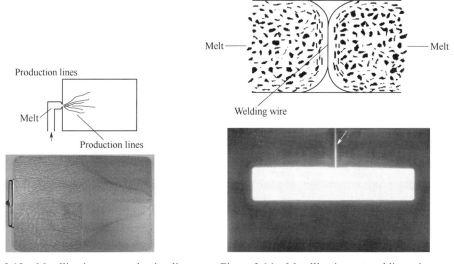

Figure 3.13   Metallic pigment production line   Figure 3.14   Metallic pigment welding wire

The impact on the environment (security)   Aluminum powder is no acute toxic. however, when it is used as color paste the toxicity of solvents should be considered, which may contain solvents and hazards and can cause toxic effects after inhalation; Copper and brass powder is also non-toxic. Powdered aluminum pigment is flammable solid and will release hydrogen through reacting with water. The most secure and rapid method is covered with dry sand after fire. In this case, the aluminum dust cloud must be avoided. It can lead to suddenly explode so that smoking, fire and static sparks must be prohibited.

The main manufacturers aboard are German ECKART, American Silberline and German Schlenk.

(1) Metallic pigment 1　Chemical composition: aluminum powders. C.I. structure No. 77000; CAS registration No. [7429-90-5].

Silver powder actually is aluminum sheet. Aluminum particles are scaly which have irregular edges and a plurality of faces for each piece, shown in Figure 3.15. The entire visible spectrum including blue light can be strongly reflected by the surfaces of aluminum sheet. Every surface is like a tiny mirror, and when light irradiate light is scattered in all directions which reveals the "little flash" effect.

Figure 3.15　Scaly aluminum particles

Aluminum pigment can generate bright blue-white specular reflection. Aluminum powder has a diameter-thickness ration of 40∶1 to 100∶1 and it parallels with substrate when dispersing in the carrier. When numerous aluminum powders are connecting to fill up each other's space, these particles as a whole would then cover the substrate and reflect lights, contributing to a unique covering power. This covering power is determined by the surface area, namely the diameter-thickness ratio. The covering power aggregates as the diameter-thickness ration increases when aluminum extends in grinding. Many factors can influence aluminum powder's covering power. For instance, poorly-dispersed aluminum powders usually have weak covering power. In addition, aluminum powders'covering power can be influenced if the powders are processed under high shear stress. In this situation, aluminum powders will fracture radially, thus decreasing the diameter-thickness ratio.

There are different varieties of aluminum powder. Fine aluminum powders have low surface-edge ratio of the pigment and more light scattering so that they have high hiding power and silky luster effect. Coarse aluminum powders have high surface- edge ratio and less light scattering so that they have strong metal texture. When the average particle diameter is 5μm they have excellent tinting strength and hiding power. When the average particle diameter is 20~30μm they can be used with the color pigment. in addition, they will have flicker effect when the average particle diameter is 330μm. The size of aluminum powder particles and color properties are respectively shown in Figure 3.16 and Table 3.62.

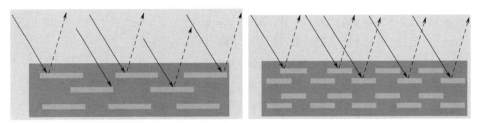

Coarse particle size of aluminum powder   Smaller particle size of aluminum powder

Figure 3.16 Light scattering phenomenon of different sizes of aluminum particles

Table 3.62 The relationship between aluminum particle size and performance

Particle size	Thin ⟶ Thick
Brightness	Low ⟶ High
Luster	Small ⟶ Large
Color saturation	Small ⟶ Large
Hiding power	Large ⟶ Small

The melting point of aluminum powder is 660 ℃, but at high temperatures the fine aluminum powders are very sensitive to oxygen, and it will form a layer of aluminum oxide in the surface. Therefore, aluminum lost its bright luster and become rough and pale. Thus on the surface of aluminum powder pigments it can form a thin layer of high-powered protective film of SiO_2 and acrylic resin to have excellent heat resistance, weather ability and acid resistance.

According to the form of aluminum powder pigments it can be divided into powder and paste. Recently the aluminum bars which are aluminum powders and polyethylene wax as main ingredients for the masterbatch process abroad.

Resist series silver powder of German ECKART Company is coated with silicon oxide on the surface, shown in Table 3.63.

(2) Metallic pigment 2 Chemical composition: copper powders and copper zinc powders (Cu, CuZn). C.I. Structure No. 77400. CAS registration No. [7440-50-8].

Table 3.63 Aluminum powder grades and performance prepared by German ECKART company

Trademark	Silver bars	Particle size /μm			Effect
		D10	D50	D90	
Chromal XV	Mastersafe 05203		5	11	High hiding power coloring
Chromal X	Mastersafe 10203	3	9	22	High hiding power coloring
Chromal IV		3	14	30	Metallic
Chromal I	Mastersafe 20203	5	20	48	Slightly metallic flashes
Reflexal 214	Mastersafe 31203	17	31	54	Fine flashes
Reflexal 212	Mastersafe 49203	27	49	74	Fine flashes
Reflexal 211	Mastersafe 60203	30	60	95	Moderate flashes
Reflexal 210	Mastersafe 75203	40	75	110	Moderate flashes
	Mastersafe 95203	56	95	130	Moderate and high flashes
Reflexal 145	Mastersafe 145203		145		Moderate and high flashes
Reflexal 245	Mastersafe 245203		245		High flashes
Reflexal 345	Mastersafe 345203		345		

Golden powder which can utilize and obtain metallic luster is actually copper and bronze powder (copper and zinc alloy powder). Along with the increase of content of zinc in copper the color change from red to glaucoma. The coloring effect of golden powder changes along with the particle size varies, fine particle size of the metal pigments (10~20μm) possess very high covering power as like satin luster. Coarse particle size of the metallic pigments (50~100μm) can obtain clear flash of bright golden. The transparency of plastics should be excellent to make the golden powder exhibit good metallic effect, therefore, trying to avoid to use it with titanium dioxide including pearl pigment coated with titanium dioxide. In addition, increasing the pressure during processing which can make the golden powder aligned can improve powder metal effect. The carbon chain of PP exists reactive tertiary carbon atoms and the metallic copper make PP degrade easily so that it is not appropriate for PP coloration.

It is same as aluminum powder that golden powder also need to be coated with silicon oxide on the surface, such as resist series golden powder and bars (Mastersafe Gold) of German ECKART Company. The performance in plastics are shown in Table 3.64. Golden powder has very excellent heat resistance, chemical stability and outdoor weather ability.

Table 3.64 Gold powder grades and performance prepared by German ECKART company

Trademark	Gold bars	Particle size/μm			Performance
		D10	D50	D90	
Resist Rotoflex		3	8	15	High hiding power coloring
	Mastersafe Gold 10103		10		High hiding power coloring
Resist AT		6	14	30	Metallic
	Mastersafe Gold 17103		17		Slightly golden flashes
Resist CT		12	28	50	Fine flashes
	Mastersafe Gold 35103		35		Fine flashes
Resist LT		17	40	73	Moderate Flashes
	Mastersafe Gold 75103		75		Moderate Flashes

3.5.3.2 Pearl Pigments

Mica-titanium pearl pigment is a kind of sheet-like inorganic pigment which possesses high refractive index, high gloss. It utilizes mica as substrate which is coated with one or more layers of high refractive transparent metal oxide films on the surface. Through optical interference it will exhibit natural soft pearly luster or metallic flash. Meanwhile, pearl pigments is widely used in the plastics industry due to light fastness, high temperature resistance (800℃), acid and alkali resistance, non-conductive, easily dispersed, fastness and non-migration, non-toxicity. Pearl pigments exist in the form of sheet and its average particle sizes range from 5 μm to 500 μm. The main raw materials of pearl pigments are natural mica, synthetic mica, glass, etc. The pearl pigments which are made up of glass have the best quality. In the market most of pearl pigments select natural mica as the substrate, and the quality of natural mica which is mainly produced in India has an important effect on the pearl effect.

After classification the mica (glass) beginning chemically react in the hydrolysis reactor. According to the requirements it should add $TiCl_4$, $FeCl_3$, $SnCl_4$ and other compounds when hydrolysis. After hydrolysis semi-finished products require calcination and its temperature is generally among 700 to 900℃. After calcination, the titanium compound and iron compound which are produced through hydrolysis generate respectively titanium oxide and ferric oxide, these oxides are very stable which can form nano-coating on the surface of mica (glass).

At first, titanium dioxide is coated on the substrate which will help form silver-white pearl pigments, shown in Figure 3.17 and Figure 3.18. With the increase of the thickness of the titanium dioxide coating the color will be from silver to interference color (Magic color), shown in Figure 3.19. The so-called magic color refers to color of pearl pigments, it is substantially white or very light color when we view from the front and show its natural color when we are from the side. With the increase of the coating thickness the color changes from golden yellow to red, purple, blue, green. For white and silver interference color series in order to improve light fastness and weather resistance some products need to be coated with a layer of tin oxide which aims at converting anatase titanium dioxide into rutile through calcination.

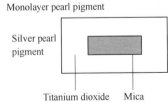

Figure 3.17 Mica coated with silver pearl pigment   Figure 3.18 Electron micrographs of silver pearl

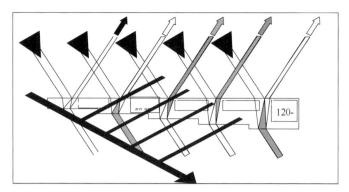

Figure 3.19 Interference color-magic color mechanism

If it is doubly coated on the substrate, that is, first coated with a layer of titanium dioxide, and then coated with a layer of iron oxide. This process can produce golden pearl pigment. In general, with the increase of the thickness of iron oxide the color changes from shallow golden to red golden which is showed in Figure 3.20. If coating iron oxide on the substrate directly the color changes from bronze, red to red-purple, brown green is showed in Figure 3.21.Therefore, the pearl pigments are generally divided into white series, symphony series, gold series, metal series according to the different chromatic light.

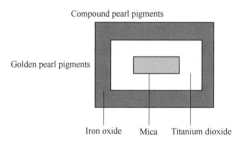

Figure 3.20 Mica doubly coated with golden pearl pigment

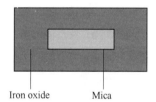

Figure 3.21 Bronze-mica coated with iron oxide directly

The impact on the environment (security) Mica-based pearl pigment is mixture. It is made up of mica and one or more kinds of metal oxides.

Acute toxicity Acute toxicity tests showed that the LD_{50} of oral dose in rat is more than 5000mg/kg and on the skin or mucous membranes did not show any irritation or sensitization. In the case of contacting pearl pigments for normal career the effects on human health is evaluated and it shows no harmful effects.

Chronic toxicity It has not been confirmed chronic effects for contacting the pearl pigments.

Major foreign manufacturers Merck Co., Ltd., Engelhard of Chemicals division of BASF.

Major domestic manufacturers Hebei Ouke Fine Chemical Co., Ltd., Wenzhou Kunwei pearl pigments Co., Ltd., Wenzhou Taizhu Co., Ltd.

(1) The silvery white series pearlescent pigments

According to the various sizes of platelets, pearlescent pigments show different effects in plastics. The flicker effect and hiding power of pearlescent pigments are related to particle size. For example, the smaller the particle size, the higher the hiding power and the worse the flicker effect (Figure 3.22). The classification and effect comparison of pearlescent pigments are shown in Table 3.65. The grades and properties of Merck 100 silver white series are shown in Table 3.66.

<15μm　　5～25μm　　10～60μm　　10～125μm　　20～180μm

Figure 3.22　Influence of pearlescent pigments of particle size on properties

Table 3.65　The relationship between particle size and property of pearlescent pigments

Marking of particle size grades	Particle size/μm	Corresponding mesh	Luster	Hiding power
M　(Mote grade)	<15	1500	Delicate luster	Best
F　(Fine grade)	2～25	1000～1200	Brocade luster	Well
N　(Normal grade)	10～60	600～800	Pearl luster	Better
S　(Shiny grade)	20～100	400～600	Flicker luster	Weaker
L　(Lightning grade)	20～200	200	Dazzling luster	Weak

Table 3.66　The grades and properties of Merck100 silver white series

Product	TiO$_2$ cladding rate / %	Particle size / μm	Quality	Application
Iriodin® 100	29 (anatase)	10～60	Better hiding power and natural pearl luster	Common
Iriodin® 111	43	15	Best hiding power and delicate luster	Film or hose extrusion molding
Iriodin® 120	38	5～25	Brocade luster	Food packaging Hose bottle
Iriodin®119-KU26	40～44 (rutile)	5～25	Brocade luster	No yellowing phenomenon in PE hose
Iriodin® 153	16	20～100	Weaker hiding power and high flicker luster	Achieve the effect of aluminum with carbon black (0.01%)
Iriodin® 163	14	20～180	Weak hiding power and ultra-dazzling luster	Achieve the effect of aluminum with carbon black (0.01%)

(2) The colored series pearlescent pigments

The colored pearlescent pigments on tiny mica platelets control the TiO_2 coating thickness to achieve interference, which is not endowed with common pigments, and the color effects depend on the angle of vision. The colored pearlescent pigments can be mixed with transparent organic pigments or the silvery white series pearlescent pigments for color matching. The grades and properties of Merck 200 colored pearlescent pigments are shown in Table 3.67.

Table 3.67 The grades and properties of Merck 200 colored pearlescent pigments

Product		Particle size / μm	Quality	
			Luster	Hiding power
Interference golden	Iriodin® 201	1000~1200	Brocade luster	Well
	Iriodin® 205	600~800	Pearl luster	Better
	Iriodin® 249	400~600	Flicker luster	Weaker
Interference blue	Iriodin® 221	1000~1200	Brocade luster	Well
	Iriodin® 225	600~800	Pearl luster	Better
	Iriodin® 289	400~600	Flicker luster	Weaker
Interference red	Iriodin® 211	1000~1200	Brocade luster	Well
	Iriodin® 215	600~800	Pearl luster	Better
	Iriodin® 259	400~600	Flicker luster	Weaker
Interference green	Iriodin® 231	1000~1200	Brocade luster	Well
	Iriodin® 235	600~800	Pearl luster	Better
	Iriodin® 299	400~600	Flicker luster	Weaker
Interference violet	Iriodin® 219	600~800	Pearl luster	Better

(3) The metal series pearlescent pigments

The metal series pearlescent pigments give golden shade on tiny mica platelets by controlling TiO_2 coating, adding a layer of iron oxide, and being mixed with an interference layer. Compared to metal pigments, the metal series pearlescent pigments have a higher thermal resistance, it can achieve 800 ℃ without discoloration. Owing to its hiding power weaker than metal pigments, the metal series pearlescent pigments can be mixed with transparent organic pigments, carbon black, ultramarine to achieve different effects. The grades and properties of Merck Iriodin® 300 golden series pearlescent pigments are shown in Table 3.68.

Table 3.68 The grades and properties of Merck Iriodin® 300 golden series pearlescent pigments

Product	TiO_2 cladding rate/%	Fe_2O_3 cladding rate/%	Particle size /μm	Quality
Iriodin® 300	38	3	10~60	Brilliant golden
Iriodin® 302	48	10	10	Brocade luster
Iriodin® 303	23	22	23	Pink brilliant golden

The metal series pearlescent pigments are coated directly by a layer of iron oxide on mica platelets, the color varies from bronze, red to red-violet, breen. The grades and properties of Merck Iriodin® 500 golden series pearlescent pigments are shown in Table 3.69.

Table 3.69　The grades and properties of Merck Iriodin® 500 golden series pearlescent pigments

Product	Fe_2O_3 cladding rate/%	Particle size /μm	Quality
Iriodin® 500	38	10~60	Brilliant bronze pearl
Iriodin® 502	42	10~60	Brilliant brownish red pearl
Iriodin® 504	46	10~60	Brilliant wine red pearl

Because the metal series pearlescent pigments are coated with iron oxide on mica platelets, it should pay attention to the use of high shearing forces on its destruction. And if the system contains chloride ions or sulfide ions, it leads to a generation of ferric chloride or iron sulfide and causes the color turning to black.

(4) Color flop (chameleon) pigments

Color flop pigments (customarily named "chameleon" pigments) use natural mica, aluminum as the substrate, provide a wet chemical coating method, and introduce a low refractive index medium SiO_2 in the chemical coating layer which is coated alternately with a high refractive index medium Fe_2O_3 or TiO_2 to cause color changes. Pure SiO_2 crystals are birefringence, the light would be divided into two rays through a birefringent crystal, a constant called "ordinary ray" and another called "extraordinary ray". The "extraordinary ray" changes depend on the angle of incidence, that is to say the "extraordinary ray" refractive index is a function of the direction in birefringent crystals. To separate the ordinary ray from the extraordinary ray we can make use of the polarized light obtained from natural light by the birefringent crystals, and this optical principle is used by the polarizing prism. Cutting the birefringent crystals into different thickness of platelets to through a monochromatic light to form a variety of gratings may cause some spectrum approach to disappearance. Using this optical phenomenon and according to the variation rules of color, controlling the thickness of SiO_2 is equivalent to inserting a polarizer into each coating layer of pearlescent pigments, and adjusting the thickness of the polarizer to meet the design requirements, which would show magical colors (Figure 3.23).

Color flop pigments should make the flaky pigment exhibit a unified orientation as much as possible to achieve the optimal effects. It can be obtained by the continuity of the processing such as thin-layer extrusion and blow molding. Taking advantage of injection molding and calendering process can also achieve the excellent effects.

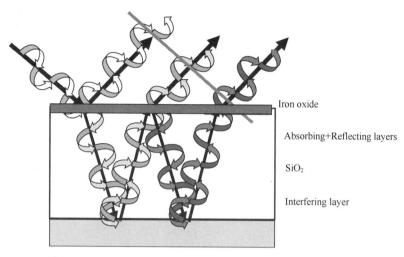

Figure 3.23　The color changing principle of color flop pigments

(5) Laser marking pigments

The use of laser for marking is widely accepted, especially in the case of traditional printing processes can't be applied because of the surface shape, size, or structure of components. Laser marking is a process without any contact, therefore the inaccessible contact surfaces and bend components can be marked without any difficulty. This is a distinct advantage of laser technology.

Laser marking refers to producing high partial energy in polymer matrix by laser radiation. The energy is absorbed by polymer matrix or additives and converted into heat energy. When the heat energy reaches a certain value, it may cause a variety of physical or chemical changes inside the polymer and result in marked effect.

Plastic products with a laser marking are very convenient. Laser beam is focused by an optical device, which control laser beam toward to the marked components by computer. Under the influence of the heat caused by the impact of the laser beam, the surface of the plastic component is changed, and the result is a marking in this area. However, the actual situation is more complicated.

Laser marking can be divided into three categories according to the effect: the light matrix marking dark, the dark matrix marking light, marking color. The marking effects depend on the physical and chemical reaction, laser parameters of the polymer, and correspond to different laser marking additives, a single laser marking additive can't simultaneously achieve three effects.

The laser marking color not only depends on the additives, but also on the plastic matrix. Some polymers are easy to carbonization at a high temperature and suitable for dark marking, such as polyethylene terephthalate (PET)/polybutylene terephthalate (PBT), polycarbonate (PC), acrylonitrile-butadiene-styrene copolymer (ABS), polystyrene-butadiene copolymer (SB), and so forth. And some polymers are easy to be decomposed into small molecules, difficult to carbonization at high temperature and suitable for light marking, such as polyoxymethylene (POM), polymethyl methacrylate (PMMA), and the polymers containing a sufficient amount of black pigments, and so forth. Some polymers can either be a dark marking or be a light marking, such as polyamide (PA), polyethylene (PE), polypropylene (PP), polystyrene (PS), styrene- acrylonitrile copolymer (SAN).

In short, the implement of laser marking effect is the interaction result of a laser beam, the resin properties and additives interaction. The color changes of compounds are induced by laser, or the control of laser energy and the absorption of additives result in resin carbonization (the light matrix marking dark) and thermal decomposition of the foam (the dark matrix marking light), and complete laser marking.

Currently, the most commonly used laser marking additive are inorganic compounds, after laser irradiation the plastic color changed by itself or by absorbing heat and result in obvious marking. Singapore scholars found the titanium dioxide (TiO_2) as a laser marking additive is reduced to TiO or Ti under laser irradiation, then cause a dark marking in plastics. Accordingly, TiO_2 is widely used in laser marking of plastics, meanwhile pearlescent pigments, silicates, silicon dioxide are also extensive use. Laser marking additives include the salts (such as the phosphates or mixtures of Cu, Fe, Sn, Sb) of color changes under the laser irradiation, and the color marked is black or dark gray. The company DSM Holland has developed the laser marking additives containing Sb_2O_3 of which the average particle size is larger than 0.5 μm, and silicate pearlescent pigments coated by metal oxides are added to provide a synergistic effect. This laser marking additive can significantly improve PA6 marking and obtain a high contrast ratio dark marking in the light. Original US Engelhard Corporation has developed an additive composed by tin oxide and a calcined Sb_2O_3 powders, and the calcined powders are blended by coprecipitation, the surface concentration of the particles is larger than the inside. The advantage of this formulation is the original plastic color has no change with the additive powder and no excessive bubbles, thus the obtained marking surface smooth. Separately added tin oxide or Sb_2O_3 can't get a good marking result.

3.5.4 Main Varieties and Properties of Effect Pigments (Organic)

The main varieties of effect pigments (organic) are daylight fluorescent pigments, optical brighteners, color changed pigments (thermal pigments and optically variable pigments), speckled effect pigments, marble batches.

3.5.4.1 Daylight Fluorescent Pigments

Fluorescence phenomenon is a photoluminescence process. In this process, after the short wavelength electromagnetic waves within the range of the ultraviolet light or visible light short are absorbed, the long wavelength electromagnetic waves are released. The latter is usually within the visible range and superimposed with a conventional reflection, thus appear bright fluorescent color.

Daylight fluorescent pigments consist of fluorescent dyes dissolved in a polymer matrix. Modified polyamide is used mainly as polymer matrix, but other polymers as carriers of daylight fluorescent dyes are also quite common. The daylight fluorescent dyes are dissolved in the polymer matrix, and the products are obtained by milling to a fine powder. Commonly used fluorescent dye is Rhodamine Red B, and the greenish yellow region is derivatives of perylene. Daylight fluorescent pigments show typical properties of pigments, such as high tinting strength, high thermal resistance, tiny particle size, and high solvent resistance. The recent development of a fluorescent pigment with the polyester carrier is also successful and the thermal resistance is up to above 285 ℃.

Daylight fluorescent pigments can be applied to partially crystalline polymers and polyolefins are the main application areas of the daylight fluorescent pigments.

Daylight fluorescent pigments are sensitive to temperature and time. Therefore, it requires special attention in the production of masterbatch and later colored final products. The processing temperature should be as low as possible with the short residence time. In practice it is often limited, since the processing of polymers requires a specific temperature to achieve the proper polymer melt. In addition, the melting of a polymer also takes some time, depending on the temperature. At too low a temperature or too short residence time a plasticizing shear force has enough power to make melting polymer incomplete through an extruder. At the same time a lot of frictional heat is generated. Results may cause thermal damage of the daylight fluorescent pigments and the color loses its partial brightness.

It is common to have the phenomenon of heading back and repetitive use in injection molding, and the sprue parts are also universal. Therefore it requires more attention because every heating can increase the degree of thermal damage.

The concentration of the daylight fluorescent pigments in the polymer matrix is rather low, which is recommended that the concentration is between 1% and 2% of the final products.

Another disadvantage of the daylight fluorescent pigments is the low light fastness. The utraviolet light of daylight is harmful. The high brightness fluorescent pigment is the result of the transformation of absorbed ultraviolet light into visible light. Therefore the use of UV absorber would inevitably reduce the high brightness.

The method to improve the light fastness of the daylight fluorescent pigments is added the nonfluorescent pigments in the same shade. Using a daylight fluorescent colorant in plastic products can cause color fading and decrease brightness, but the tone would not change significantly.

The market supply daylight fluorescent pigments in the colors yellow, green, red, blue and violet. The yellow pigments are the main product, because yellow is used worldwide as a warning color.

It is worth noting that most of daylight fluorescent pigments do not meet FDA requirements, and the varieties according with food contact are not enough. It requires careful use of daylight fluorescent pigments when the products directly contact with food. The brands and properties of ZVM daylight fluorescent pigments in NOVO-GLO British Company are shown in Table 3.70.

Table 3.70 The brands and properties of ZVM daylight fluorescent pigments in NOVO-GLO British Company

Product	Luster	Product brands	Luster
ZVM 17	Tartrazine	ZVM 16	Yellowish orange
ZVM 15	Orange	ZVM 14	Reddish orange
ZVM 13	Brilliant red	ZVM 12	Peach
ZVM 11	Pink	ZVM 21	Magenta
ZVM 19	Blue	ZVM 18	Green

3.5.4.2 Optical Brighteners

Plastic products generally have a slight absorption of the blue light (450~480 nm) in visible light (wavelength of 400~800 nm), which causes the lack of blue and increases slight yellow, then the whiteness is affected and show a dated sense. Therefore, different measures to improve the whiteness and brightness of products are taken. There are two common methods. One is blue whitening, it means that adding a small amount of blue pigments (such as ultramarine blue) to hide the yellow part of matrix by increasing the blue reflection to achieve a whiter. Although blue whitening can improve the whiteness, the effect is limited and the total amount of reflection is reduced, after that cause the decrease in brightness and luster. The optical brighteners are a kind of organic compounds absorbing ultraviolet light and exciting the blue or bluish violet fluorescence. On the one hand, the products added optical brighteners can reflect the visible light on the substance and transform the absorbed invisible ultraviolet light (wavelength of 300~400 nm) into a blue or bluish violet visible light to emit. Moreover, blue and yellow are complementary colors which results in eliminating yellow in the matrix and improving the whiteness and brightness of products. On the other hand it increases the light reflectance of the substance and the intensity of the reflected light is more than the intensity of original visible light which projected onto the treated substance, then the whiteness seems to be increased and it achieves the purpose of whitening.

The optical brighteners can be regarded as white dyes to improve the whiteness of plastics effectively. With the development of thermoplastics, optical brighteners gradually penetrate into this area. In plastics the application of optical brighteners is to increase the whiteness and brightness and change the original yellowish plastic products. For the colored plastics the products are more brilliant and for black can increase the brightness of products.

The optical brighteners have the properties of the dyes, and dissolved in the polymer to show a whitening effect. Compared with dyes, optical brighteners can be used in partially crystalline polymers. This is because of the extremely low application concentration in the polymer. In principle, optical brighteners can be used for all plastics, but in some places the choice has certain restrictions. Therefore, the selection of the type and amount of optical brighteners should be careful and the selected brightener should show a good whitening effect at low concentration. In practice, the amount of optional brighteners is only 50~500 mg/kg in thermoplastics. Only in some special cases such as thermoplastics processing cycle, the amount was required to exceed 1000 mg/kg.

According to the chemical structure, optical brighteners can be divided into five categories: stilbene, coumarin, pyrazoline, benzoxazole, naphthalimide. Owing to the rapid development of optical brighteners, the categories have been classified in more specific in recent literatures such as triazinylamino stilbene, oxazole ring, diacetyl amino substituted, coumarin, pyrazoline, naphthalimide, oxadiazole, triazole, carbon ring, furan, imidazole, and so forth. The main optical brighteners in plastics are shown in Table 3.71.

Table 3.71 The main optical brighteners in plastics

Product	Index number	CAS	Melting point /°C	Formula	Molecular weight	Application
CBS-127	C.I.FB 359		216~222	$C_{30}H_{26}O_2$	418	Applied to PVC and polyethylene products, blue fluorescence in all polymers
OB	C.I.FB 184	12224-40-7	196~203	$C_{26}H_{26}SO_2N_2$	430	Widely used in PVC, PS, ABS, PE, PP, and so forth
OB-1	C.I.FB 398		351~353	$C_{28}H_{18}N_2O_2$	414	Applied to the efficient optical brightener of polyester, and widely used in ABS, PS, HIPS, PA, PC, PP, EVA, PVC (rigid), and so forth
KCB	C.I.FB 367	63310-10-1	210~212	$C_{24}H_{14}SO_2N_2$	362	Widely used in EVA, the first variety of optical brighteners in sneakers
KSN	C.I.FB 368	17313-08	275~280	$C_{30}H_{22}N_2O_2$	442	Excellent upgrading products in optical brighteners, violet shade, excellent light fastness resistance, applied to all plastics

3.5.4.3 Color Changed Pigments

(1) Optically variable pigments

Optically variable pigments (OVP) mainly refer to the color of pigments varies with the angle changes. These pigments have have anti-counterfeiting for inks and even for automotive coatings. Optically variable pigments are manufactured through the optical films with specific optical spectrum by crushing process.

According to the design requirements of specific film structures, the optically variable films are formed under the high vacuum conditions by making many kinds of materials with different refractive index to be precipitated sequentially on the same carrier. The thickness of films meets the interference condition of light, it may cause a series of optical interference functions after films irradiated by light, and the color of pigments varies with the viewing angle of the human eyes. The structure is shown in Figure 3.24.

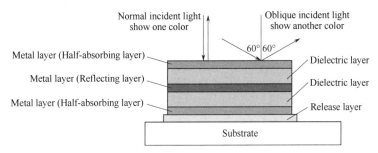

Figure 3.24 Structure of optically variable pigments

In production, PET, PVC, polyimide, glass, metal and others are generally selected to be the substrate materials. Polyvinyl ethanol, cellulose or acrylic resins such as polymethacrylate and others are commonly used in organic release layer materials, Inorganic release layer materials include sodium chloride or sodium fluoride, etc., and generally the release layers should be dissolved in organic solvent (such as acetone or methyl ethyl ketone, polypropylene, acetyl cellulose, etc.) or water. The substrates are placed in a vacuum chamber by the method of thermal evaporation, electron beam evaporation, magnetron sputtering, and so forth. Three-layer or more of the films are evaporated on the coating layer and five-layer is the best choice. For example, when the reflection color of normal incidence is red and that of oblique incidence is green, the film structures (from the bottom to the top of each layer thickness) are 5 nm (Cr film), 379 nm (Al_2O_3 film), 100 nm (Al film) 379 nm (Al_2O_3 film) and 5 nm (Cr film). When the reflection color of normal incidence is green and that of oblique incidence is blue, the film structures (from the bottom to the top of each layer thickness) are 5 nm (Cr film), 319 nm (Al_2O_3 film) 100 nm (Al film), 319 nm (Al_2O_3 film) and 5 nm (Cr film).

Changing the number of film layers can produce similar optical spectrum effect in the visible wave band and completely different effect in the ultraviolet and infrared wave band. After the releasing process, the film and substrate are separated, and then crushed to make pigment particle size to achieve the requirements. Typically, the particle size is between 5~200 μm. The larger the particle size is, the more obvious the effect becomes.

Optically variable pigments are now widely used in the field of security, cosmetics and packaging, coating, and so forth. To show a superior color changed effect, a very high degree of transparency is required in plastics such as transparent PP, PET, PVC and others.

(2) Thermal pigments

Colored plastic cups become colorless when the hot liquid is poured in, a baby spoon turns red when the food is too hot. These are a few examples of the thermochromism, commonly referred to the thermal pigments. Thermal pigments will change color at a high temperature. This process is reversible, for example, the pigment can be colorless at room temperature 20℃, and colored at 40℃. After cooling to room temperature the product became colorless again. This process can be repeated many times.

Thermal pigments are prepared by electron transfer organic compounds system. Electron transfer organic compound is an organic chromogenic system with a special chemical structure. At a specific temperature the electron transfer causes the change in molecular structure of the organics to achieve the color change. This material is not only brightness, but also can be achieved from "colored-colorless" state.

Thermal pigments are tiny particles with particular colors at a specific temperature, and the average diameter is 2~7 μm. Each particle is comprised of the tiny allochroic capsule units which containing organic acids, solvents and colorants. When the ambient temperature is below the melting point of the solvent, a colorant combining with an organic acid show a colored state. When it is above the melting point, a colorant separated from an organic acid show a colorless state. The external of thermal pigments is a transparent layer shell neither dissolution nor melt, which protects the color material from eroding by other chemicals. Thus avoiding the damage of the external shell is very important.

Thermal pigment units are available in blue, black, red, green, yellow, and so forth. Thermal pigments can be mixed with common pigments to achieve the effect that color changes from one to another. For example, the red thermal pigment mixed with common pigment is able to cause the effect from orange to yellow.

Due to the specific structure of thermal pigments, compared with organic pigments the hiding power and color matching have a certain gap. At the same time, it should avoid contacting with polar solvents and ammonia, moreover, should avoid intense ultraviolet radiation and long time at high temperature. In the production process of masterbatch and plastic products, in order to achieve fine dispersion, thermal pigments are usually provided to customer in the form of thermal masterbatches. For example, Chromazone products in Yi Jie Chemical Co. Ltd. show the color in red, blue and black, and provide two thermal pigments of low temperature (about 17 ℃) and high temperature (about 47℃).

3.5.4.4 Speckled Pigments

There are many ways to generate speckled effects, some of which are described in the previous section, such as aluminum platelets with large particle in metal effect pigments pearlescent pigments. In this section speckled effect refers to the colored chemical fibers consist of tiny particles by dissection, and it increases with the increasing size of fibers, but due to technical reasons the diameter and length of fibers are limited. On one hand, the method of fiber production limits the diameter and length, On the other hand, too coarse and long fibers may cause faults in products, particularly in thin-walled plastic parts.

Many different types of polymers are used as chemical fiber matrices. The requirements for fiber matrices are a good thermal resistance, and no softening or melting at the processing temperature of coloring. For the coloring of fibers, it should use the colorants with good thermal resistance and light fastness. Currently commercial varieties are white, black and colorful, main color blue, green and red.

Considering the plenty of fiber speckled pigments it needs to check if the fiber can be applied in the selected polymer. In the plastic molding speckled pigments are added to a white or black colored plastic sample, and compared the fibers with the raw fibers. If the fiber speckled pigments show an insufficient thermal resistance at the selected processing temperature, the shape change of the fibers can be observed by visual. Fiber softening or melting would cause a visible color shift. Black fibers with poor thermal resistance make the white sample gray, while white fiber will lighten the black sample, colorful fibers will tint white fiber sample in the color of fibers.

The concentration of fiber speckled effect pigments in the final products depends on the desired intensity of the speckled effect, therefore, the effective concentration can vary considerably.

3.5.4.5 Marble Batches

The design principle of marble batches is very simple, the melt viscosity of masterbatch is much higher than the processing temperature of the polymer, and this is the basis of the marble effect. The imitation of the marble effect can't depend on a single substance, but requires two different designed masterbatches, therefore, one is masterbatch and another is marble batch.

In theory the design of marble batches seems simple, but in fact it requires a good balance between the processing parameters of the marble batches and the colored polymers. A too large difference in properties would cause incompatibility of both components and result in defective products, however a too small difference shows an insufficient marble effect.

The normal procedure is to run an expected test with different designed marble batches on the machine, it is the only way to test the marble effect and the quality of colored final product. Moreover, the small test samples used in color matching are worthless.

The principle of marble batches is well defined. However, considering the numerous polymers and the varieties of design types, it is difficult to make a satisfactory marble effect at once. Usually to achieve the desired result should take several tests. In addition, it should be noted that marble effect can't always occur, for example, when the mold contains a gap or is rough. In this case the marble effect can be botchy or impossible to appear.

Chapter 4
Main Types and Properties of Organic Pigments

4.1 History of Organic Pigments

It is difficult to determine the exact age of the organic pigments to start using, since in ancient the organic pigments fade easily, it is difficult to retain so far. In ancient times, plant source coloring material (such as rubia, indigo grass) or animal coloring material (from the conch Tyrien) were used to alternative inorganic pigments. Colorants were called "Pigment" by biologists, because it is extracted from plants and animals. Scientific research shows that the main color components of Rubia is Alizarin (1, 2-dihydroxyanthraquinone), the indigo grass's is Indigo. These organic pigment should be classified as organic dyes rather than pigments, which have solubility, at least they are the origin of the modern organic pigments.

Organic dyes and synthetic organic pigments start synthetic, after the first synthetic dyes mauveine was produced by British chemist Perkin in 1856, the Griess diazotization of aniline was discovered by German chemist in 1858, and aniline diazonium salt coupling reaction with an aromatic amine or aromatic phenols in 1861. The rise of large-scale synthesis of dyes, laid the foundation for organic pigments industry. Organic pigments are developed gradually accompanied by the development of dye industry.

Lithol Red was the first pigment prepared by water-soluble dye, Golden Red C (Pigment Red 53:1) come out in 1903, and it is widely used in plastics until today. Monoazo and disazo benzidine yellow pigment began to put on the market in 1910, and red azo pigments into the market in 1931, since a number of patents related to yellow, orange monoazo pigments published in 1909.

Phthalocyanine blue and Phthalocyanine green pigments are the milestones in the history of organic pigments, come out respectively in 1935 and 1938. It filled the blank that the shortage of high performance blue and green organic pigments. The production of phthalocyanine pigments keep growing, because the synthesis process is simple, low production cost, bright shade, strong tinting strength, also has excellent heat resistance, light fastness, weather resistance and chemical stability. Phthalocyanine pigments were developed very rapidly in half a century. Currently there are two crystal common forms for phthalocyanine blue: α type was reddish blue tones, β-type was greenish blue tone. Phthalocyanine green was the polychlorinated matter of copper phthalocyanine, and the yellow phthalocyanine green pigment product latterly was the substituent of copper phthalocyanine chlorine or copper phthalocyanine bromine.

Phthalocyanine yellow, orange, red and purple pigment were developed gradually from the 1950s, which have the similar color fastness with green and blue pigment. Yellow and red azo condensation pigment were developed by Switzerland Ciba – Geigy Company in 1954, which have excellent heat resistance and migration resistance. In 1955, US Company Du Pont has developed quinacridone red and violet pigments. Benzimidazolone yellow, orange and red pigment were introduced to the market by Germany Hurst in 1960s. In 1970s Switzerland Ciba-Geigy and BASF has developed a yellow isoindolinone and isoindoline pigments. In 1980s Ciba Company launched a new product Diketopyrrolo-Pyrrolo (e.g. DPP) based red pigment.

4.2 Classification of Organic Pigments

There are many classifications for organic pigments. Based on the colors of chromatography organic pigments can be divided in yellow hues, orange hues, purpurin hues, green and blue hues. Based on the source can be divided in natural organic pigments and synthetic organic pigments. According to uses can be divided into organic pigments for ink(e.g. offset ink, water-based inks etc.), organic pigments for paints(e.g. solvent-based paint, waterborne coatings etc.), organic pigments for plastics(e.g. PVC, ABS, PP, PA, etc.).

In this book organic pigments is classified based on the chemical structure. The reason the light fastness, weather resistance and heat resistance properties are very similar if they have the same chemical structure. To some extent foreshadowed the typical properties of the pigment.

Organic pigments can be divided based on the chemical structure into three categories, Azo pigments, Phthalocyanine pigments and Phthalocyanine and Fused ring ketone pigments. These three types of pigment can be further subdivided based on the structural characteristics.

(1) Azo pigments can be divided monoazo, disazo, condensation azo pigments based on the number of azo group. Monoazo lakes and diclofenac benzidine series disazo pigment was considered as the classic azo pigments, which have been largely used in plastic coloring for the complete chromatogram (red, yellow, orange), bright color and reasonable price, but there are various defects in heat resistance, light fastness, migration resistance and other aspects especially in light shade because of the structure and other factors. Further, conventional benzidine yellow and orange pigments can decomposed if polymer processing temperature exceeds 200℃. Decomposition products are diclofenac-benzidine. Diclofenac-benzidine is carcinogenic in animals, and there may be carcinogenic aromatic amine to humans.

The application performance of classic azo pigments can be greatly improved, through two monoazo pigments molecule can be linked by an amide group synthetic high molecular weight condensed azo pigments, so we consider disazo condensation pigments as macromolecular pigment. In addition to increasing the molecular weight of the pigment, benzimidazolone pigments introduced some particular group that contain cyclic amide at the molecular chain to reduce the solubility of the pigment molecules. At the same time the heat resistance, weather resistance and migration resistance were improved too. These two types of pigments have excellent heat resistance, light fastness, weather resistance, and chemical stability.

(2) Phthalocyanine pigments mainly have blue and green hue, also have bright shade, high tinting strength, excellent light fastness, weather resistance, heat resistance and chemical stability. Copper phthalocyanine blue pigment has polymorphism, which means the same compound has the ability to generate a variety of different crystal structures. Crystal form affect the application performance, such as the tinting strength and color shade were different for different green halogenated copper phthalocyanine pigment.

(3) Heterocyclic and polycyclic ketonic pigments contain quinacridoue, dioxazine, isoindolinone, diketopyrrolo-pyrrolo and anthraquinone heterocyclic pigments and so on. This kind of pigment and benzimidazolone pigments, disazo condensation pigments were considered as high performance organic pigments.

Varieties of organic pigments used in plastics classification based on the chemical structure are shown in Table 4.1.

Classification of organic pigments are based on chemical structure, to a certain extent, indicates the typical properties of such pigments. The positioning of organic pigments based on structure, color area and performance is shown in Figure 4.1. From Figure 4.1 we can get a preliminary understanding and awareness of the entire organic pigments system.

Table 4.1 Varieties of organic pigments used in plastics classification based on the chemical structure

Types	Chemical structure		Color	Varieties (Index number)
Azo pigments	Monoazo	Azo salt/lakes	Yellow	Pigment Yellow 62 / 168 / 183 / 191
			Red	Pigment Red 48:1 / 48:2 / 48:3 / 48:4 Pigment Red 53:1 / 53:3 / 57:1
		Naphthol AS pigments	Red	Pigment Red 170 / 247
		Benzimidazolone pigments	Yellow	Pigment Yellow 120/180 / 181/214
			Orange	Pigment Orange 64 / 72
			Red	Pigment Red 175 / 176 / 185/208
			Violet	Pigment Violet 32
			Brown	Pigment Brown 25
	Disazo pigments		Yellow	Pigment Yellow 12 / 13 / 14 / 17 / 81 / 83
			Orange	Pigment Orange 13 / 34
	Disazo condensation pigments		Yellow	Pigment Yellow 93 / 95 / 128
			Red	Pigment Red 144 / 166 / 214 / 242
			Brown	Pigment Brown 23 / 41
Phthalocyanine pigments	Phthalocyanine pigments		Blue	Pigment Blue 15 / 15:1 / 15:3
	Chlorinated phthalocyanine pigments		Green	Pigment Green 7
	Bromine chloride phthalocyanine pigments			Pigment Green 36
Polycyclic pigments	Dioxazine pigments		Violet	Pigment Violet 23 / 37
	Quinacridone pigments		Red	Pigment Red 122 / 202; Pigment Violet 19 (γ crystal form)
			Violet	Pigment Violet 19 (β crystal form)
	Perylene pigments		Red	Pigment Red 149 / 178 / 179
			Violet	Pigment Violet 29
	Isoindolinone pigments		Yellow	Pigment Yellow 109 / 110
			Orange	Pigment Orange 61
	Isoindoline pigments		Yellow	Pigment Yellow 139
	Diketopyrrolo-pyrrolo pigments		Orange	Pigment Orange 71 / 73
			Red	Pigment Red 254 / 255 / 264 / 272
	Quinophthalone pigments		Yellow	Pigment Yellow 138
	Pteridine pigments		Yellow	Pigment Yellow 215
	Anthraquinone, anthrone pigments		Yellow - orange	Pigment Orange 43, Pigment Yellow 147
			Red - blue	Pigment Blue 60, Pigment Red 177
	Metal complex pigments		Yellow	Pigment Yellow 150
			Orange	Pigment Orange 68

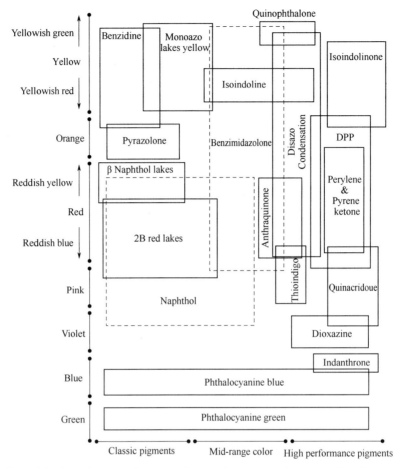

Figure 4.1　Location map of organic pigments' structure, color area and performance

4.3　Importance of Organic Pigments in Plastics Coloring

Inorganic pigments, organic pigments and solvent dyes were commonly used for plastic coloring. Organic pigments present bright color, high tinting strength and many varieties, which can't be substituted, by other pigments.

4.3.1　Differences Between Organic and Inorganic Pigments

Organic pigments chromatography is relatively wide and complete, especially tinting strength and color brightness significantly is higher than inorganic pigments, Inorganic pigments are generally somewhat dim. Compared with organic pigments, inorganic pigments have a superior mechanical strength, hiding power, heat resistance, light fastness and solvent resistance. The production of organic pigments is more complex than inorganic pigments and the price is relatively expensive. Inorganic and organic pigments have their own advantages and disadvantages respectively. In application, organic pigments and inorganic pigments often mixed together to achieve the best results. Performance comparison of organic pigments and inorganic pigments are shown in Table 4.2.

Table 4.2 Performance comparison of organic pigments and inorganic pigments

Performance	Organic pigments	Inorganic pigments
Chromatography species	Variety, chromatography wide	Fewer species, narrow chromatographic
Color characteristics	Colorful, brightness	Darker color, low brightness
Tinting strength	High	Low
Transparency	High, low hiding power	Low, high hiding power
Density	Low, most of less than 2.5 g/cm^3	High, most of less than 2.5 g/cm^3
Particle size	Small, large specific surface area	Big, Smaller surface area
Agglomeration tendency	High	Low
Dispersion	Not very good	Better
Solubility	Partially dissolved, depending on the concentration and structure	Completely insoluble
Heat resistance	Depends on varieties	Excellent
Light-fastness	Depends on varieties	Excellent
Acid and alkali resistance	Better, excellent	Some varieties change color
Solvent resistance	Medium-excellent	Excellent
Warpage	Some species (e.g. phthalocyanine) high	None
Security	Most better	Contain chromium cadmium and other heavy metals lead

4.3.1.1 Differences of Tinting Strength

The different tinting strength of organic pigments and inorganic pigments result from the different color-producing mechanisms, because of the different chemical structure. There is a conjugated double bonds system (π-electron system) chromophores in organic pigment structure, which is the reason for organic pigments to produce colors. The electrons of conjugated double bonds could transition from the ground state to an excited state by absorbing visible light selectively, so the color organic pigment present is complementary. Inorganic pigment without any double bonds or chromophores, all chemical elements were composed by positive nuclei (protons) which surrounding the (negative) electronics. The size of the nuclei and the number of electrons were decided by the position of the element in the periodic table. All the elements in the periodic table were positioned depending on the periodic cycle of atom's number, this determines the number of electrons and electron shell arrangement. Electron rotating in a fixed energy level, that called the electron orbit. The electrons of inorganic pigment could transition from the ground state to an excitation state, by absorbing energy (such as sunlight), that is to say transition from a lower energy level to a higher energy level, it requires a certain energy. Electrons transition from an excited state to a lower energy level (ground state) will produce a series of emission lines, when these emission lines in the visible wavelength range, colors will be seen. The emitted light intensity of inorganic pigments is lower than light intensity of organic pigments produced by conjugated double bonds, this is the reason that the tinting strength of inorganic pigments is lower than organic pigments.

4.3.1.2 Differences of Application

Inorganic pigments usually been calcined in high temperature (about 700~900°C), so the heat resistance is excellent. Some other inorganic pigment produced by different ways also exhibit a sufficiently high heat resistance, only very few kinds of inorganic pigments (chrome yellow) show poor heat resistance. However, the heat resistance of organic pigments depend on chemical structure and difference crystal form.

In partially crystalline polymers such as polyolefin, many organic pigment particles act as nuclei in the polymer melt, single crystals grow into spherulites, and the degree of crystallinity was increased too. The shrinkage of polymer melt depend not only on processing temperature, cooling rate, and pigment concentration factor, but also crystallinity. Because most of organic pigments used in plastic coloring are needle-like crystals, plastic products are easy to warpage. The inorganic pigments will not cause warpage for the spherical crystal. Pigment crystal structure is shown in Figure 4.2 and Figure 4.3.

Figure 4.2 Inorganic Pigment Brown 24 TEM image

Figure 4.3 Organic Pigment Blue 15∶3 TEM image

Inorganic pigments are metal oxides and salts, the density is generally in 3.5~5g / cm^3, small surface area, small gaps between the particles, so it is easy to be wetting and dispersing. Generally inorganic pigments have a relatively good dispersion than organic pigments in plastic. Especially titanium dioxide and chromium cadmium pigments in plastics is most easily dispersed.

4.3.1.3 Differences of Security

Lead, chromium, cadmium inorganic pigments are prohibited to use in toys, home appliances, and food contact packaging materials by the European Toys Directive, EU REACH SVHC (SVHC), EU ROHS legislation, EU legislation AP 89-1 and United States "Consumer Product Safety Improvement Act" (CPSIA). Due to the strengthen of environmental protection, many inorganic pigments will be increasingly restricted in use, and organic pigments in plastic applications are getting more attention. Trace impurities such as aromatic primary amines, polychlorinated biphenyls and dioxins etc. may occur in the process of organic pigment synthesis, which effect the use of plastics in food or cosmetics packing materials, therefore, many countries' legislation contains special requirements for the purity of the organic pigments.

4.3.1.4 Complementary and Synergistic Effect of Organic and Inorganic Pigments

Whether use organic pigments or inorganic pigments for plastic coloring, technical and economic as well as national regulations must be considered. The need for transparency and high tinting strength, particularly in thin articles or fiber drawing coloring organic pigments can be chosen. If the color mixing, light color, high hiding power, heat resistance, light fastness and weather resistance were needed inorganic pigments is better.

It's worth noting that BASF's PALIOTAN®, Germany HUBACK Companies' TICO® pigments is the composite of high-degree organic pigments and inorganic pigments that combines the advantages of inorganic pigments and organic pigments. Inorganic pigments can provide required hiding power, and organic pigments can improve tinting strength and the vividness of composite material. Products include different performance levels of yellow, orange and red.

4.3.2 Difference of Organic Pigments and Solvent Dyes

Colorant that can be completely dissolved in the polymer at a certain processing temperature is defined as solvent dyes. Solvent dyes are named for that it can dissolve in various organic solvents. In contrast, colorant that completely insoluble in the polymer at a certain processing temperature is defined as pigments. However, this distinction is sometimes proved ambiguous. In certain situations, some organic pigments only partially soluble in the polymer melt, DPP pigment (Pigment Red 254) is a typical example. In most of the polymers Pigment Red 254 exhibthe bright red, but used in PC coloring, when processing temperature above 320°C, it completely dissolved and showing a bright fluorescent yellow. At this time Pigment Red 254 exhibite typical dye characteristics.

Organic pigments and solvent dyes are organic compounds with color, which have the similar chemical structure, even some organic compounds can be used as both dye and organic pigments. However, organic pigments and dyes are indeed two different concepts.

(1) Solvent dyes are compounds that can absorb and transmission a certain wavelengths of light, without any kind of light-scattering. Coloring pigments used in plastics is a common effect that pigment particles (crystals) surface for light absorption, reflection and scattering (Figure 4.4). Therefore, solvent dyes and pigments are different, solvent dyes used in plastic coloring is transparent. Characteristics of solvent dyes are very high tinting strength and bright colorful.

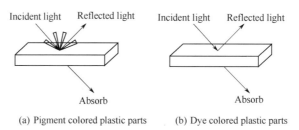

(a) Pigment colored plastic parts   (b) Dye colored plastic parts

Figure 4.4   Distinguish optical behavior of dyes and pigments

(2) Organic pigments present highly dispersed particle state in plastic coloring, so the original crystalline state always exist. The color of the pigment is on behalf of the color of the product. Because of this, the crystal state of the pigment is very important in terms of pigment, the heat resistance, light fastness, weather resistance, and dispersion of pigments in plastic coloring were affected by the crystalline state of pigment. The solvent dyes is completely dissolved in polymer keep the molecular state when coloring plastics, in this case the crystalline state of dye solvent is not so important, that means the crystalline state of dye is not closely related to the coloring behavior, all the performance are related to the chemical structure.

(3) In comparison, the molecular weight of solvent dyes is relatively low, so the heat resistance and light fastness is not as good as organic pigments.

(4) Migration can occur when solvent dyes are used in some low glass transition temperature thermoplastics, especially in polyethylene, polypropylene and plasticized polyvinyl chloride.

Difference of organic pigments and solvent dyes are shown in Table 4.3.

Table 4.3 Difference of organic pigments and solvent dyes

Application features	Organic pigments	Solvent dyes	
		Alcohol soluble	Oil-soluble
Molecular size	Larger	Smaller	Smaller
Polarity of molecules	Low	High	Low
Types of substituents	NO_2, Cl, Br, OCH_3	Lipophilic groups acid / base bonds	Lipophilic groups free base, branched chain alky
Allomorphic	Yes	None	None
Coloring patterns	Fine particles	Molecular state	Molecular state
Solvent solubility	Insoluble	Solution	Solution

4.4 Relationship between Color Performance and Pigment Chemical Structure, Crystal Structure, Application Media

There are a lot of purpose plastics coloring, the ultimate goal is to enhance the value of plastic products. For users, the effective value of the pigment depends on the color properties, processability and application performance of pigments. Firstly the properties depend on the chemical structure. Organic pigments contain monoazo pigments, azo lakes, diszao condensation pigments, phthalocyanine pigments, quinacridone pigments, dioxazine pigments, isoindolinone pigments, diketopyrrolo-pyrrolo pigments, anthraquinone heterocyclic. The luster and application performance of pigment molecule were decided by the structure, but the pigment molecules with the same class of chemical structures, the combination of the substituents varies because different atoms. Therefore, the properties is different because of the different chemical structure of pigments.

When pigment is used for plastic coloring, it is not soluble in these plastic media but in the presence of molecular form, most dispersed in a plastic medium as micro-nanometer particle morphology composed by many molecules, pigment molecule particle through absorption, reflection, transmission, refraction of light, which projected onto the surface of these application media to achieve the purpose of the plastics coloring. Therefore, application of organic pigments in the plastics coloring is not only depend on the structure of the pigment, but also the particle surface properties, particle size and particle distribution which produced in the manufacturing process. The relationship between particle size and hue, coloring strength, dispersion, light fastness, weather resistance of organic pigments has been discussed in great detail in Chapter 2 of this book, that help us to have a fully understand of the performance indicators. Only in this way can we make full use of all kinds of plastics processing conditions, in order to ultimately create value. In chapter 2 of this book, we also discussed the close relationship between application performance and application media of organic pigments. Organic pigments have different physical and chemical properties in different media. Therefore, application performance of organic pigments and the use of media should match, otherwise, in the specific application process will cause mass accidents. Relationship between application and properties of organic pigments in plastic is shown in Figure 4.5.

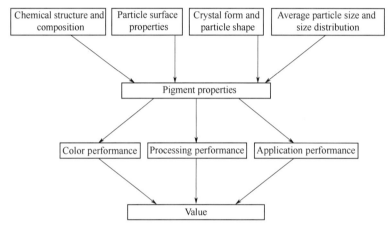

Figure 4.5　Relationship between application and properties of organic pigments in plastic

4.5　Azo Pigment

Azo pigment is a pigment that contains azo group (-N=N-) in molecule chain, the azo group of pigment molecule is imported by diazotization and coupling reaction. It is can be further classified according to the number of azo groups of pigment molecule or the structure characteristics of coupling component in the azo pigment.

4.5.1　Monoazo Pigment

Monoazo pigment refers to a pigment that contains only one azo group in molecule chain, mainly monoazo lakes pigment, naphthol AS and benzimidazolone pigments, color covered yellow, orange, red and brown. Wherein the benzimidazolone pigments is a class of high performance organic pigments, which is differ from the generic azo pigment, it has a very excellent fastness, light fastness, weather resistance, heat resistance, high solvent resistance, migration resistance, good chemical stability, good acid resistance and alkali resistance.

4.5.1.1 Monoazo Lake Pigment

The monoazo dye with molecules containing carboxylic acid group and sulfonic acid group is soluble in alkaline aqueous medium. However, after the reaction of monoazo dye molecule and alkaline earth metal compounds, they will be converted into the carboxylate or sulfonate that is not only insoluble in water but also insoluble in organic solvent, so it can be used as a pigment. The main chromatographic is yellow and red.

(1) Yellow lake pigment

In order to improve the heat resistance and migration resistance of monoazo yellow pigment, a sulfonic acid group is introduced in the molecule, and then transformed into lake pigment with a high molecular polarity, the solvent resistance, migration resistance and heat resistance in organic solvents and other media is much higher than non-lake pigment. This kind of pigment has a lower tinting strength and high hiding power, and some varieties have excellent heat resistance, which is more suitable for coloring of plastics. Specific varieties are as follows.

① **C.I. Pigment Yellow 62**   C.I. structure No 13940. Molecular formula: $C_{17}H_{15}N_4O_7S \cdot 1/2Ca$; CAS registry No [12286-66-7].

Structure formula

$$\left[\begin{array}{c} \text{structure} \end{array} \right] \cdot \tfrac{1}{2} Ca^{2+}$$

Color characterization   Pigment Yellow 62 is reddish (middle shade) yellow. Tinting strength is relatively low. With 1% titanium dioxide preparation 1 / 3SD HDPE need 0.5% of pigment. But to achieve the same effect, slightly reddish benzidine Pigment Yellow 13 just require 0.17% of pigment.

Main properties   Show in Table 4.4, Table4.5 and Figure 4.6.

Table 4.4 The application properties of Pigment Yellow 62 used in PVC

Project		Pigment	Titanium dioxide	Light fastness degree	Weather resistance degree (2000h)	Migration resistance degree	Heat resistance degree	
							180°C/30min	200°C/10min
PVC	Full shade	0.1%		7	3~4	5	5	5
	Reduction	0.2%	2%	6~7	2~3	5	5	5

Table 4.5 The application properties of Pigment Yellow 62 used in HDPE

Project		Pigment	Titanium dioxide	Light fastness degree	Weather resistance degree (3000h)	Migration resistance degree
HDPE	Full shade	0.1%		7	2~3	5
	Reduction	0.1%	1%	7	1	5

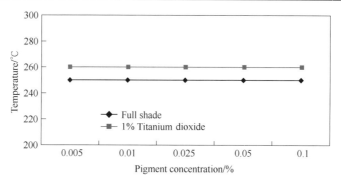

Figure 4.6 The heat resistance of Pigment Yellow 62 in HDPE

Application range   Shown in Table 4.6.

Table 4.6 Application range of Pigment Yellow 62

General plastics		Engineering plastics		Spinning	
LL/LDPE	●	PS/SAN	○	PP	○
HDPE	●	ABS	○	PET	×
PP	●	PC		PA6	×
PVC(soft)	●	PBT	×		
PVC(rigid)	●	PA	×		
Rubber	●	POM			

● Recommended to use, ○ Conditional use, × Not recommended to use.

Variety characteristic   Pigment Yellow 62 has relatively low tinting strength, it is suitable for light color varieties coloring. The Light fastness and heat resistance of Pigment Yellow 62 are superior to dichloro benzidine pigment, nor security issues. The quantity demand in the coloring of polyolefin is increasing because of the price and performance advantage. Pigment Yellow 62 is suitable for PVC coloring and general polyolefin plastics, but used in HDPE and other crystallization plastics will affect warpage variant.

② **C.I. Pigment Yellow 168**   C.I. structure No 13960. Molecular formula: $C_{16}H_{12}ClN_4O_7S \cdot 1/2Ca$; CAS registry No [71832-85-4].

Structure formula

$$\left[\begin{array}{c} \text{Cl} \quad \overset{\text{CH}_3}{\underset{}{\text{C}=\text{O}}} \\ \text{C}_6\text{H}_4\text{--NHC--CH--N=N--C}_6\text{H}_3(\text{NO}_2)\text{--SO}_3 \\ \overset{}{\underset{\text{O}}{}} \end{array} \right] \cdot \frac{1}{2}\text{Ca}^{2+}$$

Color characterization Pigment Yellow 168 is greenish yellow, with relatively low tinting strength.

Main properties Shown in Table 4.7, Table4.8 and Figure 4.7.

Table 4.7 Application properties of Pigment Yellow 168 used in PVC

Project		Pigment	Titanium dioxide	Light fastness degree	Weather resistance degree (2000h)	Migration resistance degree	Heat resistance degree	
							180°C/30min	200°C/10min
PVC	Full shade	0.1%		7~8	4	5	5	5
	Reduction	0.2%	2%	7	3~4	5	5	5

Table 4.8 Application properties of Pigment Yellow 168 used in HDPE

Project		Pigment	Titanium dioxide	Light fastness degree	Weather resistance degree (3000h)	Migration resistance degree
HDPE	Full shade	0.1%		7	3	5
	Reduction	0.1%	1%	7	1	5

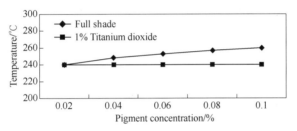

Figure 4.7 The heat resistance of Pigment Yellow 16 in HDPE

Application range Shown in Table 4.9.

Table 4.9 Application range of Pigment Yellow 168

General plastics		Engineering plastics		Spinning	
LL/LDPE	●	PS/SAN	○	PP	○
HDPE	●	ABS	○	PET	×
PP	●	PC		PA	×
PVC(soft)	●	PBT		PAN	
PVC(rigid)	●	PA			
Rubber	●	POM			

● Recommended to use, ○ Conditional use, × Not recommended to use.

Variety characteristic The application performance of Pigment Yellow 168 is close to Pigment Yellow 62. Pigment Yellow 168 has a good migration resistance, is suitable for PVC and general polyolefin plastics coloring, but used in HDPE and other crystallization plastics will affect warpage and variant slightly.

③ **C.I. Pigment Yellow 183** C.I. structure number: 18792. Molecular formula: $C_{16}H_{10}C_{12}N_4O_7S_2Ca$; CAS registry number: [65212-77-3]

Structure formula

$$\left[\begin{array}{c} \text{Cl}\underset{\text{Cl}}{\bigcirc}\overset{\text{SO}_3^-}{\underset{}{\,}}\text{N}=\text{N}-\underset{\text{HO}}{\overset{\text{CH}_3}{\bigcirc}}\underset{\text{SO}_3^-}{\overset{\text{N}-\text{N}}{\bigcirc}} \end{array} \right] \cdot \text{Ca}^{2+}$$

Color characterization Pigment Yellow 183 is reddish yellow, and a special variety for coloring of plastic. The tinting strength is relatively low, formulated 1/3SD PVC with 1% of titanium dioxide required 0.34% of pigment, and formulated 1/3SD HDPE with 1% of titanium dioxide required 0.43% of pigment.

Pigment Yellow 183 is polymorphism. There are two commodity forms classified by the difference of transparency. Transparency variety has a high tinting strength with slightly green. Another variety with a low tinting strength and biased red.

Main properties Shown in Table 4.10~Table 4.15, Figure 4.8 and Figure 4.9.

Table 4.10 Application properties of Pigment Yellow 183(transparent) used in PVC

	Project	Pigment	Titanium dioxide	Light fastness degree	Migration resistance degree
PVC	Full shade	0.1%		8	5
	Reduction	0.1%	1%	7~8	

Table 4.11 Application properties of Pigment Yellow 183(transparent) used in HDPE

	Project	Pigment	Titanium dioxide	Light fastness degree	Migration resistance degree
HDPE	Full shade	0.1%		7	5
	Reduction	0.1%	1%	6~7	

Table 4.12 Application range of Pigment Yellow 183(transparent)

General plastics		Engineering plastics		Spinning	
LL/LDPE	●	PS/SAN	●	PP	●
HDPE	●	ABS	●	PET	○
PP	●	PC	○	PA6	×
PVC(soft)	●	PBT	○		
PVC(rigid)	●	PA6	○		
Rubber	●	PMMA	●		

● Recommended to use, ○ Conditional use, × Not recommended to use.

Table 4.13 Application properties of Pigment Yellow 183(covered) used in PVC

	Project	Pigment	Titanium dioxide	Light fastness degree	Migration resistance degree
PVC	Full shade	0.1%		8	5
	Reduction	0.1%	1%	7	

Table 4.14 Application properties of Pigment Yellow 183(cover) used in HDPE

Project		Pigment	Titanium dioxide	Light fastness degree	Weather resistance degree (3000h)	Migration resistance degree
HDPE	Full shade	0.1%		7	3~4	5
	Reduction	0.1%	1%	6~7		

Table 4.15 Application range of Pigment Yellow 183(cover)

General plastics		Engineering plastics		Spinning	
LL/LDPE	●	PS/SAN	●	PP	●
HDPE	●	ABS	●	PET	×
PP	●	PC	○	PA6	×
PVC(Soft)	●	PBT	○		
PVC(Rigid)	●	PA6	○		
Rubber	●	PMMA	●		

● Recommended to use, ○ Conditional use, × Not recommended to use.

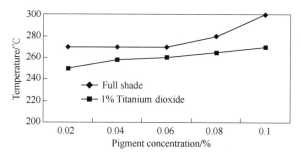

Figure 4.8 Heat resistance of Pigment Yellow 183(transparent) in HDPE

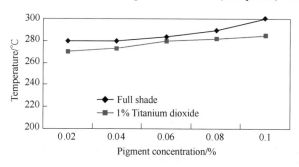

Figure 4.9 Heat resistance of Pigment Yellow 183(covered) in HDPE

Variety characteristic Pigment Yellow 183 has relatively low tinting strength, it is suitable for color matching of light-colored varieties. The Full shade of Pigment Yellow 183 (cover) can still meet requirements of product that exposed outdoors for a long time. Pigment Yellow 183 has excellent heat resistance in a wide concentration range, and used for ABS is up to 300°C. In addition to using for coloring of general purpose polyolefin also can be used for coloring of styrene engineering plastic. Pigment Yellow

183 used in HDPE and other crystallization plastics will affect warpage variant.

④ **C.I. Pigment Yellow 191** C.I. structure number: 18795. Molecular formula: $C_{17}H_{13}ClN_4O_7S_2Ca$. CAS registry number: [129423-54-7].

Structure formula

$$\left[H_3C \underset{Cl}{\overset{SO_3^-}{\bigcirc}} -N=N- \underset{HO}{\overset{CH_3}{\bigcirc}} \underset{SO_3^-}{\overset{N-N}{\bigcirc}} \right] \cdot Ca^{2+}$$

Color characterization Pigment Yellow 191 is reddish yellow, and the color is close to Pigment Yellow 83. The tinting strength is much lower than Pigment Yellow 83. Formulated 1/3SD HDPE with 1% of titanium dioxide required 0.34% of pigment, and only 0.08% of pigment is needed for Pigment Yellow 83 to achieve the same effect.

The difference of Pigment Yellow 191∶1 and Pigment Yellow 191 is the laking salt, Pigment Yellow 191∶1 is aluminum salt.

Main properties Shown in Table 4.16~Table 4.21, Figure 4.10 and Figure 4.11.

Table 4.16 Application properties of Pigment Yellow 191 used in PVC

	Project	Pigment	Titanium dioxide	Light fastness degree	Migration resistance degree
PVC	Full shade	0.1%		7	
	Reduction	0.1%	0.5%	6	5

Table 4.17 Application properties of Pigment Yellow 191 used in HDPE

	Project	Pigment	Titanium dioxide	Light fastness degree	Weather resistance degree (3000h, 0.2%)
HDPE	Full shade	0.1%		7	3
	Reduction	0.1%	1%	6~7	

Table 4.18 Application range of Pigment Yellow 191

General plastics		Engineering plastics		Spinning	
LL/LDPE	●	PS/SAN	●	PP	●
HDPE	●	ABS	●	PET	×
PP	●	PC	●	PA6	×
PVC(soft)	●	PBT	●	PAN	×
PVC(rigid)	●	PA	×		
Rubber	●	POM	○		

● Recommended to use, ○ Conditional use, × Not recommended to use.

Table 4.19 Application properties of Pigment Yellow 191:1 used in PVC

Project		Pigment	Titanium dioxide	Light fastness degree	Weather resistance degree (2000h)	Migration resistance degree	Heat resistance degree	
							180°C/30min	200°C/10min
PVC	Full shade	0.1%		7~8	3~4	5	5	5
	Reduction	0.2%	2%	6~7	2~3	5	5	5

Table 4.20 Application properties of Pigment Yellow 191:1 used in HDPE

Project		Pigment	Titanium dioxide	Light fastness degree	Weather resistance degree (3000h)	Migration resistance degree
HDPE	Full shade	0.1%		7~8	3~4	5
	Reduction	0.1%	1%	7		5

Table 4.21 Application range of Pigment Yellow 191:1

General plastics		Engineering plastics		Spinning	
LL/LDPE	●	PS/SAN	●	PP	●
HDPE	●	ABS	●	PET	×
PP	●	PC	○	PA6	×
PVC(soft)	●	PBT	○	PAN	
PVC(rigid)	●	PA6	○		
Rubber	●	PMMA	●		

● Recommended to use, ○ Conditional use, × Not recommended to use.

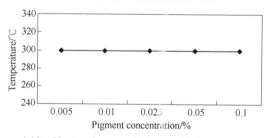

Figure 4.10 Heat resistance of Pigment Yellow 191 in HDPE

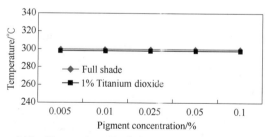

Figure 4.11 Heat resistance of Pigment Yellow 191:1 in HDPE

Variety characteristic Pigment Yellow 191 has a relatively low tinting strength, it is suitable for color matching of light-colored varieties. The full shade of Pigment Yellow 191 can still meet requirements of exposing outdoors for a long time. Pigment Yellow 191 has excellent heat resistance, especially for coloring light product can still maintain a good heat resistance, and used for PC is up to 330°C. In addition to using for coloring of PVC and general purpose polyolefin, also can be used for coloring of engineering plastics.

(2) Red lake pigment

① **Naphthol red lake pigment**

2-naphthol as a coupling component diazonium salt reaction with sulfonic acid containing aromatic amine, and then laking with metal salt (calcium salt, barium salt, strontium salt, etc.), a variety of bluish red and yellowish red can be prepared. This kind of lake pigments are more ancient species of lake pigments, the most famous variety golden red C began to be produced in Germany in 1902. Laking pigment has a better solvent resistance and migration resistance, the main varieties used in plastics are as follows.

C.I. Pigment Red 53 : 1 C.I. structure number: 15585. Molecular formula: $C_{34}H_{24}C_{12}N_4O_8S_2Ba$. CAS registry number: [5160-02-1].

Structure formula

$$\left[\begin{array}{c} H_3C \\ Cl- \end{array} \! \begin{array}{c} OH \\ -N\!\!=\!\!N- \\ SO_3^- \end{array} \right] \cdot \frac{1}{2} Ba^{2+}$$

Color characterization Pigment Red 53 : 1 is an ancient lake variety and shows very high color saturation. The shade is the most yellowish in azo lake pigments. Compared with other similar pigments, Pigment Red 53 : 1 is high tinting strength and present good shade in plastics.

Main properties Shown in Table 4.22~Table 4.24, Figure 4.12.

Table 4.22 Application properties of Pigment Red 53 : 1 used in PVC

Project		Pigment	Titanium dioxide	Light fastness degree	Migration resistance degree
PVC	Full shade	0.1%		6	4~5
	Reduction	0.1%	0.5%	2	

Table 4.23 Application properties of Pigment Red 53 : 1 used in HDPE

Project		Pigment	Titanium dioxide	Light fastness degree
HDPE	Full shade	0.22%		3
	1/3SD	0.22%	1%	2-3

Table 4.24 Application range of Pigment Red 53 : 1

General plastics		Engineering plastics		Spinning	
LL/LDPE	●	PS/SAN	●	PP	○
HDPE	●	ABS	●	PET	×
PP	●	PC	●	PA6	×
PVC(soft)	●	PBT	×	PAN	×
PVC(rigid)	●	PA	×		
Rubber	●	POM	×		

● Recommended to use, ○ Conditional use, × Not recommended to use.

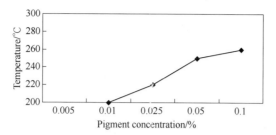

Figure 4.12 Heat resistance of Pigment Red 53∶1 in HDPE

Variety characteristic Pigment Red 53∶1 shows good color performance, low price, excellent heat resistance and poor light fastness. In addition to suitable for general polyolefins coloring, it also can be used in styrenic engineering plastics.

② **2-hydroxy-3-naphthoic acid red lake pigments**

2-hydroxy-3-naphthoic acid (2,3 acid), also known as bona acid, regarded as coupling components reacts with aromatic amine diazonium salt containing sulfonic acid group, and then laking with metal, which can prepare red pigments with a variety of red spectrums. For example 2B red and ruby red 4B are the most used in plastics. 2B red series in the medium red zone show high tinting strength, excellent color performance, good migration fastness and heat resistance, moderate light fastness and good dispersibility. The main varieties used in plastics are as follows.

a) C.I. Pigment Red 48∶1 C.I. structure number: 15865∶1. Molecular formula: $C_{18}H_{11}ClN_2O_8SCa$. CAS registry number: [7585-41-3].

Structure formula

$$\left[H_3C-\underset{Cl}{\underset{|}{\bigcirc}}\underset{SO_3^-}{\overset{|}{-}}N=N-\underset{}{\bigcirc}\underset{OH}{\overset{|}{-}}\underset{COO^-}{\overset{|}{-}} \right] \cdot Ba^{2+}$$

Color representation Pigment Red 48∶1 is a barium salt lake, which belongs to 2B red series. It shows yellowish red and the most yellow phase red pigment in Pigment Red 48 series. Pigment Red 48∶1 has medium tinting strength and only needs 0.38% pigment to prepare 1/3 SD of HDPE with 1% titanium dioxide.

Main properties Shown in Table 4.25～Table 4.27, Figure 4.13.

Table 4.25 The application properties of Pigment Red 48∶1 used in PVC

Project		Pigment	Titanium dioxide	Light fastness degree	Migration resistance degree	Heat resistance degree	
						180°C/30min	200°C/10min
PVC	Full shade	0.1%		4	4～5	5	5
	Reduction	0.2%	2%	2-3	4～5	5	5

Table 4.26 The application properties of Pigment Red 48∶1 used in HDPE

Project		Pigment	Titanium dioxide	Light fastness degree	Migration resistance degree
HDPE	Full shade	0.1%		6	5
	Reduction	0.1%	1%	3-4	5

Table 4.27 The application range of Pigment Red 48∶1

General plastics		Engineering plastics		Spinning	
LL/LDPE	●	PS/SAN	○	PP	○
HDPE	●	ABS	×	PET	×
PP	○	PC	×	PA6	×
PVC(Soft)	●	PBT	×	PAN	×
PVC(Rigid)	○	PA	×		
Rubber	●	POM			

● Recommended to use, ○ Conditional use, × Not recommended to use.

Figure 4.13 The heat resistance of Pigment Red 48∶1 in HDPE

Variety characteristic The heat resistance of Pigment Red 48∶1 is poor in transparent PP, but it can be up to 200~230℃/5min under 1/3 SD. At higher temperatures it will quickly become dark red. Pigment Red 48∶1 can be used in the coloring of PVC and shows good migration fastness, bleeding resistance and does not migrate out in flexible PVC.

b) C.I. Pigment Red 48∶2 C.I. structure number: 15865∶2. Molecular formula: $C_{18}H_{11}ClN_2O_6SCa$. CAS registry number: [7023-61-2].

Structure formula

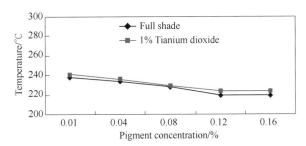

Color characterization Pigment Red 48∶2 is a barium salt lake, belongs to 2B red series. The shade is bluish red. It has high tinting strength and only needs 0.21% pigment to prepare 1/3 SD of HDPE with 1% titanium dioxide.

Main properties Shown in Table 4.28~Table 4.30, Figure 4.14.

Table 4.28 Application properties of Pigment Red 48:2 used in PVC

Project		Pigment	Titanium dioxide	Light fastness degree	Weather resistance degree (2000h)	Migration resistance degree	Heat resistance degree	
							180°C/30min	200°C/10min
PVC	Full shade	0.1%		7	2-3	5	4~5	4~5
	Reduction	0.2%	2%	6	1	4.9	4~5	4~5

Table 4.29 Application properties of Pigment Red 48:2 used in HDPE

Project		Pigment	Titanium dioxide	Light fastness degree	Weather resistance degree (3000h)	Migration resistance degree
HDPE	Full shade	0.1%		7	1~2	5
	Reduction	0.1%	1%	6	1	5

Table 4.30 Application range of Pigment Red 48:2

General plastics		Engineering plastics		Spinning	
LL/LDPE	●	PS/SAN	○	PP	●
HDPE	●	ABS	×	PET	×
PP	●	PC	×	PA6	×
PVC(soft)	●	PBT	×	PAN	×
PVC(rigid)	○	PA	×		
Rubber	●	POM			

● Recommended to use, ○ Conditional use, × Not recommended to use.

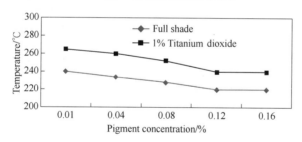

Figure 4.14 Heat resistance of Pigment Red 48:2 in HDPE

Variety characteristic Pigment Red 48:2 is moderate fastness, good performance-price ratio and the light fastness is significantly higher than Pigment Red 48:1. It is an important variety in the coloring of polyolefin plastics. Pigment Red 48:2 is suitable for the pre-coloring of spinning of polypropylene fiber. The beautiful bright red formulated with dark concentration is extensively used in carpets. It has no effect on the warpage in crystalline plastics such as HDPE.

c) C.I. Pigment Red 48:3 C.I. structure number: 15865:3. Molecular formula: $C_{18}H_{11}ClN_2O_6SSr$. CAS registry number: [15782-05-5].

Structure formula

$$\left[H_3C - \underset{Cl}{\underset{|}{\bigcirc}} \underset{SO_3^-}{\overset{}{-}} N = N - \underset{}{\bigcirc\!\bigcirc} \overset{OH}{\underset{}{-}} \overset{COO^-}{} \right] \cdot Sr^{2+}$$

Color characterization Pigment Red 48∶3 is a kind of strontium salt, belongs to 2B red series. The shade is yellower than Pigment Red 48∶2. It has medium tinting strength and only needs 0.25% pigment to prepare 1/3 SD of HDPE with 1% titanium dioxide.

Main properties Shown in Table 4.31~Table 4.33, Figure 4.15.

Table 4.31 Application properties of Pigment Red 48∶3 used in PVC

Project		Pigment	Titanium dioxide	Light fastness degree	Weather resistance degree (2000h)	Migration resistance degree	Heat resistance degree	
							180°C/30min	200°C/10min
PVC	Full shade	0.1%		6	2	5	5	5
	Reduction	0.2%	2%	5~6	1	5	5	5

Table 4.32 Application properties of Pigment Red 48∶3 used in HDPE

Project		Pigment	Titanium dioxide	Light fastness degree	Migration resistance degree
HDPE	Full shade	0.1%		6	5
	Reduction	0.1%	1%	4	5

Table 4.33 The application range of Pigment Red 48∶3

General plastics		Engineering plastics		Spinning	
LL/LDPE	●	PS/SAN	○	PP	●
HDPE	●	ABS	○	PET	×
PP	●	PC	×	PA6	×
PVC(soft)	●	PBT	×	PAN	×
PVC(rigid)	●	PA	×		
Rubber	●	PMMA	×		

● Recommended to use, ○ Conditional use, × Not recommended to use

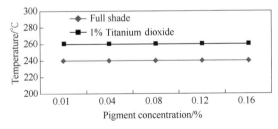

Figure 4.15 Heat resistance of Pigment Red 48∶3 in HDPE

Variety characteristic The light fastness of Pigment Red 48∶3 is best in 2B red series, and the heat resistance can be up to 240℃/5min in polyolefin plastics. It would rapidly turn blue and dark above 240℃. The properties are excellent in 2B red series and can be used in the coloring of PVC and general polyolefin. It has no effect on the warpage in crystalline plastics such as HDPE.

d) C.I. Pigment Red 57∶1   C.I. structure number: 15850∶1. Molecular formula: $C_{18}H_{12}N_2O_6SCa$. CAS registry number: [5281-04-9].

Structure formula

$$\left[H_3C-\underset{}{\bigcirc}-\underset{SO_3^-}{\underset{|}{\bigcirc}}\overset{OH}{\underset{}{\bigcirc}}\overset{COO^-}{\underset{}{\bigcirc\bigcirc}}-N=N-\right] \cdot Ca^{2+}$$

Color characterization   Pigment Red 57∶1 is a calcium salt lake, bluish red and bluest in azo lake pigments. It has high tinting strength and only needs 0.14% pigment to prepare 1/3 SD of HDPE with 1% titanium dioxide.

Main properties   Shown in Table 4.34~Table 4.36, Figure 4.16.

Table 4.34   Application properties of Pigment Red 57∶1 used in PVC

Project		Pigment	Titanium dioxide	Light fastness degree	Weather resistance degree (2000h)	Migration resistance degree	Heat resistance degree	
							180°C/30min	200°C/10min
PVC	Full shade	0.1%		6	1	5	5	5
	Reduction	0.2%	2%	4	1	4.9	5	5

Table 4.35   Application properties of Pigment Red 57∶1 used in HDPE

Project		Pigment	Titanium dioxide	Light fastness degree	Weather resistance degree (3000h)	Migration resistance degree
HDPE	Full shade	0.1%		6~7	1	5
	Reduction	0.1%	1%	4~5	1	5

Table 4.36   Application range of Pigment Red 57∶1

General plastics		Engineering plastics		Spinning	
LL/LDPE	●	PS/SAN	○	PP	●
HDPE	●	ABS	×	PET	×
PP	●	PC	×	PA6	×
PVC(soft)	●	PBT	×	PAN	×
PVC(rigid)	○	PA	×		
Rubber	●	POM			

● Recommended to use, ○ Conditional use, × Not recommended to use.

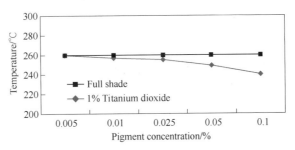

Figure 4.16 Heat resistance of Pigment Red 57∶1 in HDPE

Variety characteristic Pigment Red 57∶1 is cheap and the light fastness is only 2 in flexible PVC with 0.1% Pigment Red 57∶1. The light fastness is worse when compatibility with titanium dioxide which can affect the application in plastics. The heat resistance of Pigment Red 57∶1 can be up to 220~240℃/5min in polyolefin plastics. The heat resistance would sharply decrease with the concentration less than 0.1%, so the pigment is only used for deep shade. It has effect on the warpage in crystalline plastics such as HDPE but small.

(3) 2-hydroxy-3 naphthamide (Naphthol AS) red lake pigments

Naphthol AS derivatives are condensated between substituted aniline and 2, 3 acids, which are valuable in organic pigments. The common feature of these pigments is the molecule caining one or two sulfonic acid groups which not only make pigments becoming insoluble lake salt but also significantly improve the solvent resistance and migration fastness.

C.I. Pigment Red 247 C.I. structure number: 15915. Molecular formula: $C_{32}H_{26}N_4O_7S \cdot 1/2Ca$. CAS registry number: [43035-18-3].

Structure formula

$$\left[{}^-O_3S-\!\!\!\!\bigcirc\!\!\!\!-HNOC-\!\!\!\!\bigcirc\!\!\!\!\overset{CH_3}{\underset{N=N}{}}-\!\!\!\!\bigcirc\!\!\!\!\overset{OH}{\underset{}{}}\!\!-CONH-\!\!\!\!\bigcirc\!\!\!\!-OCH_3 \right] \cdot \tfrac{1}{2}Ca^{2+}$$

Color characterization Pigment Red 247 is positive red with bright bluish. It has good tinting strength and only needs 0.28% pigment to prepare 1/3 SD of HDPE with 1% titanium dioxide.

Main properties Shown in Table 4.37~Table 4.39, Figure 4.17.

Table 4.37 Application properties of Pigment Red 247 used in PVC

Project		Pigment	Titanium dioxide	Light fastness degree	Migration resistance degree
PVC	Full shade	0.1%		6~7	5
	Reduction	0.1%	0.5%	6	5

Table 4.38 Application properties of Pigment Red 247 used in HDPE

Project		Pigment	Titanium dioxide	Light fastness degree
HDPE	Full shade	0.27%		6~7
	1/3 SD	0.27%	1%	5~6

Table 4.39 Application range of Pigment Red 247

General plastics		Engineering plastics		Spinning	
LL/LDPE	●	PS/SAN	●	PP	●
HDPE	●	ABS	●	PET	×
PP	●	PC	●	PA6	×
PVC(soft)	●	PBT	●	PAN	×
PVC(rigid)	●	PA	×		
Rubber	○	POM	×		

● Recommended to use, ○ Conditional use, × Not recommended to use.

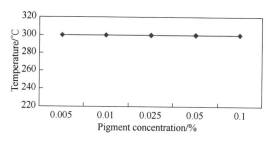

Figure 4.17 Heat resistance of Pigment Red 247 in HDPE

Variety characteristic Pigment Red 247 shows excellent light fastness and heat resistance of which can be up to 300℃. It has no effect on the warpage in crystalline plastics such as HDPE and no bleeding in PVC.

4.5.1.2 Naphthol AS pigment derivatives

Naphthol AS derivatives are condensated between substituted aniline and 2, 3 acids, which have crystal polymorphism. The crystal structure of these pigments has the following characteristics. Molecular chains are almost parallel, azo groups present as hydrazone structure not azo, molecule exist hydrogen bonds. So AS pigment derivatives have excellent properties which depend on substituent groups and would be better with the increasing of amide groups in molecules, for example, Pigment Red 187 contains three amide groups. They are slap-up varieties in the coloring of plastics. Compared with similar chromatographic azo pigments, as pigment derivatives show high tinting strength and the heat resistance has great difference because of the variety. Naphthol AS red pigment chromatographies are between bluish red and yellowish red.

① **C.I. Pigment Red 170** C.I. structure number: 12475. Molecular formula: $C_{26}H_{22}N_4O_4$. CAS registry number: [2786-76-7].

Structure formula

[Structure of Pigment Red 170 showing H₂NC(O)—C₆H₄—N=N—naphthalene with HO, C(O)NH—C₆H₄—OCH₂CH₃ substituents]

Color characterization Pigment Red 170 is positive red, high saturation and is the leader in similar pigments. It has good tinting strength and only needs 0.22% pigment to prepare 1/3 SD of HDPE with 1% titanium dioxide. Pigment Red 170 is crystal pleomorphism and has two crystal forms which is different in transparency. Transparency varieties with slight blue are commonly known as Permanent Red F5RK; High opacity varieties with yellow are commonly known as Permanent Red F3RK.

Main properties Shown in Table 4.40~Table 4.44, Figure 4.18 and Figure 4.19.

Table 4.40 Application properties of Pigment Red 170 (F3RK) used in PVC

Project		Pigment	Titanium dioxide	Light fastness degree	Migration resistance degree
PVC	Full shade	0.1%		7~8	
	Reduction	0.1%	0.5%	6~7	2

Table 4.41 Application properties of Pigment Red 170 (F3RK) used in HDPE

Project		Pigment	Titanium dioxide	Light fastness degree	Weather resistance degree (3000h, 0.2%)
HDPE	Full shade	0.22%		8	3
	1/3 SD	0.22%	1%	7~8	

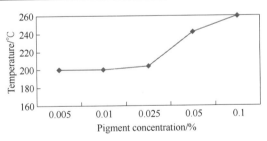

Figure 4.18 Heat resistance of Pigment Red 170 (F3RK) in HDPE

Table 4.42 Application range of Pigment Red 170 (F3RK)

General plastics		Engineering plastics		Spinning	
LL/LDPE	●	PS/SAN	×	PP	●
HDPE	●	ABS	×	PET	×
PP	●	PC	×	PA6	×
PVC(soft)	○	PBT	×	PAN	●
PVC(rigid)	○	PA	×		
Rubber	×	POM			

● Recommended to use, ○ Conditional use, × Not recommended to use.

Table 4.43 Application properties of Pigment Red 170 (F5RK) used in HDPE plastics

Project		Pigment	Titanium dioxide	Light fastness degree
HDPE	Full shade	0.21%		8
	1/3 SD	0.21%	1%	7~8

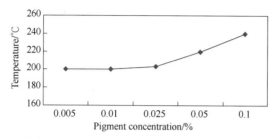

Figure 4.19 Heat resistance of Pigment Red 170 (F5RK) in HDPE

Table 4.44 Application range of Pigment Red 170 (F3RK)

General plastics		Engineering plastics		Spinning	
LL/LDPE	●	PS/SAN	×	PP	●
HDPE	●	ABS	×	PET	×
PP	●	PC	×	PA6	×
PVC(soft)	×	PBT	×	PAN	●
PVC(rigid)	×	PA	×		
Rubber	×	POM	×		

● Recommended to use, × Not recommended to use.

Variety characteristic Pigment Red 170 has excellent light fastness, F3RK can basically meet the requirements for long-term exposure. It shows excellent heat resistance in deep shade and conversely in light. Pigment Red 170 is good performance-price ratio and suitable for general purpose polyolefin plastics of moderate fastness. Pigment Red 170 (F3RK) can be used in the coloring of PVC, but it would bleed in flexible PVC. Pigment Red 170 (F5RK) can't be used in the coloring of PVC.

Pigment Red 170 is suitable for the pre-coloring of spinning of polypropylene fiber and has no effect on the warpage in crystalline plastics such as HDPE.

② **C.I. Pigment Red 187** C.I. structure number: 12486. Molecular formula: $C_{34}H_{28}ClN_5O_7$. CAS registry number: [59487-23-9].

Structure formula

Color characterization Pigment Red 187 is high transparency bluish red and is the bluest in similar pigments. It has good tinting strength and only needs 0.22% pigment to prepare 1/3 SD of HDPE with 1% titanium dioxide. Pigment Red 187 is crystal pleomorphism and has two crystal forms of which bluish red of high specific surface area showing commercial value.

Main properties Show in Table 4.45~Table 4.47, Figure 4.20.

Table 4.45 Application properties of Pigment Red 187 used in PVC plastics

Project		Pigment	Titanium dioxide	Light fastness degree	Weather resistance degree (3000h)	Migration resistance degree
PVC	Full shade	0.1%		7	3	
	Reduction	0.1%	0.5%	7		5

Table 4.46 Application properties of Pigment Red 187 used in HDPE plastics

Project		Pigment	Titanium dioxide	Light fastness degree
HDPE	Full shade	0.25%		8
	1/3 SD	0.25%	1%	8

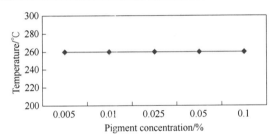

Figure 4.20 Heat resistance of Pigment Red 187 in HDPE (full shade)

Table 4.47 Application range of Pigment Red 187

General plastics		Engineering plastics		Spinning	
LL/LDPE	●	ABS	●	PP	●
HDPE	●	PC	●	PET	×
PP	●	PBT	●	PA6	×
PVC(soft)	●	PA	×	PAN	×
PVC(rigid)	●	POM	×		
Rubber	●				

● Recommended to use, × Not recommended to use.

Variety characteristic Pigment Red 187 has excellent light fastness and heat resistance in the coloring of polyolefins. In addition to the application in the coloring of PVC and general polyolefins, Pigment Red 187 also can be used in styrenic engineering plastics. It is suitable for the pre-coloring of spinning of polypropylene fiber.

4.5.1.3 Benzimidazolone Pigments

Benzimidazolone pigments get name from 5-acylamino benzimidazole ketone groups exist in the molecule. Because of introducing imide groups, benzimidazolone pigments are a kind of organic pigments with high performance, which is different from the azo pigments. They show excellent fastness in plastics and high heat resistance in polyolefins, some of which are the best heat-resistant varieties of organic pigments. They also have excellent light fastness, and the yellow series have high weather resistance, excellent solvent resistance, migration fastness, chemical stability and good acid or alkali resistance.

It is concerned that most benzimidazolone pigments do not cause warpage of polyolefin injection molding products, which can be used in large-scale and asymmetric injection products.

(1) Yellow, orange benzimidazolone pigments

Commercial pigments that substituent aromatic amines as diazo component coupled with 5-acetoacetoxyethyl benzimidazole ketone under the pigment process, the general structure is as follows:

Main varieties in the application of yellow and orange benzimidazolone pigments in plastics are shown in Table 4.48.

Table 4.48 Main varieties in the application of yellow and orange benzimidazolone pigments in plastics

C.I. pigment	C.I. structure number	X	Y	Z	W	Color light
Pigment Yellow 120	[29920-31-8]	H	COOCH$_3$	H	H	Yellow
Pigment Yellow 151	[31837-42-0]	COOH	H	H	H	Greenish yellow
Pigment Yellow 180	[77804-81-0]			A		Greenish yellow
Pigment Yellow 181	[74441-05-7]			B		Reddish yellow
Pigment Yellow 194	[82199-12-0]	COH$_3$	H	COCH$_3$	H	Yellow
Pigment Yellow 214	[25430-12-5]					Greenish yellow
Pigment Yellow 64	[72102-84-2]					Orange
Pigment Yellow 72	[78245-94-0]					Yellowish orange

A= (2-amino-substituted benzene with NH$_2$, OCH$_2$CH$_2$O, NH$_2$ groups)

B= H$_2$NOC—⟨⟩—HNOC—⟨⟩—NH$_2$

The chemical structure of Pigment Yellow 214 and Pigment Orange 72 have not been announced, although Pigment Orange 64 is benzimidazolone organic pigment, the structure is different from Table 4.48. Yellow 180 is greenish yellow, high tinting strength and saturation. It is stable in a wide range of concentration and can be widely used in polyolefin, which has the potential to be standard color.

① **C.I. Pigment Yellow 120**   C.I. structure number: 11783. Formula: $C_{21}H_{19}N_5O_7$. CAS registry number: [29920-31-8].

Structure formula

$$\underset{H_3COOC}{\overset{H_3COOC}{\diagup}}\!\!-\!\!N\!\!=\!\!N\!\!-\!\!CH\!\!-\!\!CNH\!\!-\!\!\underset{}{\overset{O}{\diagup}}$$

Color characterization   Positive(middle shade) yellow, good tinting strength, only need 0.38% pigment to prepare 1/3 SD of PE with 1% titanium dioxide.

Main properties   Shown in Table 4.49~Table 4.51 and Figure 4.21.

Table 4.49 Application properties of Pigment Yellow 120 in PVC

Project		Pigment	Titanium dioxide	Light fastness degree	Weather resistance degree (3000h)	Migration resistance degree
PVC	Full shade	0.1%		8	4	5
	Reduction	0.1%	0.5%	8		

Table 4.50 Application properties of Pigment Yellow 120 in PE

Project		Pigment	Titanium dioxide	Light fastness degree	Weather resistance degree (3000h, 0.2%)
PE	Full shade	0.38%		8	3~4
	1/3 SD	0.38%	1%	8	

Table 4.51 Application range of Pigment Yellow 120

General plastics		Engineering plastics		Spinning	
LL/LDPE	●	PS/SAN	○	PP	●
HDPE	●	ABS	●	PET	×
PP	●	PC	×	PA6	×
PVC (soft)	●	PBT	×	PAN	●
PVC (rigid)	●	PA	×		
Rubber	●	POM	○		

● Recommended to use, ○ Conditional use, × Not recommended to use.

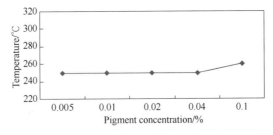

Figure 4.21 Heat resistance of Pigment Yellow 120 in HDPE (full shade)

Variety characteristics   Excellent in light fastness, 8 (1/25 SD) in flexible PVC. However, weather resistance of rigid PVC with lead / cadmium as a stabilizer is much lower than the same series of other products. It is suitable for PVC with special preparation supplies, and applicable to pre-coloring of spinning of PP.

② **C.I. Pigment Yellow 151**   C.I. structure number: 13980. Formula: $C_{18}H_{15}N_5O_5$. CAS registry number: [31837-42-0].

Structure formula

$$\underset{\text{COOH}}{\bigcirc}-N=N-CH-\underset{\underset{CH_3}{C=O}}{\overset{O}{\overset{\|}{C}}}NH-\underset{H}{\overset{H}{\underset{N}{\bigcirc}}}=O$$

Color characterization   Greenish yellow, good tinting strength, only need 0.38% pigment to prepare 1/3 SD of PE with 1% titanium dioxide.

Main properties   Shown in Table 4.52~Table 4.54 and Figure 4.22.

Table 4.52   Application properties of Pigment Yellow 151 in PVC

	Project	Pigment	Titanium dioxide	Light fastness degree	Migration resistance degree
PVC	Full shade	0.1%		7~8	5
	Reduction	0.1%	0.5%	7~8	

Table 4.53   Application properties of Pigment Yellow 151 in PE

	Project	Pigment	Titanium dioxide	Light fastness degree	Weather resistance degree (3000h, 0.2%)
PE	Full shade	0.38%		8	3~4
	1/3 SD	0.38%	1%	8	

Table 4.54   Application range of Pigment Yellow 151

General plastics		Engineering plastics		Spinning	
LL/LDPE	●	PS/SAN	○	PP	○
HDPE	●	ABS	○	PET	×
PP	●	PC	×	PA6	×
PVC (soft)	●	PBT	×	PAN	×
PVC (rigid)	●	PA	×		
Rubber	●	POM	×		

● Recommended to use, ○ Conditional use, × Not recommended to use.

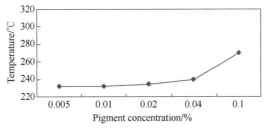

Figure 4.22   Heat resistance of Pigment Yellow 151 in HDPE (full shade)

Variety characteristics   Excellent in light fastness, 8 (1/25 SD) in flexible PVC and PS, good weather resistance in rigid PVC. It has excellent heat resistance which is up to 260 ℃ in PE (1/3 SD), beyond this temperature the shade turns red and the saturation reduced. It is suitable for pre-coloring of spinning of PP. It is also the ideal alternative to chrome yellow varieties.

③ **C.I. Pigment Yellow 180** C.I. structure number: 21290; Formula: $C_{36}H_{32}N_{10}O_8$; CAS registry number: [77804-81-0].

Structure formula

Color characterization Greenish yellow, the only disazo pigment of benzimidazolone yellow series, good tinting strength, only need 0.3% pigment to prepare 1/3 SD of HDPE with 1% titanium dioxide.

Main properties Shown in Table 4.55～Table 4.57 and Figure 4.23.

Table 4.55 Application properties of Pigment Yellow 180 in PVC

	Project	Pigment	Titanium dioxide	Light fastness degree	Migration resistance degree
PVC	Full shade	0.1%		6～7	5
	Reduction	0.1%	0.5%	6～7	5

Table 4.56 Application properties of Pigment Yellow 180 in PE

	Project	Pigment	Titanium dioxide	Light fastness degree
PE	Full shade	0.16%		6～7
	1/3 SD	0.16%	1%	6～7

Table 4.57 Application range of Pigment Yellow 180

General plastics		Engineering plastics		Spinning	
LL/LDPE	●	PS/SAN	●	PP	●
HDPE	●	ABS	●	PET	×
PP	●	PC	●	PA6	×
PVC (soft)	●	PBT	●	PAN	●
PVC (rigid)	●	PA	×		
Rubber	●	POM	●		

● Recommended to use, × Not recommended to use.

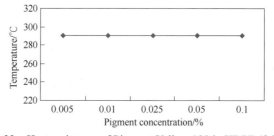

Figure 4.23 Heat resistance of Pigment Yellow 180 in HDPE (full shade)

Variety characteristics The heat resistance is excellent in HDPE and independent of the concentration within a wide range, and decreases only less than 0.005%. Pigment Yellow 180 is not only applied to general polyolefin plastics but also to general engineering plastics such as polycarbonate, polystyrene and polyester. It is suitable for pre-coloring of spinning of PP. It may become a standard color potential variety of medium yellow.

④ **C.I. Pigment Yellow 181** C.I. structure number: 11777. Formula: $C_{25}H_{21}N_7O_5$; CAS registry number: [74441-05-7].

Structure formula

$$H_2NC(=O)-C_6H_4-NHC(=O)-C_6H_4-N=N-CH-C(=O)NH-\text{benzimidazolone}$$
$$\text{with } C=O, CH_3 \text{ substituent}$$

Color characterization Reddish yellow, lower tinting strength than Pigment Yellow 180, only need 0.4% pigment to prepare 1/3 SD of HDPE with 1% titanium dioxide.

Main properties Shown in Table 4.58～Table 4.60 and Figure 4.24.

Table 4.58 Application properties of Pigment Yellow 181 in PVC

Project		Pigment	Titanium dioxide	Light fastness degree	Weather resistance degree (3000h)	Migration resistance degree
PVC	Full shade	0.1%		8		5
	Reduction	0.1%	0.5%	8		5

Table 4.59 Application properties of Pigment Yellow 181 in PE

Project		Pigment	Titanium dioxide	Light fastness degree	Weather resistance degree (3000h, 0.2%)
PE	Full shade	0.44%		8	3～4
	1/3 SD	0.44%	1%	8	

Table 4.60 Application range of Pigment Yellow 181

General plastics		Engineering plastics		Spinning	
LL/LDPE	●	PS/SAN	●	PP	●
HDPE	●	ABS	●	PET	×
PP	●	PC	×	PA6	×
PVC (soft)	●	PBT	●	PAN	●
PVC (rigid)	●	PA	×		
Rubber	●	POM	●		

● Recommended to use, ○ Conditional use, × Not recommended to use.

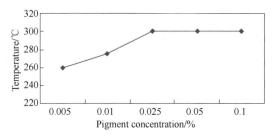

Figure 4.24 Heat resistance of Pigment Yellow 181 in HDPE (full shade)

Variety characteristics   Excellent in light fastness, 8 in different plastics. Good heat resistance in a very low coloring concentration. Pigment yellow 181 is not only applied to general polyolefin plastics but also to general engineering plastics such as polystyrene. It is suitable for pre-coloring of spinning of PP.

⑤ **C.I. Pigment Yellow 214**

Color characterization   Bright greenish yellow. Currently the best variety of yellow organic pigments in most green shade and color saturation, high tinting strength, only need 0.19% pigment to prepare 1/3 SD of HDPE with 1% titanium dioxide.

Main properties   Shown in Table 4.61~Table 4.63 and Figure 4.25.

Table 4.61 Application properties of Pigment Yellow 214 in PVC

Project		Pigment	Titanium dioxide	Light fastness degree	Weather resistance degree (3000h)	Migration resistance degree
PVC	Full shade	0.1%		7		5
	Reduction	0.1%	0.5%	6~7		5

Table 4.62 Application properties of Pigment Yellow 214 in PE

Project		Pigment	Titanium dioxide	Light fastness degree
PE	Full shade	0.19%		7
	1/3 SD	0.19%	1%	6~7

Table 4.63 Application range of Pigment Yellow 214

General plastics		Engineering plastics		Spinning	
LL/LDPE	●	PS/SAN	●	PP	●
HDPE	●	ABS	●	PET	×
PP	●	PC	×	PA6	×
PVC (soft)	●	PBT	○	PAN	○
PVC (rigid)	●	PA	×		
Rubber	●	POM	●		

● Recommended to use, ○ Conditional use, × Not recommended to use.

Variety characteristics   The light fastness of Pigment Yellow 214 is lower than Pigment Yellow 180 and 6~7 in HDPE (1/3 SD with 1% titanium dioxide). The heat resistance is up to 280℃/5min in polyolefin plastics and engineering plastics. It is applied not only to the coloring of polyolefin but also to the coloring of styrenic engineering plastics.

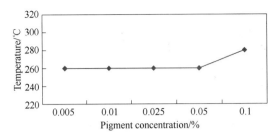

Figure 4.25　Heat resistance of Pigment Yellow 214 in HDPE (full shade)

⑥ **C.I. Pigment Orange 64**　C.I. structure number: 12760. Formula: $C_{12}H_{10}N_6O_4$. CAS registry number: [72102-84-2].

Structure formula

Color characterization　Bright reddish orange, high tinting strength, need 0.42% pigment to prepare 1/3 SD of HDPE with 2% titanium dioxide.

Main properties　Shown in Table 4.64～Table 4.66 and Figure 4.26.

Table 4.64　Application properties of Pigment Orange 64 in PVC

Project		Pigment	Titanium dioxide	Light fastness degree	Weather resistance degree (3000h)	Migration resistance	Heat resistance	
							180℃/30min	200℃/10min
PVC	Full shade	0.1%		7～8	3	5	4～5	4～5
	Reduction	0.2%	2%	7～8	1	5	5	5

Table 4.65　Application properties of Pigment Orange 64 in HDPE

Project		Pigment	Titanium dioxide	Light fastness degree	Weather resistance degree (3000h)	Migration resistance degree
HDPE	Full shade	0.1%		8	3～4	5
	Reduction	0.1%	1%	7～8	1	5

Table 4.66　Application range of Pigment Orange 64

General plastics		Engineering plastics		Spinning	
LL/LDPE	●	PS/SAN	●	PP	●
HDPE	●	ABS	●	PET	×
PP	●	PC	○	PA6	×
PVC (soft)	●	PET	×	PAN	○
PVC (rigid)	●	PA	×		
Rubber	●	POM	●		

● Recommended to use, ○ Conditional use, × Not recommended to use.

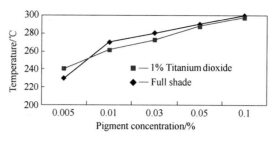

Figure 4.26 Heat resistance of Pigment Orange 64 in HDPE (full shade)

Variety characteristics   The light fastness of Pigment Orange 64 is good and 7~8 in PVC (1/3 SD with 2% titanium dioxide). The heat resistance get up to 290℃/5min. It is applied not only to the coloring of polyolefin but also to the coloring of styrenic engineering plastics. It is the logo color of HDPE pressure pipes. Owing to the high tinting strength, high color saturation, excellent overall properties, Pigment Orange 64 may become a standard color potential variety of orange after a certain level of price decreasing.

⑦ **C.I. Pigment Orange 72**   Benzimidazolone pigment structure.

Color characterization   Bright yellowish orange, acceptable color saturation, good tinting strength, need 0.24% pigment to prepare 1/3 SD of HDPE with 1% titanium dioxide.

Main properties   Shown in Table 4.67~Table 4.69 and Figure 4.27.

Table 4.67 Application properties of Pigment Orange 72 in PVC

Project		Pigment	Titanium dioxide	Light fastness degree	Weather resistance degree (3000h)	Migration resistance degree
PVC	Full shade	0.1%		8	4	
	Reduction	0.1%	0.5%	7~8		5

Table 4.68 Application properties of Pigment Orange 72 in PE

Project		Pigment	Titanium dioxide	Light fastness degree	Weather resistance degree (3000h, 0.2%)
PE	Full shade	0.24%		8	4~5
	1/3 SD	0.24%	1%	8	

Table 4.69 Application range of Pigment Orange 72

General plastics		Engineering plastics		Spinning	
LL/LDPE	●	PS/SAN	●	PP	●
HDPE	●	ABS	●	PET	×
PP	●	PC	×	PA6	×
PVC (soft)	●	PBT	×	PAN	●
PVC (rigid)	●	PA	×		
Rubber	●	POM	●		

● Recommended to use, × Not recommended to use.

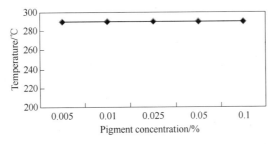

Figure 4.27 Heat resistance of Pigment Orange 72 in HDPE (full shade)

Variety characteristics The light fastness of Pigment Orange 72 is 8 in HDPE (1/3 SD with 2% titanium dioxide), the full shade meets the requirements for long-term exposure. In addition to the coloring of general polyolefin it can also be used to the coloring of styrene engineering plastics.

(2) Red, violet, brown benzimidazolone pigments

Red, violet, brown benzimidazolone pigments are a coupling component prepared by condensation between 5-amino-benzimidazolone and 2-hydroxy-3-naphthoic acid to couple with substituted aromatic amine diazonium salt. The general structure is as follows:

Main varieties of red, brown benzimidazolone pigments are shown in Table 4.70.

Table 4.70 Main varieties of red, brown benzimidazolone pigments

C.I. Pigment	CAS	X	Y	Z	Shade
Red 175	[6985-92-8]	COOCH$_3$	H	H	Bluish red
Red 176	[12225-06-08]	OCH$_3$	H	CONHC$_6$H$_5$	Carmine
Red 185	[51920-12-8]	OCH$_3$	SO$_2$NHCH$_3$	CH3	Carmine
Red 208	[31778-10-6]	COOC$_4$H$_9$	H	H	Red
Violet 32	[12225-08-6]	OCH$_3$	SO$_2$NHCH$_3$	OCH3	Reddish violet
Brown 25	[6992-11-6]	Cl	H	Cl	Reddish brown

① **C.I. Pigment Red 176** C.I. structure number: 12515. Formula: C$_{32}$H$_{24}$N$_6$O$_5$. CAS registry number: [12225-06-8].

Structure formula

Color characterization   Bluish red, acceptable color saturation, bluer than Pigment Red 185, high tinting strength, need 0.53% pigment to prepare 1/3 SD of PVC with 5% titanium dioxide. The reduction is beautiful pink.

Main properties   Shown in Table 4.71～Table 4.73 and Figure 4.28.

Table 4.71   Application properties of Pigment Red 176 in PVC

Project		Pigment	Titanium dioxide	Light fastness degree	Migration resistance degree
PVC	Full shade	0.1%		7	5
	Reduction	0.1%	0.5%	6～7	5

Table 4.72   Application properties of Pigment Red 176 in HDPE

Project		Pigment	Titanium dioxide	Light fastness degree
HDPE	Full shade	0.21%		7
	1/3 SD	0.21%	1%	7

Table 4.73   Application range of Pigment Red 176

General plastics		Engineering plastics		Spinning	
LL/LDPE	●	PS/SAN	○	PP	●
HDPE	●	ABS	×	PET	×
PP	●	PC	●	PA6	×
PVC (soft)	●	PBT	×	PAN	●
PVC (rigid)	●	PA	×		
Rubber	●	POM	×		

● Recommended to use, ○ Conditional use, × Not recommended to use.

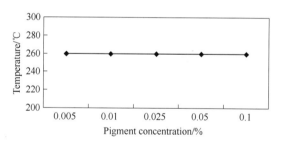

Figure 4.28   Heat resistance of Pigment Red 176 in HDPE (full shade)

Variety characteristics   The light fastness and heat resistance of Pigment Red 176 are excellent, but the heat resistance is decreased greatly after adding titanium dioxide. It is applied to the coloring of PVC and general polyolefin plastics, medium fastness, high performance-price ratio. It is suitable for foaming technology of EVA at 160℃ and 30min. It can also be used to the pre-coloring of spinning of polypropylene fibers.

② **C.I. Pigment Red 185**   C.I. structure number: 12516. Formula: $C_{27}H_{24}N_6O_6S$. CAS registry number: [51920-12-8].

Structure formula

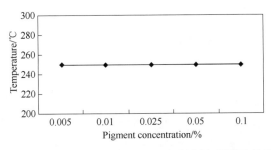

Color characterization Crystal polymorphism, bright bluish red, high tinting strength, need 0.45% pigment to prepare 1/3 SD of PVC with 5% titanium dioxide.

Main properties Shown in Table 4.74～Table 4.76 and Figure 4.29.

Table 4.74 Application properties of Pigment Red 185 in PVC

	Project	Pigment	Titanium dioxide	Light fastness degree	Migration resistance degree
PVC	Full shade	0.1%		7～8	5
	Reduction	0.1%	0.5%	7	5

Table 4.75 Application properties of Pigment Red 185 in PE

	Project	Pigment	Titanium dioxide	Light fastness degree
PE	Full shade	0.2%		6
	1/3 SD		1%	5～6

Table 4.76 Application range of Pigment Red 185

General plastics		Engineering plastics		Spinning	
LL/LDPE	●	PS/SAN	●	PP	×
HDPE	●	ABS	×	PET	×
PP	●	PC	×	PA6	×
PVC (soft)	●	PBT	×	PAN	○
PVC (rigid)	●	PA	×		
Rubber	●	POM	×		

● Recommended to use, ○ Conditional use, × Not recommended to use.

Figure 4.29 Heat resistance of Pigment Red 185 in HDPE (full shade)

Variety characteristics The light fastness and heat resistance of Pigment Red 185 is excellent, and the heat resistance is independent of the concentration within a range between 0.1% and 0.005%, but the heat resistance decreases with increasing amounts of titanium dioxide. It is applied to the coloring of PVC and general polyolefin plastics.

③ **C.I. Pigment Red 175**　C.I. structure number: 12513. Formula: $C_{26}H_{19}N_5O_5$. CAS registry number: [6985-92-8].

Structure formula

Color characterization　Dull red (yellowish), high transparency, acceptable tinting strength, need 0.68% pigment to prepare 1/3 SD of PVC with 5% titanium dioxide.

Main properties　Shown in Table 4.77～Table 4.79 and Figure 4.30.

Table 4.77　Application properties of Pigment Red 175 in PVC

Project		Pigment	Titanium dioxide	Light fastness degree	Weather resistance degree (3000h)	Migration resistance degree
PVC	Full shade	0.1%		7	4	5
	Reduction	0.1%	0.5%	7～8		5

Table 4.78　Application properties of Pigment Red 175 in PE

Project		Pigment	Titanium dioxide	Light fastness degree	Weather resistance degree (3000h, 0.2%)
PE	Full shade	0.22%		8	4
	1/3 SD	0.22%	1%	8	

Table 4.79　Application range of Pigment Red 175

General plastics		Engineering plastics		Spinning	
LL/LDPE	●	PS/SAN	●	PP	●
HDPE	●	ABS	●	PET	×
PP	●	PC	×	PA6	×
PVC (soft)	●	PBT	×	PAN	●
PVC (rigid)	●	PA	×		
Rubber	●	POM	×		

● Recommended to use, × Not recommended to use.

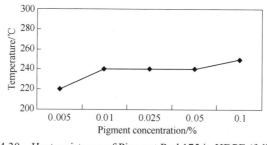

Figure 4.30　Heat resistance of Pigment Red 175 in HDPE (full shade)

Variety characteristics　Excellent in light fastness, meets the requirements for long-term exposure. The heat resistance up to 250℃, but decreases with decreasing of concentration. In addition to the coloring of PVC and general polyolefins it can also be used to the coloring of styrenic engineering plastics.

④ **C.I. Pigment Red 208**　C.I. structure number: 12514. Formula: $C_{29}H_{25}N_5O_5$. CAS registry number: [31778-10-6].

Structure formula

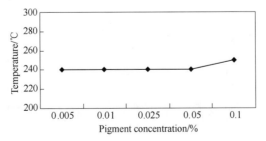

Color characterization　Positive (middle shade) red, high tinting strength, need 0.6% pigment to prepare 1/3 SD of PVC with 1% titanium dioxide.

Main properties　Shown in Table 4.80~Table 4.82 and Figure 4.31.

Table 4.80　Application properties of Pigment Red 208 in PVC

Project		Pigment	Titanium dioxide	Light fastness degree	Weather resistance degree (3000h)	Migration resistance degree
PVC	Full shade	0.1%		7	3	4~5
	Reduction	0.1%	0.5%	6~7		

Table 4.81　Application properties of Pigment Red 208 in HDPE

Project		Pigment	Titanium dioxide	Light fastness degree
HDPE	Full shade	0.13%		7
	1/3 SD	0.13%	1%	6~7

Table 4.82　Application range of Pigment Red 208

General plastics		Engineering plastics		Spinning	
LL/LDPE	●	PS/SAN	×	PP	×
HDPE	●	ABS	×	PET	×
PP	●	PC	×	PA6	×
PVC (soft)	●	PBT	×	PAN	●
PVC (rigid)	●	PA	×		
Rubber	●	POM	×		

● Recommended to use, × Not recommended to use.

Figure 4.31　Heat resistance of Pigment Red 208 in HDPE (full shade)

Variety characteristics The heat resistance up to 240 ℃ in HDPE, but the temperature drops below 200 ℃ with titanium dioxide adding. It is applied to the coloring of PVC and general polyolefin plastics. It is a standard red for the coloring of PU.

⑤ **C.I. Pigment Violet 32** C.I. structure number: 12517. Formula: $C_{27}H_{34}N_6O_7S$. CAS registry Number: [12225-08-0].

Structure formula

$$CH_3NHO_2S-\underset{CH_3O}{\underset{|}{\bigcirc}}\!\!\!\overset{OCH_3\ OH}{\underset{}{}}\!\!\!-N\!=\!N-\underset{}{\bigcirc}\!\!\!\overset{\overset{O}{\|}}{\underset{}{CNH}}-\underset{H}{\overset{H}{\underset{N}{\bigcirc}}}\!\!=\!O$$

Color characterization Bluish red, high transparency, high tinting strength, need 0.36% pigment to prepare 1/3 SD of PVC with 1% titanium dioxide.

Main properties Shown in Table 4.83~Table 4.85 and Figure 4.32.

Table 4.83 Application properties of Pigment Violet 32 in PVC

Project		Pigment	Titanium dioxide	Light fastness degree	Weather resistance degree (3000h)	Migration resistance degree
PVC	Full shade	0.1%		7~8		5
	Reduction	0.1%	0.5%	7		

Table 4.84 Application properties of Pigment Violet 32 in HDPE

Project		Pigment	Titanium dioxide	Light fastness degree
HDPE	Full shade	0.11%		7
	1/3 SD	0.11%	1%	6~7

Table 4.85 Application range of Pigment Violet 32

General plastics		Engineering plastics		Spinning	
LL/LDPE	●	PS/SAN	×	PP	×
HDPE	●	ABS	×	PET	×
PP	●	PC	×	PA6	×
PVC (soft)	●	PBT	×		
PVC (rigid)	●	PA	×		
Rubber	○	POM	×		

● Recommended to use, ○ Conditional use, × Not recommended to use.

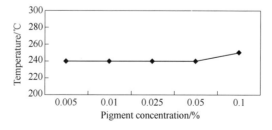

Figure 4.32 Heat resistance of Pigment Violet 32 in HDPE (full shade)

Variety characteristics Pigment Violet 32 shows excellent light fastness and is applied to the coloring of PVC. Owing to the low heat resistance in polyolefins, it is used in the coloring of films and cables. Pigment violet 32 is suitable for the coloring of chemical fiber stock solution such as cellulose acetate, viscose rayon. It can replace thioindigo pigments (Pigment Red 88), and the shade is almost identical in flexible PVC.

⑥ **C.I. Pigment brown 25**   C.I. structure number: 12510. Formula: $C_{24}H_{15}Cl_2N_5O_3$. CAS registry number: [6992-11-6].

Structure formula

Color characterization   Reddish brown, good transparency, high tinting strength in PVC, need 0.77% pigment to prepare 1/3 SD of PVC with 5% titanium dioxide.

Main properties   Shown in Table 4.86～Table 4.88 and Figure 4.33.

Table 4.86   Application properties of Pigment Brown 25 in PVC

Project		Pigment	Titanium dioxide	Light fastness degree	Weather resistance degree (3000h)	Migration resistance degree
PVC	Full shade	0.1%		8	5	4～5
	Reduction	0.1%	0.5%	8		4～5

Table 4.87   Application properties of Pigment Brown 25 in HDPE

Project		Pigment	Titanium dioxide	Light fastness degree	Weather resistance degree (3000h, 0.2%)
HDPE	Full shade	0.22%		8	4～5
	1/3 SD	0.22%	1%	8	

Table 4.88   Application range of Pigment Brown 25

General plastics		Engineering plastics		Spinning	
LL/LDPE	●	PS/SAN	×	PP	●
HDPE	●	ABS	×	PET	×
PP	●	PC	×	PA6	×
PVC (soft)	●	PBT	×	PAN	●
PVC (rigid)	●	PA	×		
Rubber	●	POM	●		

● Recommended to use, × Not recommended to use.

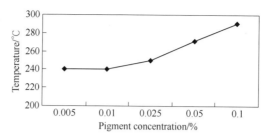

Figure 4.33 Heat resistance of Pigment Brown 25 in HDPE (full shade)

Variety characteristics Pigment Brown 25 shows excellent light fastness and meets the requirements for long-term exposure. Owing to the good heat resistance, in addition to the coloring of general polyolefins it can also be used to the coloring of styrenic engineering plastics. It may cause a slight effect on the shrinkage of HDPE, but the warpage is significant when the processing temperature is below 220 ℃. It is a standard brown for the coloring of PU.

4.5.2 Disazo Pigments

Disazo pigments refer to the structure of pigment molecule containing two azo groups, generally, it is coupled between diarylamine diazonium salt (3,3-dichlorobenzidine) and coupling components (acetoacetanilide and its derivatives or disazopyrazolone and its derivatives) which is the famous benzidine pigments. The chromatography is among the strong greenish yellow, strong reddish yellow and orange. Benzidine azo pigments came out in 1935 and the first variety was C.I. Pigment Yellow 13.

Dichlorobenzidine azo pigments show excellent color properties and high tinting strength, the tinting strength is more than twice as high as that of the monoazo yellow pigments with the same color, but the light fastness (except Pigment Yellow 83) is general. The dispersion is good during processing.

Dichlorobenzidine azo pigments will be decomposed to dichlorobenzidine in plastic processing when the temperature exceeds 200 ℃. As consequence of these studies they are not suitable for the plastics of the processing temperature above 200 ℃.

(1) Dichlorobenzidine yellow azo pigments Dichlorobenzidine yellow azo pigments currently used in the plastics have the following general structure:

$$\underset{Z}{\overset{X}{Y-\bigcirc}}-HNCOCH-N=N-\underset{R_2}{\overset{R_1\ \ R_2}{\bigcirc-\bigcirc}}-N=N-CHCONH-\underset{Z}{\overset{X}{\bigcirc}}-Y$$
$$\ COCH_3\ COCH_3$$

Varieties of dichlorobenzidine yellow azo pigments currently used in the plastics are shown in Table 4.89.

Table 4.89 Varieties of dichlorobenzidine yellow azo pigments currently used in the plastics

C.I. Pigment	CAS	R_1	R_2	X	Y	Z	Shade
Yellow 12	[6358-85-6]	Cl	H	H	H	H	Yellow
Yellow 13	[5102-83-0]	Cl	H	CH_3	CH_3	H	Yellow
Yellow 14	[5468-75-7]	Cl	H	CH_3	H	H	Yellow
Yellow 17	[4531-49-1]	Cl	H	OCH_3	H	H	Greenish yellow
Yellow 81	[22094-93-5]	Cl	Cl	CH_3	CH_3	H	Greenish yellow
Yellow 83	[5567-15-7]	Cl	H	OCH_3	Cl	OCH_3	Reddish yellow

① **C.I. Pigment Yellow 12**   C.I. structure number: 21100. Formula: $C_{32}H_{26}Cl_2N_6O_4$. CAS registry number:[5102-83-0].

Structure formula

Color characterization   Positive yellow, high tinting strength.

Main properties   Mainly used in the ink area, rarely used in plastics. Low migration resistance in PVC.

Variety characteristics   Pigment Yellow 12 is used extensively in EVA foam plastics, the bright yellow is welcome by customers. It is also applied to films and flat fibers. But migration resistance, heat resistance and light fastness should be pay attention.

② **C.I. Pigment Yellow 13**   C.I. structure number: 21100. Formula: $C_{36}H_{34}Cl_2N_6O_4$. CAS registry number: [5102-83-0].

Structure formula

Color characterization   Positive (middle shade) yellow, greener than Pigment Yellow 13, redder than Pigment Yellow 17, high tinting strength, need 0.3% pigment to prepare 1/3 SD of PVC with 5% titanium dioxide, need 0.12% pigment to prepare 1/3 SD of HDPE with 1% titanium dioxide.

Main properties   Shown in Table 4.90～Table 4.92.

Table 4.90 Application properties of Pigment Yellow 13 in PVC

Project		Pigment	Titanium dioxide	Light fastness degree	Weather resistance degree (2000h)	Migration resistance	Heat resistance	
							180℃/30min	200℃/30min
PVC	Full shade	0.1%		7	3	5	4~5	4~5
	Reduction	0.2%	2%	6	1	5	5	4~5

Table 4.91 Application properties of Pigment Yellow 13 in HDPE

Project		Pigment	Titanium dioxide	Light fastness degree	Migration resistance degree
HDPE	Full shade	0.1%		7~8	4.6
	Reduction	0.1%	1%	6~7	4.6

Table 4.92 Application range of Pigment Yellow 13

General plastics		Engineering plastics		Spinning	
LL/LDPE	●	PS/SAN	×	PP	×
HDPE	○	ABS	×	PET	×
PP	○	PC	×	PA6	×
PVC (soft)	●	PBT	×	PAN	○
PVC (rigid)	●	PA	×		
Rubber	●	POM	×		

● Recommended to use, ○ Conditional use, × Not recommended to use.

Variety characteristics Pigment Yellow 13 is low-cost, limited security, cautious using! The Migration resistance in flexible PVC is much better than Pigment Yellow 12, but it may cause bloom when the pigment concentration is less than 0.05%. The migration resistance is also well in rigid PVC. If the processing temperature is too low, Pigment Yellow 13 may affect the warpage in HDPE.

③ **C.I. Pigment Yellow 14** C.I. structure number: 21095. Formula: $C_{34}H_{30}Cl_2N_6O_4$. CAS registry number: [5468-75-7].

Structure formula

Color characterization Pure shade, greenish yellow, greener than Pigment Yellow 13, high tinting strength, need 0.31% pigment to prepare 1/3 SD of flexible PVC with 5% titanium dioxide, need 0.1% pigment to prepare 1/3 SD of HDPE with 1% titanium dioxide.

Main properties Shown in Table 4.93~Table 4.95.

Table 4.93 Application properties of Pigment Yellow 14 in PVC

	Project	Pigment	Titanium dioxide	Light fastness degree	Migration resistance degree
PVC	Full shade	0.1%		6~7	
	Reduction	0.1%	0.5%	6	3

Table 4.94 Application properties of Pigment Yellow 14 in HDPE

	Project	Pigment	Titanium dioxide	Light fastness degree
PE	Full shade	0.1%		5
	1/3 SD	0.1%	1%	4~5

Table 4.95 Application range of Pigment Yellow 14

General plastics		Engineering plastics		Spinning	
LL/LDPE	●	PS/SAN	×	PP	×
HDPE	×	ABS	×	PET	×
PP	×	PC	×	PA6	×
PVC (soft)	●	PBT	×	PAN	×
PVC (rigid)	●	PA	×		
Rubber	●	POM	×		

● Recommended to use, × Not recommended to use.

Variety characteristics Pigment Yellow 14 is low-cost, limited security, cautious using! Depending on the different countries, the application of Pigment Yellow 14 is widely different in plastic processing. It is almost not recommended for plastics in Europe, but recommended in Japan. Because of the lower price than Pigment Yellow 13 and 17, it is extensively applied to polyolefin plastics. But the migration resistance often brings great trouble for customer. Pigment Yellow 14 can be used in the coloring of rubber and viscose fiber stock solution. It should be noted that the light fastness is bad for viscose fibers.

④ **C.I. Pigment Yellow 17** C.I. structure number: 21105. Formula: $C_{34}H_{30}Cl_2N_6O_6$. CAS registry number: [4531-49-1].

Structure formula

Color characterization Greener than Pigment Yellow 14 and 12, high transparency. It has unexpected bright fluorescent effects in the coloring of plastics. The tinting strength is lower than Pigment Yellow 14, and the light fastness degree is 1~2, higher than Pigment Yellow 14 at the same concentration.

Main properties Shown in Table 4.96～Table 4.98.

Table 4.96 Application properties of Pigment Yellow 17 in PVC

Project		Pigment	Titanium dioxide	Light fastness degree	Migration resistance degree
PVC	Full shade	0.1%		7	
	Reduction	0.1%	0.5%	6～7	3

Table 4.97 Application properties of Pigment Yellow 17 in HDPE

Project		Pigment	Titanium dioxide	Light fastness degree
HDPE	Full shade	0.13%		7
	1/3 SD	0.13%	1%	6～7

Table 4.98 Application range of Pigment Yellow 17

General plastics		Engineering plastics		Spinning	
LL/LDPE	●	PS/SAN	×	PP	●
HDPE	○	ABS	×	PET	×
PP	○	PC	×	PA6	×
PVC (soft)	●	PBT	×	PAN	●
PVC (rigid)	●	PA	×		
Rubber	●	POM	×		

● Recommended to use, ○ Conditional use, × Not recommended to use.

Variety characteristics Pigment Yellow 17 is low-cost, limited security, cautious using! Because of the excellent insulation it could be used in PVC insulated cables. Pigment Yellow 17 is suitable for pre-coloring of spinning of polypropylene fibers.

⑤ **C.I. Pigment Yellow 81** C.I. structure No. 21127. Molecular formula: $C_{36}H_{32}Cl_4N_6O_6$. CAS registry number: [22094-93-5].

Structure formula

Color characterization Pigment Yellow 81 has a strong greenish shade, and the shade is more greenish after adding titanium dioxide. The tinting strength of pigment yellow 81 is low, the required concentration of pigment is 1.15% when blending with 5% of titanium dioxide to achieve 1/3 SD in flexible PVC, while the required concentration of pigment is only 0.27% when blending with 1% of titanium dioxide to achieve 1/3 SD in HDPE.

Main properties See Table 4.99～Table 4.101.

Table 4.99 Application properties of Pigment Yellow 81 in PVC

Project		Pigment	Titanium dioxide	Light fastness degree
PVC	Full shade	0.1%		7
	Reduction	0.1%	0.5%	7

Table 4.100 Application properties of Pigment Yellow 81 in HDPE

Project		Pigment	Titanium dioxide	Light fastness degree
PE	Full shade	0.27%		7
	1/3 SD	0.27%	1%	7

Table 4.101 Application of Pigment Yellow 81

General plastics		Engineering plastics		Spinning	
LL/LDPE	●	PS/SAN	×	PP	×
HDPE	○	ABS	×	PET	×
PP	○	PC	×	PA6	×
PVC (soft)	●	PBT	×	PAN	○
PVC (rigid)	●	PA	×		
Rubber	●	POM	×		

● Recommended to use, ○ Conditional use, × No recommended to use.

Varieties characteristics Pigment Yellow 81 is inexpensive and limited to safety, use with caution! The blooming may be occur in the coloring of flexible PVC according to the formulation and processing conditions if the concentration of pigment is too low. In addition, there is no effect on the warpage of HDPE.

⑥ **C.I. Pigment Yellow 83** C.I. structure No. 21108. Molecular formula: $C_{36}H_{32}Cl_4N_6O_8$. CAS registry number: [5567-15-7].

Structure formula

$$H_3CO\text{-}\underset{\underset{OCH_3}{|}}{\overset{\overset{Cl}{|}}{C_6H_2}}\text{-}NHC(O)\text{-}CH(CH_3)\text{-}N=N\text{-}C_6H_3(Cl)\text{-}C_6H_3(Cl)\text{-}N=N\text{-}CH(CH_3)\text{-}CNH(O)\text{-}\underset{\underset{OCH_3}{|}}{\overset{\overset{H_3CO}{|}}{C_6H_2}}\text{-}Cl$$

Color characterization Pigment Yellow 83 is a reddish yellow pigment, the shade is redder than Pigment Yellow 13, and the tinting strength is also stronger. The required concentration of pigment is only 0.08% when blending with 1% of titanium dioxide to achieve 1/3 SD in HDPE.

Main properties See Table 4.102~Table 4.104.

Table 4.102 Application properties of Pigment Yellow 83 in PVC

Project		Pigment	Titanium dioxide	Light fastness degree	Weather resistance degree (3000h)	Migration fastness degree
PVC	Full shade	0.1%		7~8	4-5	
	Reduction	0.1%	0.5%	7~8		5

Table 4.103 Application properties of Pigment Yellow 83 in HDPE

Project		Pigment	Titanium dioxide	Light fastness degree
HDPE	Full shade	0.8%		7
	1/3 SD	0.8%	1%	6~7

Table 4.104 Application of Pigment Yellow 83

General plastics		Engineering plastics		Spinning	
LL/LDPE	●	PS/SAN	●	PP	●
HDPE	●	ABS	○	PET	×
PP	●	PC	×	PA6	×
PVC (soft)	●	PBT		PAN	×
PVC (rigid)	●	PA	×		
Rubber	●	POM			

● Recommended to use, ○ Conditional use, × No recommended to use.

Varieties characteristics Pigment Yellow 83 is inexpensive and limited to safety, use with caution! It possesses good solvent resistance. There is no migration in PVC, even the concentration of pigment is low. It is often applied in the form of pigment preparations in the coloring of polyolefin plastics. And it is suitable for coloring polypropylene fibers during spinning.

(2) Orange disazo pigments

Currently the basic formula of dichlorobenzidine orange disazo pigments used frequently in the plastics industry is as follows:

Currently the variety of dichlorobenzidine orange disazo pigments used frequently in the plastics industry is list in Table 4.105.

Table 4.105 The variety of dichlorobenzidine orange disazo pigments used frequently in the plastics industry

C.I. pigment	CAS No.	X	Y	A	B	Shade
Pigment Orange 13	[3520-72-7]	Cl	H	CH_3	H	Orange
Pigment Orange 34	[15793-73-4]	Cl	H	CH_3	CH_3	Orange

① **C.I. Pigment Orange 13**   C.I. structure No. 21110. Molecular formula: $C_{32}H_{24}Cl_2N_8O_2$. CAS registry number: [3520-72-7].

Structure formula

$$\text{H}_3\text{C}-\underset{\underset{\text{Ph}}{|}}{\underset{N}{N}}=\underset{\text{OH}}{C}-N=N-\underset{}{C_6H_3Cl}-\underset{}{C_6H_3Cl}-N=N-\underset{\text{OH}}{C}=\underset{\underset{\text{Ph}}{|}}{\underset{N}{N}}-\text{CH}_3$$

Color characterization Pigment Orange 13 is a bright yellowish orange pigment, the shade is slightly yellower than Pigment Orange 34 and the tinting strength is also slightly stronger. Moreover, the required concentration of pigment is only 0.12% when blending with 1% of titanium dioxide to achieve 1/3 SD in HDPE.

Main properties See Table 4.106～Table 4.108.

Table 4.106 Application properties of Pigment Orange 13 in PVC

	Project	Pigment	Titanium dioxide	Light fastness degree	Migration fastness degree
PVC	Full shade	0.1%		6	
	Reduction	0.1%	0.5%	4～5	2

Table 4.107 Application properties of Pigment Orange 13 in HDPE

	Project	Pigment	Titanium dioxide	Light fastness degree
PE	Full shade	0.12%		5
	1/3 SD	0.12%	1%	4

Table 4.108 Application of Pigment Orange 13

General plastics		Engineering plastics		Spinning	
LL/LDPE	●	PS/SAN	×	PP	○
HDPE	○	ABS	×	PET	×
PP	○	PC	×	PA6	×
PVC (soft)	●	PBT	×	PAN	●
PVC (rigid)	●	PA	×		
Rubber	●	POM	×		

● Recommended to use, ○ Conditional use, × No recommended to use.

Varieties characteristics Pigment Orange 13 is inexpensive and limited to safety, use with caution! In addition, there is no effect on the warpage of HDPE.

② **C.I. Pigment Orange 34** C.I. structure No. 21115. Molecular formula: $C_{34}H_{28}Cl_2N_8O_2$. CAS registry number: [15793-72-4].

Structure formula

$$\text{H}_3\text{C}-\underset{\underset{C_6H_4CH_3}{|}}{\underset{N}{N}}=\underset{\text{OH}}{C}-N=N-C_6H_3Cl-C_6H_3Cl-N=N-\underset{\text{OH}}{C}=\underset{\underset{C_6H_4CH_3}{|}}{\underset{N}{N}}-\text{CH}_3$$

Color characterization Pigment Orange 34 is a bright yellowish orange pigment. It has high tinting strength. And the required concentration of pigment is only 0.14% when blending with 1% of titanium dioxide to achieve 1/3 SD in HDPE.

Main properties See Table 4.109～Table 4.111.

Table 4.109 Application properties of Pigment Orange 34 in PVC

Project		Pigment	Titanium dioxide	Light fastness degree	Weather resistance degree (1000 h)	Migration fastness degree	Heat resistance degree	
							180℃ /30min	200℃ /10min
PVC	Full shade	0.1%		6～7	1-2	4.7	5	5
	Reduction	0.2%	2%	5～6	1	4.7	5	5

Table 4.110 Application properties of Pigment Orange 34 in HDPE

Project		Pigment	Titanium dioxide	Light fastness degree	Migration fastness degree
HDPE	Full shade	0.1%		6～7	3～4
	Reduction	0.1%	1%	6	4

Table 4.111 Application of Pigment Orange 34

General plastics		Engineering plastics		Spinning	
LL/LDPE	●	PS/SAN	×	PP	○
HDPE	○	ABS	×	PET	×
PP	○	PC	×	PA6	×
PVC (soft)	●	PBT	×	PAN	
PVC (rigid)	●	PA	×		
Rubber	●	POM	×		

● Recommended to use, ○ Conditional use, × No recommended to use.

Varieties characteristics Pigment Orange 34 is inexpensive and limited to safety, use with caution! The blooming will occur in the coloring of flexible PVC when the concentration of pigment is lower than 0.1%, while the bleeding will occur when the concentration of pigment is high. In addition, there is no effect on the warpage and deformation of HDPE.

(3) Red disazo pigments

C.I. Pigment Red 38 C.I. structure No. 21120. Molecular formula: $C_{36}H_{28}Cl_2N_8O_6$. CAS registry number: [6358-87-8].

Structure formula

Color characterization Pigment Red 38 is a positive red pigment with a high transparency. The tinting strength is high in coloring for PVC. The required concentration of pigment is 0.33% when blending with 5% of titanium dioxide to achieve 1/3 SD in flexible PVC. The required concentration of pigment is only 0.13% when blending with 1% of titanium dioxide to achieve 1/3 SD in HDPE.

Main properties See Table 4.112~Table 4.114.

Table 4.112 Application properties of Pigment Red 38 in PVC

Project		Pigment	Titanium dioxide	Light fastness degree	Weather resistance degree (3000 h)	Migration fastness degree
PVC	Full shade	0.1%		6~7		
	Reduction	0.1%	0.5%	5~6		3

Table 4.113 Application properties of Pigment Red 38 in HDPE

Project		Pigment	Titanium dioxide	Light fastness degree
HDPE	Full shade	0.13%		6
	1/3 SD	0.13%	1%	4~5

Table 4.114 Application of Pigment Red 38

General plastics		Engineering plastics		Spinning	
LL/LDPE	●	PS/SAN	×	PP	×
HDPE	○	ABS	×	PET	×
PP	○	PC	×	PA6	×
PVC (soft)	●	PBT	×	PAN	●
PVC (rigid)	●	PA	×		
Rubber	●	POM	×		

● Recommended to use, ○ Conditional use, × No recommended to use.

Varieties characteristics The blooming will occur in the coloring of flexible PVC when the concentration of pigment is too low. And the blooming will occur in the coloring of rigid PVC when the concentration of pigment is low. It is also suitable for coloring of cables to keep good electrical properties. In addition, it is certified by FDA and especially suitable for coloring of rubber.

4.5.3 Disazo Condensation Pigments

Although the manufacturing process of monoazo pigment is simple, the performance is undesirable in coloring of plastics because of relatively small molecular weight and other reasons. And yellow disazo pigments mostly have disazodiarylide as coupling component and relatively large molecular weight, but the performance of yellow disazo pigments is substantially similar to monoazo pigments and it is also limited to safety. Therefore, Ciba Specialty Chemicals developed disazo pigments with large molecular weight and multiple amide groups in 1950s, which is got by bridging two monazo pigments with aromatic diamide. So it is called disazo condensation pigments.

The molecular structure of disazo condensation pigments contains multiple amide groups to increase the molecular weight, which greatly improved heat resistance, light fastness, weather resistance, solvent resistance and migration fastness. Moreover, the color atlas is wide, from a strong greenish yellow to bluish red or purple even brown. In addition, the dispersion of disazo condensation pigments in plastics is good (except pigment red 220). However, the production cost of disazo condensation pigments is high, and the color is not brighter than monazo pigments'.

(1) Yellow disazo condensation pigments

Currently the basic formula of yellow disazo condensation pigments used frequently in the plastics industry is as follows.

$$BHNOC\text{-}Ar(Cl)\text{-}N=N\text{-}CH(COCH_3)\text{-}CONH\text{-}A\text{-}NHCO\text{-}CH(COCH_3)\text{-}N=N\text{-}Ar(Cl)\text{-}CONHB$$

Main varieties of yellow disazo condensation pigments for plastics coloring are list in the following table.

C.I. pigment	A	B	Shade
Yellow 93	CH$_3$-Ar-Cl	Cl-Ar-CH$_3$	Yellow
Yellow 95	CH$_3$-Ar-CH$_3$	Cl-Ar-CH$_3$	Reddish yellow
Yellow 128	CH$_3$-Ar-Cl	CH$_3$-Ar-O-Ar-Cl	Greenish yellow

① **C.I. Pigment Yellow 93** C.I. structure No. 20710. Molecular formula: $C_{43}H_{35}Cl_5N_8O_6$. CAS registry number: [5580-57-4].

Structure formula

Color characterization Pigment Yellow 93 is a greenish yellow pigment and the tinting strength is in medium level. The required concentration of pigment is only 0.85% when blending with 5% of titanium dioxide to achieve 1/3 SD in flexible PVC.

Main properties See Table 4.115~Table 4.117 and Figure 4.34.

Table 4.115 Application properties of Pigment Yellow 93 in PVC

Project		Pigment	Titanium dioxide	Light fastness degree	Weather resistance degree (2000 h)	Migration fastness degree	Heat resistance degree	
							180℃ /30min	200 ℃ /10min
PVC	Full shade	0.1%		8	5	5	5	5
	Reduction	0.2%	2%	7~8	4~5	5	5	5

Table 4.116 Application properties of Pigment Yellow 93 in HDPE

Project		Pigment	Titanium dioxide	Light fastness degree	Weather resistance degree (3000h)	Migration fastness degree
HDPE	Full shade	0.1%		8	3~4	5
	Reduction	0.1%	1%	6~7	2	5

Table 4.117 Application of Pigment Yellow 93

General plastics		Engineering plastics		Spinning	
LL/LDPE	●	PS/SAN	●	PP	●
HDPE	●	ABS	○	PET	×
PP	●	PC	×	PA6	×
PVC (soft)	●	PET	×	PAN	
PVC (rigid)	●	PA	×		
Rubber	●	POM	●		

● Recommended to use, ○ Conditional use, × No recommended to use.

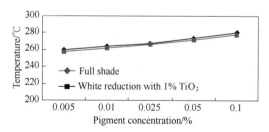

Figure 4.34 Heat resistance of Pigment Yellow 93 in HDPE

Varieties characteristics Pigment Yellow 93 possesses excellent light fastness and weather resistance. It can basically meet the requirements when exposed to the outdoors for a long time. In addition, it has excellent heat resistance in the coloring of polyolefin, and the heat-resistance temperature is up to 270℃ when blending with 1% of titanium dioxide to achieve 1/3 SD in HDPE. It will not migrate in the coloring of flexible PVC. Not only can it be used for the coloring of general-purpose polyolefin, but also can be applied to the coloring of styrenic engineering plastics. Moreover, it is suitable for coloring polypropylene fibers before spinning and is recognized as standard color in the yellow zone. There is a large effect on the warpage and a minimal effect on the deformation of HDPE.

② **C.I. Pigment Yellow 95**   C.I. structure No. 20034. Molecular formula: $C_{44}H_{38}Cl_4N_8O_6$. CAS registry number: [5280-80-8].

Structure formula

Color characterization   Pigment Yellow 95 is a reddish yellow pigment and the shade is between pigment yellow 13 and Pigment Yellow 83 in the standard depth of shade. In addition, the tinting strength is extremely high and is the highest in yellow disazo condensation pigments. The required concentration of this pigment is only 0.7% to achieve 1/3 SD in flexible PVC, while Pigment Yellow 94 is 2.1%.

Main properties   See Table 4.118~Table 4.120 and Figure 4.35.

Table 4.118   Application properties of Pigment Yellow 95 in PVC

Project		Pigment	Titanium dioxide	Light fastness degree	Weather resistance degree (2000 h)	Migration fastness degree	Heat resistance degree	
							180℃/30min	200 ℃/10min
PVC	Full shade	0.1%		8	4~5	5	5	5
	Reduction	0.2%	2%	7~8	3~4	5	5	5

Table 4.119   Application properties of Pigment Yellow 95 in HDPE

Project		Pigment	Titanium dioxide	Light fastness degree	Weather resistance degree (3000h)	Migration fastness degree
HDPE	Full shade	0.1%		7~8	3	5
	Reduction	0.1%	1%	6~7	1~2	5

Table 4.120   Application of Pigment Yellow 95

General plastics		Engineering plastics		Spinning	
LL/LDPE	●	PS/SAN	●	PP	●
HDPE	●	ABS	○	PET	×
PP	●	PC	×	PA6	×
PVC (soft)	●	PET	×	PAN	
PVC (rigid)	●	PA	×		
Rubber	●	POM	●		

● Recommended to use, ○ Conditional use, × No recommended to use.

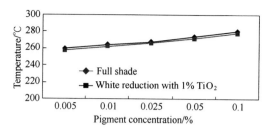

Figure 4.35   The heat resistance of Pigment Yellow 95 in HDPE

Varieties characteristics   Pigment Yellow 95 has excellent light and heat resistance, and suitable for the coloring of general-purpose polyolefin plastics. Moreover, it will not migrate in the coloring of flexible PVC. There is a minimal effect on the warpage of HDPE.

③ **C.I. Pigment Yellow 128**   C.I. structure No. 20037. Molecular formula: $C_{55}H_{37}Cl_5N_8O_6$. CAS registry number: [79953-85-8].

Structure formula

Color characterization   Pigment Yellow 128 is a greenish yellow pigment and the tinting strength is in medium level. The required concentration of pigment is only 1.35% when blending with 5% of titanium dioxide to achieve 1/3 SD in flexible PVC.

Main properties   See Table 4.121~Table 4.123 and Figure 4.36.

Table 4.121   Application properties of Pigment Yellow 128 in PVC

Project		Pigment	Titanium dioxide	Light fastness degree	Weather resistance degree (2000 h)	Migration fastness degree	Heat resistance degree	
							180℃ /30min	200 ℃ /10min
PVC	Full shade	0.1%		8	4~5	5	5	5
	Reduction	0.2%	2%	7~8	4	5	5	5

Table 4.122   Application properties of Pigment Yellow 128 in HDPE

Project		Pigment	Titanium dioxide	Light fastness degree	Weather resistance degree (3000h)	Migration fastness degree
HDPE	Full shade	0.1%		8	4~5	5
	Reduction	0.1%	1%	7~8	3	5

Table 4.123 Application of Pigment Yellow 128

General plastics		Engineering plastics		Spinning	
LL/LDPE	●	PS/SAN	○	PP	●
HDPE	●	ABS	○	PET	×
PP	●	PC	×	PA6	×
PVC (soft)	●	PET	×	PAN	
PVC (rigid)	●	PA6	×		
Rubber	●	POM	●		

● Recommended to use, ○ Conditional use, × No recommended to use.

Figure 4.36 The heat resistance of Pigment Yellow 128 in HDPE

Varieties characteristics The notable feature about Pigment Yellow 128 is the excellent weather resistance to meet the requirements when exposed to the outdoors for a long time, but the heat resistance is general. It is mainly used in the coloring of PVC, and it will not migrate in the coloring of flexible PVC. It is also suitable for the coloring of general-purpose polyolefin plastics. In addition, there is a minimal effect on the warpage of HDPE.

④ **C.I. Pigment Yellow 155** C.I. structure No. 200310. Molecular formula: $C_{34}H_{32}N_6O_{12}$. CAS registry number: [68516-73-4].

Structure formula

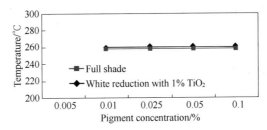

Color characterization Pigment Yellow 155 is a greenish yellow pigment and the tinting strength is in medium level. The required concentration of pigment is only 0.609% when blending with 5% of titanium dioxide to achieve 1/3 SD in flexible PVC, and the required concentration of pigment is only 0.19% when blending with 1% of titanium dioxide to achieve 1/3 SD in HDPE.

Main properties See Table 4.124~Table 4.126 and Figure 4.37.

Table 4.124 Application properties of Pigment Yellow 155 in PVC

Project		Pigment	Titanium dioxide	Light fastness degree	Weather resistance degree (3000h)	Migration fastness degree
PVC	Full shade	0.1%		8	3	
	Reduction	0.1%	0.5%	7~8		3~4

Table 4.125 Application properties of Pigment Yellow 155 in HDPE

Project		Pigment	Titanium dioxide	Light fastness degree	Weather resistance degree (3000h, 0.2%)
PE	Full shade	0.18%		8	3
	1/3SD	0.18%	1%	7~8	

Table 4.126 Application of Pigment Yellow 155

General plastics		Engineering plastics		Spinning	
LL/LDPE	●	PS/SAN	●	PP	●
HDPE	●	ABS	×	PET	×
PP	●	PC	×	PA6	×
PVC (soft)	●	PBT	×	PAN	×
PVC (rigid)	●	PA	×		
Rubber	●	POM	○		

● Recommended to use, ○ Conditional use, × No recommended to use.

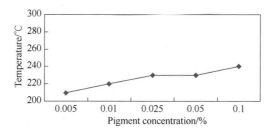

Figure 4.37 Heat resistance of Pigment Yellow 155 in HDPE (full shade)

Varieties characteristics The light fastness of Pigment Yellow 155 is excellent in the coloring of polyolefin. Not only can it be used for the coloring of general-purpose polyolefin, but also can be applied to the coloring of styrenic engineering plastics. In addition, it is not suitable for the coloring of flexible PVC because of migration. And it is mainly suitable for coloring polypropylene fibers during spinning. Moreover, it is the ideal product to replace dichlorobenzidine yellow pigment.

(2) Red disazo condensation pigments

Currently the basic formula of red disazo condensation pigments used frequently in the plastics industry are as follows:

Main varieties of red disazo condensation pigments for plastics coloring are list in Table 4.127.

Table 4.127 Main varieties of red disazo condensation pigments for plastics coloring

C.I. pigment	CAS No.	R_1	R_2	X	Y	Shade
Red 144	[5280-78-4]	Cl	H	Cl	Cl	Bluish red
Red 166	[3905-19-9]	H	H	Cl	Cl	Yellowish red
Red 214	[40618-31-3]	Cl	Cl	Cl	Cl	Bluish red
Red 242	[52238-93-3]	Cl	Cl	Cl	CF3	Red
Red 262	[211502-19-1]	CN	H	Cl	CF3	Bluish red

① **C.I. Pigment Yellow 144** C.I. structure No. 20735. Molecular formula: $C_{40}H_{23}Cl_5N_6O_4$. CAS registry number: [40716-47-0].

Structure formula

Color characterization Pigment Red 144 is a slight bluish red pigment and the tinting strength is very high. The required concentration of pigment is only 0.7% when blending with 5% of titanium dioxide to achieve 1/3 SD in PVC, and the required concentration of pigment is only 0.13% when blending with 1% of titanium dioxide to achieve 1/3 SD in HDPE. The tinting strength of Pigment Red 144 is second in the red disazo condensation pigments.

Main properties See Table 4.128~Table 4.130 and Figure 4.38.

Table 4.128 Application properties of Pigment Red 144 in PVC

Project		Pigment	Titanium dioxide	Light fastness degree	Weather resistance degree (2000 h)	Migration fastness degree	Heat resistance degree	
							180℃/30min	200 ℃/10min
PVC	Full shade	0.1%		8	4~5	5	5	5
	Reduction	0.2%	2%	7~8	2~3	4.9	5	5

Table 4.129 Application properties of Pigment Red 144 in HDPE

Project		Pigment	Titanium dioxide	Light fastness degree	Weather resistance degree (3000h)	Migration fastness degree
HDPE	Full shade	0.1%		7~8	3~4	5
	Reduction	0.1%	1%	7~8	1	5

Table 4.130 Application of Pigment Red 144

General plastics		Engineering plastics		Spinning	
LL/LDPE	●	PS/SAN	●	PP	●
HDPE	●	ABS	●	PET	×
PP	●	PC	○	PA6	×
PVC (soft)	●	PBT	○	PAN	
PVC (rigid)	○	PA	×		
Rubber	●	POM	●		

● Recommended to use, ○ Conditional use, × No recommended to use.

Figure 4.38 The heat resistance of Pigment Red 144 in HDPE

Varieties characteristics   In the 1/3 standard depth of shade the heat resistance of Pigment Red 144 is 300℃/5min in HDPE. Moreover, the heat resistance of full shade and reduction is the same. In addition, this pigment has excellent light fastness, but poor weather resistance in reduction. So it can't be used to color the products that exposed to the outdoors for a long time. And it is suitable for coloring polypropylene fibers. Not only can it be used for the coloring of PVC and general-purpose polyolefin, but also can be applied to the coloring of styrenic engineering plastics. The migration fastness of this pigment is very outstanding in flexible PVC. However, there is a large effect on the shrinkage of HDPE.

② **C.I. Pigment Yellow 166**   C.I. structure No. 20730. Molecular formula: $C_{40}H_{24}Cl_4N_6O_4$. CAS registry number: [3905-19-9].

Structure formula

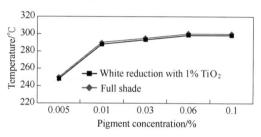

Color characterization Pigment Red 166 is a very bright yellowish red pigment and the tinting strength is high. And the required concentration of pigment is only 0.9%~1.2% (different concentration of pigment depending on companies) when blending with 5% of titanium dioxide to achieve 1/3 SD in PVC.

Main properties See Table 4.131~Table 4.133 and Figure 4.39.

Table 4.131 Application properties of Pigment Red 166 in PVC

Project		Pigment	Titanium dioxide	Light fastness degree	Weather resistance degree (2000 h)	Migration fastness degree	Heat resistance degree	
							180℃ /30min	200 ℃ /10min
PVC	Full shade	0.1%		8	4~5	5	5	5
	Reduction	0.1%	2%	7	3~4	5	5	5

Table 4.132 Application properties of Pigment Red 166 in HDPE

Project		Pigment	Titanium dioxide	Light fastness degree	Weather resistance degree (3000h)	Migration fastness degree
HDPE	Full shade	0.1%		7~8	3~4	5
	Reduction	0.1%	1%	7~8	1	5

Table 4.133 Application of Pigment Red 166

General plastics		Engineering plastics		Spinning	
LL/LDPE	●	PS/SAN	●	PP	●
HDPE	●	ABS	○	PET	×
PP	●	PET	×	PA6	×
PVC (soft)	●	PA	×	PAN	×
PVC (rigid)	●	POM	●		
Rubber	●				

● Recommended to use, ○ Conditional use, × No recommended to use.

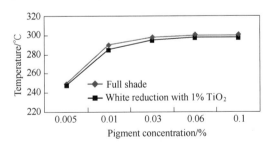

Figure 4.39 The heat resistance of Pigment Red 166 in HDPE

Varieties characteristics The heat and light fastness of pigment red 166 is very excellent in the coloring of polyolefin. And the weather resistance of pigment that in full shade for spin dying basically meets the requirements when exposed to the outdoors for a long time, but the weather resistance of pigment that in reduction with titanium dioxide becomes very poor. In addition, not only can it be used for the coloring of PVC and general-purpose polyolefin, but also can be applied to the coloring of styrenic engineering plastics. The bleeding and migration fastness is very outstanding in the coloring of PVC. However, there is a large effect on the shrinkage of HDPE. This pigment is the important variety whose comprehensive performance is excellent in the yellowish red zone.

③ **C.I. Pigment Red 214** C.I. structure No. 200660. Molecular formula: $C_{40}H_{22}Cl_6N_6O_4$. CAS registry number: [4068-31-3].

Structure formula

Color characterization Pigment Red 214 is a bluish red pigment and the shade is brighter than Pigment Red 144. The tinting strength of this pigment is high, and the required concentration of pigment is only 0.56% when blending with 5% of titanium dioxide to achieve 1/3 SD in PVC, the required concentration of pigment is only 0.13% when blending with 1% of titanium dioxide to achieve 1/3 SD in HDPE.

Main properties See Table 4.134～Table 4.136 and Figure 4.40.

Table 4.134　Application properties of Pigment Red 214 in PVC

Project		Pigment	Titanium dioxide	Light fastness degree	Weather resistance degree (3000h)	Migration fastness degree
PVC	Full shade	0.1%		7～8	3～4	
	Reduction	0.1%	0.5%	7～8		5

Table 4.135　Application properties of Pigment Red 214 in HDPE

Project		Pigment	Titanium dioxide	Light fastness degree	Weather resistance degree (3000h,0.2%)
HDPE	Full shade	0.16%		8	3
	1/3SD	0.16%	1%	7～8	

Table 4.136　Application of Pigment Red 214

General plastics		Engineering plastics		Spinning	
LL/LDPE	●	PS/SAN	●	PP	●
HDPE	●	ABS	●	PET	○
PP	●	PC	●	PA6	×
PVC (soft)	●	PBT	●	PAN	●
PVC (rigid)	●	PA	×		
Rubber	●	POM	●		

● Recommended to use, ○ Conditional use, × No recommended to use.

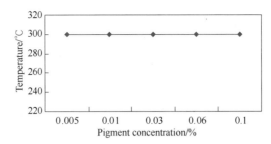

Figure 4.40 Heat resistance of Pigment Red 214 in HDPE (full shade)

Varieties characteristics The heat and light fastness of Pigment Red 214 is excellent in the coloring of polyolefin. Not only can it be used for the coloring of general-purpose polyolefin, but also can be applied to the coloring of styrenic engineering plastics. However, there is a large effect on the warpage of HDPE. In addition, it is suitable for coloring PP and polyester fibers, and the fastness properties of textiles that colored with this pigment can meet the demands of users.

④ **C.I. Pigment Red 242** C.I. structure No. 20067. Molecular formula: $C_{42}H_{22}Cl_4F_6N_6O_4$. CAS registry number: [52238-92-3].

Structure formula

Color characterization Pigment Red 242 is a very bright yellowish red pigment, and the shade is still very bright yellowish after adding titanium dioxide. And the tinting strength of this pigment is general, the required concentration of pigment is 0.884% when blending with 5% of titanium dioxide to achieve 1/3 SD in flexible PVC, and the required concentration of pigment is only 0.2% when blending with 1% of titanium dioxide to achieve 1/3 SD in HDPE.

Main properties See Table 4.137～Table 4.139 and Figure 4.41.

Table 4.137 Application properties of Pigment Red 242 in PVC

	Project	Pigment	Titanium dioxide	Light fastness degree	Migration fastness degree
PVC	Full shade	0.1%		8	
	Reduction	0.1%	0.5%	7～8	5

Table 4.138 Application properties of Pigment Red 242 in HDPE

	Project	Pigment	Titanium dioxide	Light fastness degree
PE	Full shade	0.23%		8
	1/3 SD	0.23%	1%	7～8

Table 4.139 Application of Pigment Red 242

General plastics		Engineering plastics		Spinning	
LL/LDPE	●	PS/SAN	●	PP	○
HDPE	●	ABS	●	PET	●
PP	●	PC	●	PA6	×
PVC (soft)	●	PBT	●	PAN	●
PVC (rigid)	●	PA	×		
Rubber	●	POM	●		

● Recommended to use, ○ Conditional use, × No recommended to use.

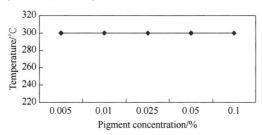

Figure 4.41 Heat resistance of Pigment Red 242 in HDPE (full shade)

Varieties characteristics The heat and light fastness of Pigment Red 242 is quite excellent in the coloring of polyolefin. Not only can it be used for the coloring of general-purpose polyolefin, but also can be applied to the coloring of styrenic engineering plastics. In addition, it will not migrate in the coloring of flexible PVC. However, there is a large effect on the warpage of HDPE. It is irreplaceable to prepare bright pink pigment with this pigment.

⑤ **C.I. Pigment Red 262** Molecular formula: $C_{43}H_{23}Cl_2F_6N_7O_4$. CAS registry number: [211502-19-1].

Structure formula

Color characterization Pigment Red 262 is bluish red. It has a high tinting strength, so when formulated with 5% titanium dioxide to make the PVC with 1/3 standard depth, the amount is only 0.405%. And when formulated with 1% titanium dioxide to make HDPE with 1/3 standard depth, the amount is 0.115%.

Main properties As shown in Table 4.140~Table 4.142 and Figure 4.42.

Table 4.140 Application properties of Pigment Red 262 in PVC plastics

	Project	Pigment	Titanium dioxide	Light fastness degree	Migration resistance degree
PVC	Full shade	0.1%		7~8	
	Reduction	0.1%	0.5%	7~8	4

Table 4.141 Application properties of Pigment Red 262 in HDPE plastics

Project		Pigment	Titanium dioxide	Light fastness degree	Weather resistance degree (3000h,0.2%)
PE	Full shade	0.12%		7~8	3
	1/3SD	0.12%	1%	8	

Table 4.142 Applications of Pigment Red 262

General plastics		Engineering plastics		Spinning	
LL/LDPE	●	PS/SAN	●	PP	●
HDPE	●	ABS	●	PET	×
PP	●	PC	●	PA6	×
PVC(soft)	●	PBT	○	PAN	×
PVC(rigid)	●	PA	×		
Rubber	●	POM	×		

● Recommended to use, ○ Conditional use, × Not recommended to use.

Figure 4.42 Heat resistance of Pigment Red 262 in HDPE

Varieties characteristics In the polyolefin, the heat and light fastness of Pigment Red 262 are excellent. Besides coloring the general-purpose polyolefin, the Pigment Red 262 can also be used in styrene-based engineering plastics coloring.

(3) Brown disazo condensation pigments

At present, the generic structure of brown disazo condensation pigments which commonly used in the plastics industry is as follows.

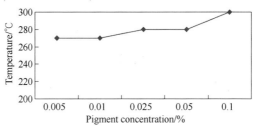

The main varieties of brown disazo condensation pigments which used in coloring plastics are shown in Table 4.143.

Table 4.143 Main varieties of brown condensed azo pigments for plastics coloring

Species	CAS No	A	R	Shade
Brown 23	[35869-64-8]	Cl-C6H3-Cl	NO2-C6H3-	Brownish yellow
Brown 41	[68516-75-5]	naphthalene	Cl2-C6H3-	Brownish yellow

198 Coloring of Plastics: Fundamental-Application-Masterbatch

① **C.I. Pigment Brown 23** C.I. structure number: 20060. Molecular formula: $C_{40}H_{23}Cl_3N_8O_8$. CAS registry number: [35869-64-8].

Structure formula

Color characterization Pigment Brown 23 is brownish red, and the tinting strength is ordinary. When formulated with 1% titanium dioxide to make flexible PVC with 1/3 standard depth, the amount is 0.75%.

Main properties As shown in Table 4.144～Table 4.146 and Figure 4.43.

Table 4.144 Application performance of Pigment Brown 23 in PVC

Project		Pigments	Titanium dioxide	Light fastness degree	Weather resistance degree (2000h)	Migration resistance degree	Heat resistance degree	
							180℃/30min	200℃/10min
PVC	Full shade	0.1%		8	5	4.9	5	5
	Reduction	0.2%	2%	7～8	4	4.8	5	5

Table 4.145 Application performance of Pigment Brown 23 in HDPE

Project		Pigments	Titanium dioxide	Light fastness degree	Migration resistance degree
HDPE	Full shade	0.1 %		6～7	5
	Reduction	0.1 %	1%		5

Table 4.146 Applications of Pigment Brown 23

General plastics		Engineering plastics		Spinning	
LL/LDPE	○	PS/SAN	○	PP	×
HDPE	○	ABS	○	PET	×
PP	○	PC	×	PA6	×
PVC(soft)	●	PET	×	PAN	
PVC(rigid)	●	PA	×		
Rubber	●	POM	●		

● Recommended to use, ○ Conditional use, × Not recommended to use.

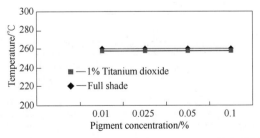

Figure 4.43 Heat resistance of Pigment Brown 23 in HDPE

Varieties characteristics In the polyolefin, the heat and light fastness of Pigment Brown 23 are excellent, and it can meet the requirements of long-term exposed outdoors when used in rigid PVC. Besides coloring the general-purpose polyolefin, the Pigment Brown 23 can also be used in styrene-based engineering plastics coloring. Pigment Brown 23 at a processing temperature of 220℃ can significantly affect the warp of HDPE injection products, but the effects reduce when temperature rises. Pigment Brown 23 is suitable for spinning polypropylene fibers and coloring the acrylic raw liquor, but the weather resistance can't meet the requirements that to be placed in long-term exposed outdoor.

② **C.I. Pigment Brown 41**   Molecular formula: $C_{44}H_{26}Cl_4N_6O_4$. CAS registry number: [68516-75-5].

Structure formula

Color characterization   Pigment Brown 41 is yellowish brown. It is specially designed to replace the brown varieties of benzimidazolone. The tinting strength of Pigment Brown 41 is not very high, when formulated with 1% titanium dioxide to make the flexible PVC with 1/3 standard depth, the amount is only 0.41%, while the hiding power is good.

Main properties   As shown in Table 4.147~Table 4.149 and Figure 4.44.

Table 4.147   Application performance of Pigment Brown 41 in PVC

Project		Pigments	Titanium dioxide	Light fastness degree	Weather resistance degree (3000h)	Migration resistance degree
PVC	Full shade	0.1%		8	5	4
	Reduction	0.1%	0.5%	8		

Table 4.148   Application performance of Pigment Brown 41 in HDPE

Project		Pigments	Titanium dioxide	Light fastness degree	Weather resistance degree (3000h)
PE	Full shade	0.2%		8	4~5
	1/3SD	0.2%	1%	8	

Table 4.149 Applications of Pigment Brown 41

General plastics		Engineering plastics		Spinning	
LL/LDPE	●	PS/SAN	●	PP	●
HDPE	●	ABS	×	PET	×
PP	●	PC	×	PA6	×
PVC(soft)	●	PBT	×	PAN	●
PVC(rigid)	●	PA	×		
Rubber	●	POM	●		

● Recommended to use, ○ Conditional use, × Not recommended to use.

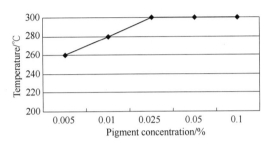

Figure 4.44 Heat resistance of Pigment Brown 41 in HDPE (full shade)

Varieties characteristics The heat and light fastness of Pigment Brown 41 are excellent, and it can meet the requirements of long-term exposed outdoors. Pigment Brown 41 is suitable for coloring the general-purpose polyolefin and polyvinyl chloride.

4.6 Phthalocyanine Pigments

The phthalocyanine pigment has been one of the largest species of organic pigments, it not only has excellent comprehensive application performances, like heat resistance, light fastness, migration resistance and weather resistance, but also has advantages of convenience, low cost, etc. Currently, none of the chemicals can replace it, and its production accounts is a quarter of organic pigments production. The shades of phthalocyanine pigment in plastics coloring are blue and green. Phthalocyanine pigment including Pigment Blue 15∶1, Pigment Blue 15∶3 and Pigment Green 7 have become the standard color in the coloring of plastics..

The phthalocyanine pigments and their metal phthalocyanine are polymorphism, which means a compound can generate a number of different crystal structures. The polymorphism phenomenon exists with varying degrees in many kinds of compounds, but more typical in phthalocyanine pigments. Different crystals of phthalocyanine pigments have different melting points, and also have difference in the crystal shape, density, surface color and reflection of light. These physical properties are closely related with the phthalocyanine pigment application properties in plastics. So studying and understanding the knowledge have a practical significance. Now, we know the phthalocyanine pigments have eight modifications (α, β, γ, δ, ε, Π, χ, R). However, only crystal modification of α, β, γ, ε used as the pigments, especially α, β are most commonly used. The stability of the phthalocyanine pigment depends on the crystal form, wherein the β-modification is strongest, α-modification is weakest. The α-modification is unstable and it can't withstand a processing temperature in plastic molding and then it will be converted to stable β-modification. To avoid this phenomenon, α-modification can be stable by adding 0.5~1 atom chlorine and can also get a crystalline surface treatment by adding special stable substance.

The blue and green varieties used in plastic coloring are shown in Table 4.150.

Table 4.150   The blue and green varieties used in plastic coloring

Species	CAS registry number	Structure types	Application characteristics
Pigment Blue 15	[147-14-8]	CuPc	Unstable α-modification, reddish blue
Pigment Blue 15∶1	[147-14-8]	Low chlorinated CuPc	Stable α-modification, reddish blue
Pigment Blue 15∶3	[147-14-8]	CuPc	Stable β-modification, greenish blue
Pigment Green 7	[1328-53-6]	(CuPc) Cl15-16	Polychlorinated copper phthalocyanine
Pigment Green 36	[14302-13-7]	(CuPc) Cl10Br6	Bromo-chloro-copper phthalocyanine

4.6.1   Blue Phthalocyanine Pigments

① **C.I. Pigment Blue 15**   C.I. structure number: 74160. Molecular formula: $C_{32}H_{16}CuN_8$. CAS registry number: [147-14-8].

Structure formula

Color characterization   Pigment Blue 15 is a copper phthalocyanine with a metastable α-modification. The stability of the pigment is poor, and the crystal form will transform into a stable β-modification (greenish) when it is under high temperature or aromatic hydrocarbon solvents. As compared to Pigment Blue 15∶3, the Pigment Blue 15 is reddish blue, the color is more vivid and tinting strength is higher.

Main properties   As shown in Table 4.151~Table 4.152 and Figure 4.45.

Table 4.151　Application performance of Pigment Blue 15 in PVC plastics

Project		Pigments	Titanium dioxide	Light fastness degree	Weather resistance degree (5000h)	Migration resistance degree
PVC	Full shade	0.1%		8	4~5	5
	Reduction	0.1%	1%	8	3~4	

Table 4.152　Applications of Pigment Blue 15

General plastics		Engineering plastics		Spinning	
LL/LDPE	●	PS/SAN	×	PP	×
HDPE	×	ABS	×	PET	×
PP	×	PC	×	PA6	×
PVC(soft)	●	PBT	×	PAN	×
PVC(rigid)	●	PA	×		
Rubber	●	POM			

● Recommended to use, × Not recommended to use.

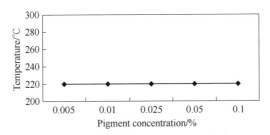

Figure 4.45　Heat resistance of Pigment Blue 15 in HDPE (full shade)

Varieties characteristics　When Pigment Blue 15 is applied in plastics processing, the temperature should not be higher than 200℃. As the temperature increases, Pigment Blue 15 will be transformed from α type to β-modification. Pigment Blue 15 can be used for coloring PVC, the migration resistance is good, and it has a high light fastness. Pigment Blue 15 can also be used for coloring EVA.

② **C.I. Pigment Blue 15∶1**　C.I. structure number: 74160. Molecular formula: $C_{32}H_{16}CuN_8$. CAS registry number: [147-14-8].

Structure formula　The same as Pigment Blue 15.

Color characterization　The crystal structure type of Pigment Blue 15∶1 is a stable α- crystalline form. The chemical composition is a mixture of phthalocyanine blue and a small amount of one-chlorinated copper phthalocyanine. Compared with Pigment Blue 15, the shade became partial green because of the presence of chlorinated copper phthalocyanine. And the transparency and tinting strength are slightly lower. As for polyolefins, Pigment Blue 15∶1 has a high tinting strength, and when formulated with 1% titanium dioxide to make the HDPE with 1/3 standard depth, the amount is less than 0.1%.

Main properties　As shown in Table 4.153~Table 4.155 and Figure 4.46.

Table 4.153 Application performance of Pigment Blue 15∶1 in PVC

Project		Pigments	Titanium dioxide	Light fastness degree	Weather resistance degree (5000h)	Migration resistance degree
PVC	Full shade	0.1%		8	5	5
	Reduction	0.1%	1%	8	4	5

Table 4.154 Application performance of Pigment Blue 15∶1 in HDPE

Project		Pigments	Titanium dioxide	Light fastness degree	Weather resistance degree (3000h)
HDPE	Full shade	0.1%		8	5
	Reduction	0.1%	1%	8	5

Table 4.155 Applications of Pigment Blue 15∶1

General plastics		Engineering plastics		Spinning	
LL/LDPE	●	PS/SAN	●	PP	●
HDPE	●	ABS	●	PET	●
PP	●	PC	●	PA	●
PVC(soft)	●	PBT	●	PAN	●
PVC(rigid)	●	PA6	●		
Rubber	●	POM	●		

● Recommended to use.

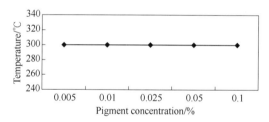

Figure 4.46 Heat resistance of Pigment Blue 15∶1 in HDPE (full shade)

Varieties characteristics The comprehensive fastness of Pigment Blue 15∶1 is excellent, and it is suitable for coloring general purpose polyolefin plastics, general engineering plastics and specialty plastics. Pigment Blue 15∶1 is the standard color for coloring plastics. Pigment Blue 15∶1 is suitable for coloring polypropylene, polyester and fibers of polyamide 6. The obtained textile has a very good light fastness and application fastness. The dispersibility of Pigment Blue 15∶1 is not very good, especially not suitable for plastic film products. When Pigment Blue 15∶1 is used in HDPE and other crystalline plastics, it can seriously affect the warpage and deformation of plastics.

③ **C.I. Pigment Blue 15∶3** C.I. structure number: 74160. Molecular formula: $C_{32}H_{16}CuN_8$. CAS registry number: [147-14-8].

Structure formula The same as Pigment Blue 15.

Color characterization Pigment Blue 15∶3 is a single copper phthalocyanine, the crystal structure type is a stable β-modification. By ball milling method the β-modification copper phthalocyanine is rod-like, as compared with α-modification phthalocyanine pigments, its shade is obviously partial green. While by kneading method the crystal is equiaxial, and its shade is greener. Pigment Blue 15∶3 is greenish blue, the shade is pure, and the transparency is also very good. The tinting strength of Pigment Blue 15∶3 is 15%~20% lower than α-modification phthalocyanine pigments.

Main properties　　As shown in Table 4.156~Table 4.158 and Figure 4.47.

Table 4.156　Application properties of Pigment Blue 15∶3 in PVC

Project		Pigments	Titanium dioxide	Light fastness degree	Weather resistance degree (5000h)	Migration resistance degree
PVC	Full shade	0.1%		8	4~5	5
	Reduction	0.1%	1%	8	4~5	

Table 4.157　Application performance of Pigment Blue 15∶3 in HDPE

Project		Pigments	Titanium dioxide	Light fastness degree	Weather resistance degree (3000h)
HDPE	Full shade	0.1%		8	5
	Reduction	0.1%	1%	8	5

Table 4.158　Applications of Pigment Blue 15∶3

General plastics		Engineering plastics		Spinning	
LL/LDPE	●	PS/SAN	●	PP	●
HDPE	●	ABS	●	PET	●
PP	●	PC	●	PA	●
PVC(soft)	●	PBT	●	PAN	●
PVC(rigid)	●	PA6	●		
Rubber	●	POM	●		

● Recommended to use.

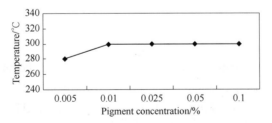

Figure 4.47　Heat resistance of Pigment Blue 15∶3 in HDPE (full shade)

Varieties characteristics　　For Pigment Blue 15∶3, the heat resistance is better than the α- crystalline form phthalocyanine pigment. But in the PE, the heat resistance will reduce when Pigment Blue 15∶3 was reduction with titanium dioxide, such as the PE containing 0.1% Pigment Blue 15∶3 and 0.5% of titanium dioxide, the heat temperature is only 250℃. As for Pigment Blue 15∶3, the comprehensive fastness is excellent, and it is suitable for coloring general purpose polyolefin plastics, general engineering plastics and specialty plastics. Pigment Blue 15∶3 is the standard color for coloring plastics. Pigment Blue 15∶3 is mostly supplied in the form of pigment preparations when it is used in the plastics. Pigment Blue 15∶3 is widely used in coloring the fiber solution of polypropylene, polyester, acrylic, nylon, rayon and so on. The obtained textile has a very good light fastness and application fastness. The Pigment Blue 15∶3 has a nucleation phenomenon, so it can seriously affect the warpage and deformation of plastics.

4.6.2 Green Phthalocyanine Pigments

In the molecule of metal phthalocyanine, when the 16 hydrogen atoms are replaced by halogen, we obtained green phthalocyanine pigments. In fact, halogenated copper phthalocyanine as a pigment is chlorine-substituted or chloro-bromo mixing substituted. The quantity and quality of chlorine-substituted or chloro-bromo mixing substituted directly affect the shade of green phthalocyanine.

① **C.I. Pigment Green 7** C.I. structure number: 74260. Molecular formula: $C_{32}Cl_{12-16}Cu\ N_8$. CAS registry number: [1328-53-6].

Structure formula

Color characterization Pigment Green 7 is perchlorinated copper phthalocyanine, one molecule has an average of 14~15 chlorine atoms and the shade is bluish green. The tinting strength of Pigment Green 7 is much lower than that of phthalocyanine blue pigment. For example, when formulated with 1% titanium dioxide to make HDPE with 1/3 standard depth, the amount is 0.2%, while the amount of β-modification phthalocyanine blue pigment is only 0.1%. The reason is that after the chlorine atom is introduced into copper phthalocyanine, the molecular weight is increased while the tinting strength is decreased.

Main properties As shown in Table 4.159~Table 4.161 and Figure 4.48.

Table 4.159 Application properties of Pigment Green 7 in PVC

Project		Pigments	Titanium dioxide	Light fastness degree	Migration resistance degree	Weather resistance degree (5000h)
PVC	Full shade	0.1%		8	5	5
	Reduction	0.1%	1%	8		4~5

Table 4.160 Application properties of Pigment Green 7 in HDPE

Project		Pigments	Titanium dioxide	Light fastness degree	Weather resistance degree (3000h)
HDPE	Full shade	0.1%		8	5
	Reduction	0.1%	1%	8	5

Table 4.161 Applications of Pigment Green 7

General plastics		Engineering plastics		Spinning	
LL/LDPE	●	PS/SAN	●	PP	●
HDPE	●	ABS	●	PET	●
PP	●	PC	○	PA	●
PVC(soft)	●	PET	○	PAN	●
PVC(rigid)	●	PA6	●		
Rubber	●	POM	●		

● Recommended to use, ○ Conditional use.

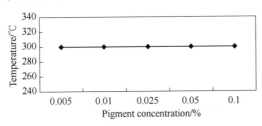

Figure 4.48 Heat resistance of Pigment Green 7 in HDPE (full shade)

Varieties characteristics As for Pigment Green 7, the comprehensive fastness is excellent when it is used in plastics, and it is suitable for coloring general purpose polyolefin plastics, general engineering plastics and specialty plastics. Pigment Green 7 is the standard color for coloring plastics. Pigment Green 7 is mostly supplied in the form of pigment preparations when it is used in the plastics. Pigment Green 7 is widely used in coloring the fiber solution of polypropylene, polyester, acrylic, nylon, rayon and so on. The obtained textile has a very good light fastness and application fastness. It is suitable for making outdoor products like outdoor awning and so on. When Pigment Green 7 is used in HDPE and other crystalline plastics, it can affect the warpage and deformation of plastics, but much smaller than that of blue phthalocyanine pigments.

② **C.I. Pigment Green 36** C.I. structure number: 74265. Molecular formula: $C_{32}Cl_{12}CuN_8Br_6$. CAS registry number: [14302-13-7].

Structure formula

$$\left[\text{Cu phthalocyanine} \right] Cl_xBr_y \quad x+y=12\sim15$$

Color characterization Pigment Green 36 is chloro-bromo mixing substituted copper phthalocyanine, the average number of substituted chlorine and bromine atoms are 14~15. And the shade depends on the number of substituted bromine and chlorine atoms in molecule of the copper phthalocyanine. The yellowish tint increases with increasing number of bromine atoms. The pigments can be divided into two categories. One is the number of substituted bromine atoms<6, which named the yellowish phthalocyanine green 3G. The other is the number of substituted bromine atoms⩾6, which named yellowish phthalocyanine Green 6G. The yellowish tint increases with increasing number before G. Therefore Pigment Green 36 is more vivid than Pigment Green 7. Because of the atomic weight of bromine is relatively larger than that of chlorine, the molecular weight of Pigment Green 36 is larger than that of Pigment Green 7 and the tinting strength is lower.

Main properties As shown in Table 4.162 to Table 4.164 and Figure 4.49.

Table 4.162 Application properties of Pigment Green 36 in PVC

Project		Pigments	Titanium dioxide	Light fastness degree	Migration resistance degree	Weather resistance degree (5000h)
PVC	Full shade	0.1%		8	5	5
	Reduction	0.1%	1%	8		4~5

Table 4.163 Application properties of Pigment Green 36 in HDPE

Project		Pigments	Titanium dioxide	Light fastness degree	Weather resistance degree (3000h)	Migration resistance degree
HDPE	Full shade	0.1%		8	5	5
	Reduction	0.1%	1%	8	5	

Table 4.164 Applications of Pigment Green 36

General plastics		Engineering plastics		Spinning	
LL/LDPE	●	PS/SAN	●	PP	○
HDPE	●	ABS	○	PET	○
PP	●	PC	●	PA	○
PVC(soft)	●	PBT	●	PAN	
PVC(rigid)	●	PA6	○		
Rubber	●	POM	●		

● Recommended to use, ○ Conditional use.

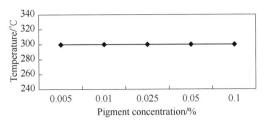

Figure 4.49　Heat resistance of Pigment Green 36 in HDPE (full shade)

Varieties characteristics　The plastic applications fastness of Pigment Green 36 and Pigment Green 7 are almost the same. As for Pigment Green 36, the comprehensive fastness is excellent, and it is suitable for coloring general purpose polyolefin plastics, general engineering plastics and specialty plastics. Pigment Green 36 is mostly supplied in the form of pigment preparations when it is used in the plastics. Pigment Green 36 is widely used in coloring the fiber solution of polypropylene, polyester, acrylic, nylon, rayon and so on. The obtained textile has a very good light fastness and application fastness. It is suitable for making outdoor products like outdoor awning and so on.

When Pigment Green 36 is used in HDPE and other crystalline plastics, it can affect the warpage and deformation of plastics, but much smaller than that of blue phthalocyanine pigments. The tinting strength of Pigment Green 36 is lower than that of Pigment Green 7 while the price is much higher than Pigment Green 7, so the application in plastics is much less than Pigment Green 7.

4.7　Heterocyclic and Polycyclic Ketonic Pigments

Classic azo organic pigments have been widely used in plastics coloring due to the complete chromatography, bright color, reasonable price, but because of the chemical structure and other factors, it has shortcomings in heat resistance, light fastness, migration resistance, etc. The gap is greater especially when the light color is used. In addition, the conventional yellow and orange benzidine pigments the conventional yellow and orange benzidine pigments will be decomposed into dicloxacillin benzidine when the processing temperature of polymer exceeds 200℃. The effects of pigment decomposition on human body and the environment are more and more attract people's attention. In the face of today's fierce market competition, the appearance and performance of products have never been like this to affect the consumer's purchasing desire, especially in the markets of automotive, household appliances, consumer goods, etc. Therefore, in order to meet market demands, all the countries are focus on the development of heterocyclic and polycyclic ketonic pigments used in coloring plastics.

4.7.1 Dioxazine Pigments

The pigment molecules contain the parent structure of triphenyl dioxazine which is orange. When the different substituents and heterocyclic were introduced to the both sides of the benzene ring, the pigment will present a bright purple and its molecule has plane symmetry. In the current market dioxazine pigments are carbazole dioxazine violet. They have beautiful, bright and pure bluish purple tones and a particularly high tinting strength and gloss. They also have excellent light fastness, good heat resistance, weather resistance and solvent resistance, the performance can be rivaled with phthalocyanine pigments, but the migration resistance is not very good. Dioxazine violet is the standard color for coloring plastics. But the disadvantages and advantages are both obvious. Because of the high tinting strength, the performance decreases at the low-concentration. To be special attention, when titanium dioxide was added to the dioxazine violet, the performance decreases even more.

(1) **C.I. Pigment Violet 23**   C.I. structure number: 51319. Molecular formula: $C_{34}H_{22}Cl_2N_4O_2$. CAS registry number: [6358-30-1].

Structure formula

Color characterization   The basic color of Pigment Violet 23 is reddish purple, another variety with bluish purple color can also be obtained through special treatment. Pigment Violet 23 is a general purple species. Its production is a large number. Pigment Violet 23 has a particularly high tinting strength, when formulated with 1% titanium dioxide to make HDPE with 1/3 standard depth, the amount is only 0.07%. In flexible PVC, the tinting strength is very high while the migration resistance is not very good when it is applied in light color.

Main properties   As shown in Table 4.165～Table 4.167 and Figure 4.50.

Table 4.165   Application properties of Pigment Violet 23 in PVC

Project		Pigments	Titanium dioxide	Light fastness degree	Weather resistance degree (3000h)	Migration resistance degree
PVC	Full shade	0.1%		7~8	5	4
	Reduction	0.1%	0.5%	7~8		

Table 4.166 Application performance of Pigment Violet 23 in HDPE

Project		Pigments	Titanium dioxide	Light fastness degree	Weather resistance degree (3000h, Natural 0.2%)
HDPE	Full shade	0.07%		7~8	4~5
	1/3SD	0.07%	1%	7~8	5

Table 4.167 Applications of Pigment Violet 23

General plastics		Engineering plastics		Spinning	
LL/LDPE	●	PS/SAN	●	PP	●
HDPE	●	ABS	○	PET	×
PP	●	PC	×	PA6	○
PVC(soft)	●	PBT	×	PAN	●
PVC(rigid)	●	PA	○		
Rubber	●	POM	×		

● Recommended to use, ○ Conditional use, × Not recommended to use.

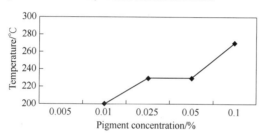

Figure 4.50 Heat resistance of Pigment Violet 23 in HDPE (full shade)

Varieties characteristics Pigment Violet 23 can be used for coloring polyolefin, the heat-resistance temperature of 1/3SD polyolefin is up to 280℃. If the temperature exceeds this limit, the shade will shift to the red phase, 1/25SD polyolefin still resistant to 200℃. Pigment Violet 23 can be used for coloring transparent polystyrene, but it can't resistant to a higher temperature above 220℃ in this medium while the Pigment Violet 23 will decompose above this temperature. Pigment Violet 23 also can be used for coloring polyester plastics and it can withstand 280℃/6h without decomposition. If the concentration is too low, it will be partially dissolved to make its shade reddish at this temperature.

The light fastness of Pigment Violet 23 is excellent, the degree is up to eight, but the degree of light fastness will be reduced sharply to 2 when it is diluted to 1/25SD with titanium dioxide. Therefore the concentration for Pigment Violet 23 used in transparent products should not be less than 0.05%.

Pigment Violet 23 is suitable for coloring general purpose polyolefin plastics and general engineering plastics. Pigment Violet 23 is not suitable for coloring soft polyvinylchloride due to the poor migration. Pigment Violet 23 is suitable for coloring polypropylene, polyester and fibers of polyamide 6 before spinning, its concentration can't be too low or there will be a chromatic aberration. When Pigment Violet 23 is used in HDPE and other crystalline plastics, it can seriously affect the warpage and deformation of plastics.

A very small amount of Pigment Violet 23 added into titanium dioxide can cover the yellow shade, and resulting in a very pleasing white color. About 100g of titanium dioxide only need 0.0005-0.05g Pigment Violet 23.

(2) **C.I. Pigment Violet 37** C.I. structure number: 51345. Molecular formula: $C_{40}H_{34}N_6O_8$. CAS registry number: [57971-98-9].

Structure formula

Color characterization The shade of Pigment Violet 37 is more reddish than Pigment Violet 23. The tinting strength of Pigment Violet 37 in most medium is lower than Pigment Violet 23. When used to make products with 1/3 standard depth in coloring polyolefin, the amount of Pigment Violet 37 is 0.09%, while that of Pigment Violet 23 is 0.07%.

Main properties As shown in Table 4.168~Table 4.170 and Figure 4.51.

Table 4.168 Application properties of Pigment Violet 37 in PVC

Project		Pigments	Titanium dioxide	Light fastness degree	Weather resistance degree (2000h)	Migration resistance degree	Heat resistance degree	
							180℃/30min	200℃/10min
PVC	Full shade	0.1%		7~8	4~5	5	5	5
	Reduction	0.2%	2%	6~7	2~3	5	5	5

Table 4.169 Application performance of Pigment Violet 37 in HDPE

Project		Pigments	Titanium dioxide	Light fastness degree	Weather resistance degree (3000h, full shade.2%)	Migration resistance degree
HDPE	Full shade	0.1%		8	4~5	4.9
	Reduction	0.1%	1%	7~8	2~3	4.9

Table 4.170 Applications of Pigment Violet 37

General plastics		Engineering plastics		Spinning	
LL/LDPE	●	PS/SAN	●	PP	●
HDPE	●	ABS	×	PET	×
PP	●	PC	×	PA	×
PVC(soft)	●	PET	○	PAN	
PVC(rigid)	●	PA6	×		
Rubber	●	POM			

● Recommended to use, ○ Conditional use, × Not recommended to use.

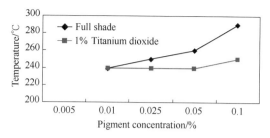

Figure 4.51　Heat resistance of Pigment Violet 37 in HDPE

Varieties characteristics　The overall performance of Pigment Violet 37 is slightly better than that of Pigment Violet 23, especially light fastness and migration resistance at the light color. Pigment Violet 37 can be used for coloring polyolefin, the heat-resistance temperature of 1/3SD polyolefin is up to 290℃. If the concentration of pigment is too low, the heat resistance will be sharply declined.

The Light fastness of Pigment Violet 37 in PE is excellent, the degree is up to eight, and the degree of reduction is 7～8. Pigment Violet 23 is suitable for coloring general purpose polyolefin plastics and general engineering plastics. For coloring flexible polyvinylchloride, the application of Pigment Violet 37 is better than that of Pigment Violet 23 and the migration resistance is excellent. Pigment Violet 37 is suitable for coloring polypropylene, polyester and fibers of polyamide 6 before spinning. The concentration can't be too low or there will be a chromatic aberration. And as the same of Pigment Violet 23, when Pigment Violet 37 is used in HDPE and other crystalline plastics, it can seriously affect the warpage and deformation of plastics.

4.7.2　Quinacridone Pigments

Quinacridone pigments are diketone derivatives of quinolin acridinium or quinacridinyl, their color area covered from bluish red to purplish red. The molecular weight of quinacridone pigments is small, the molecular structure is simple, but they have excellent heat resistance, light fastness, migration resistance, weather resistance and excellent overall performance at the low concentrations or reduction. The quinacridone pigments are difficult to disperse in plastics, therefore the use of an effective dispersing agent and sufficient shear during the manufacture of color masterbatch are essential. Otherwise incomplete dispersed pigments will produce visually disturbing color specks in the final product.

The common structure of quinacridone pigments is as follows. The varieties of quinacridone pigments used in plastic coloring are shown in Table 4.171.

Table 4.171 Varieties of quinacridone pigment used in plastic coloring

Species	CAS registry number	R_2	R_3	R_4	R_2^*	R_3^*	R_4^*	Shade	Crystal modification number
C.I. Pigment Violet 19	73900	H	H	H	H	H	H	Purplish—bluish red	5
C.I. Pigment Red 122	73915	CH_3	H	H	CH_3	H	H	pinkish red	4
C.I. Pigment Red 202	73907	Cl	H	H	Cl	H	H	Bluish red —purple	3

(1) **C.I. Pigment Violet** 19(β-modification) C.I. structure number: 73900. Molecular formula: $C_{20}H_{12}N_2O_2$. CAS registry number: [1047-16-1].

Structure formula

Color characterization Pigment Violet 19 (β-modification) is reddish—purple. The shade is very close to Pigment Red 88 (four chlorinated thioindigo pigment), but more vivid. The tinting strength of Pigment Violet 19 (β-modification) is high. When formulated with 1% titanium dioxide to make HDPE with 1/3 standard depth, the amount is 0.19%.

Pigment Violet 19 (β-modification) is rarely using alone, it is often mixed with other pigments. The mainly mixing inorganic pigments are iron oxide red or molybdate red. The mainly mixing organic pigments are Pigment Orange 36, after blending the resulting pigment exhibits a deep red. The light and weather resistance of mixing with iron oxide red or molybdate red are higher than that of pigment mixed with titanium dioxide.

Main properties As shown in Table 4.172~Table 4.174 and Figure 4.52.

Table 4.172 Application properties of Pigment Violet 19 (β-modification) in PVC

Project		Pigments	Titanium dioxide	Light fastness degree	Weather resistance degree (2000h)	Migration resistance degree	Heat resistance degree	
							180℃/30min	200℃/10min
PVC	Full shade	0.1%		7~8	4~5	5	5	5
	Reduction	0.2%	2%	7~8	4~5	5	5	5

Table 4.173　Application performance of Pigment Violet 19 (β-modification) in HDPE

Project		Pigments	Titanium dioxide	Light fastness degree	Weather resistance degree (3000h)	Migration resistance degree
HDPE	Full shade	0.1%		8	4	5
	Reduction	0.1%	1%	8	2	5

Table 4.174　Applications of Pigment Violet 19 (β-modification)

General plastics		Engineering plastics		Spinning	
LL/LDPE	●	PS/SAN	●	PP	●
HDPE	●	ABS	○	PET	○
PP	●	PC	×	PA	×
PVC (soft)	●	PET	○	PAN	
PVC (rigid)	●	PA	○		
Rubber	●	POM	○		

● Recommended, ○ Conditional use, × Not recommended to use.

Figure 4.52　Heat resistance of Pigment Violet 19 (β-modification) in HDPE (full shade)

Varieties characteristics　The light fastness and weather resistance of Pigment Violet 19 (β-modification) are much higher than those of Pigment Red 88, so Pigment Violet 19 (β-modification) can meet the requirements of long-term exposed outdoors. The heat resistance of Pigment Violet 19 (β-modification) is excellent and not relative to pigment concentration in a wide range. The comprehensive fastness of Pigment Violet 19 (β-modification) is excellent, and it is suitable for coloring general purpose polyolefin plastics and general engineering plastics. When it is used in flexible PVC, the bleeding resistance and durability are excellent. When Pigment Violet 19(β-modification) is used in HDPE and other crystalline plastics, it can make a medium warpage and deformation of plastics.

(2) **C.I. Pigment Violet** 19(γ-modification)　C.I. structure number: 73900. Molecular formula: $C_{20}H_{12}N_2O_2$. CAS registry number: [1047-16-1].

Structure formula

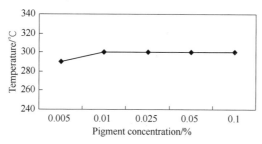

Color characterization Pigment Violet 19 (γ-modification) is bluish red. The Pigment Violet 19 (γ-modification) has a yellower shade than Pigment Violet 19 (β-modification). The commercial Pigment Violet 19(γ-modification) has a wide particle size distribution, and the distribution has a great impact on the coloring performance. If the average particle size is small, it will hare a high tinting strength, bright and bluish shade, but its light and weather resistance is relatively poor. If the average particle size is large, it will have a low tinting strength, good relatively light fastness and weather resistance.

The tinting strength of Pigment Violet 19(γ-modification) is high, when formulated with 1% titanium dioxide to make HDPE with 1/3 standard depth, the amount is only 0.27% (small particle).

Main properties As shown in Table 4.175～Table 4.180 and Figure 4.53; Figure 4.54.

Table 4.175 Application properties of Pigment Violet 19(γ-modification, small particle size) in PVC

Project		Pigments	Titanium dioxide	Light fastness degree	Weather resistance degree (3000h)	Migration resistance degree
PVC	Full shade	0.1%		7～8	4～5	5
	Reduction	0.2%	2%	8		

Table 4.176 Application performance of Pigment Violet 19(γ-modification, small particle size) in HDPE

Project		Pigments	Titanium dioxide	Light fastness degree	Weather resistance degree (3000h, full shade 10.2%)
HDPE	Full shade	0.27%		8	4～5
	1/3SD	0.27%	1%	8	

Table 4.177 Applications of Pigment Violet 19(γ-modification, small particle size)

General plastics		Engineering plastics		Spinning	
LL/LDPE	●	PS/SAN	○	PP	●
HDPE	●	ABS	●	PET	○
PP	●	PC	●	PA6	○
PVC(soft)	●	PBT	●	PAN	
PVC(rigid)	●	PA	○		
Rubber	●	POM	○		

● Recommended to use, ○ Conditional use.

Table 4.178 Application performance of Pigment Violet 19(γ-modification, large particle size) in PVC

Project		Pigments	Titanium dioxide	Light fastness degree	Weather resistance degree (3000h)	Migration resistance degree
PVC	Full shade	0.1%		8	5	5
	Reduction	0.1%	0.5%	8		

Table 4.179 Application performance of Pigment Violet 19(γ-modification, large particle size) in HDPE

Project		Pigments	Titanium dioxide	Light fastness degree	Weather resistance degree (3000h, Natural0.2%)
HDPE	Full shade	0.3%		8	4～5
	1/3SD	0.3%	1%	8	

Table 4.180 Applications of Pigment Violet 19(γ-modification, large particle size)

General plastics		Engineering plastics		Spinning	
LL/LDPE	●	PS/SAN	●	PP	●
HDPE	●	ABS	●	PET	○
PP	●	PC	●	PA6	○
PVC(soft)	●	PBT	●	PAN	
PVC(rigid)	●	PA	○		
Rubber	●	POM	○		

● Recommended to use, ○ Conditional use.

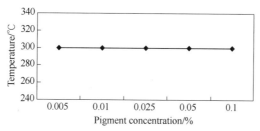

Figure 4.53 Heat resistance of Pigment Violet 19(γ-modification, small particle size) in HDPE (full shade)

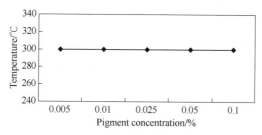

Figure 4.54 Heat resistance of Pigment Violet 19(γ-modification, large particle size) in HDPE (full shade)

Varieties characteristics The light fastness and weather resistance of Pigment Violet 19 (γ-modification) are better than those of Pigment Violet 19 (β-modification), so Pigment Violet 19 (γ-modification) can meet the requirements of long-term exposed outdoors.

For colored polyester, Pigment Violet 19 (γ-modification) can withstand a high temperature of 240～290℃ for 5～6h, even in the processing temperature of 320℃ in polycarbonate. As for Pigment Violet 19(γ-modification), the comprehensive fastness is excellent and it is suitable for coloring general purpose polyolefin plastics. When it is used in flexible PVC, the bleeding resistance and durability are excellent. Pigment Violet 19 (γ-modification) is suitable for coloring general engineering plastics. Pigment Violet

19 (γ-modification) can be used for coloring the polyester, but there is a chromatic aberration often appear when used at low concentrations. That is the reason that the Pigment Violet 19 (γ-modification) is partially dissolved in the polyester. Pigment Violet 19 (γ-modification) is also applicable for the injection molding and extrusion molding of nylon, not only because it is resistant to high processing temperature of nylon, but also because it is entirely chemically inert in nylon. Pigment Violet 19 (γ-modification) is suitable for polypropylene, polyester, nylon 6 fiber before spinning and spinning. When Pigment Violet 19(γ-modification) is used in HDPE and other crystalline plastics, it can make a medium warpage and deformation of plastics.

(3) **C.I. Pigment Red 122** C.I. structure number: 73915. Formula: $C_{22}H_{16}N_2O_2$. CAS registry number: [980-26-7]/[16043-40-6].

Structure formula

Color characterization Pigment Red 122 is a very bright bluish red pigment, and the shade is close to magenta. The tinting strength is higher than Pigment Puple19 (γ-modification). For example, to modulate the same chrominance sample, the amount of Pigment Red 122 is 80% of γ-modification Pigment Puple19.

Main properties Shown in Table 4.181~Table 4.183 and Figure 4.55.

Table 4.181 Application properties of Pigment Red 122 in PVC

Project		Pigment	Titanium dioxide	Light fastness degree	Weather resistance degree (3000h)	Migration fastness degree
PVC	Full shade	0.1%		8	5	5
	Reduction	0.1%	0.5%	8		

Table 4.182 Application properties of Pigment Red 122 in HDPE

Project		Pigment	Titanium dioxide	Light fastness degree	Weather resistance degree (3000h, full shade 0.2%)
HDPE	Full shade	0.22 %		8	5
	1/3 SD	0.22 %	1%	8	

Table 4.183 Application range of Pigment Red 122

General plastics		Engineering plastics		Spinning	
LL/LDPE	●	PS/SAN	●	PP	●
HDPE	●	ABS	●	PET	○
PP	●	PC	●	PA6	○
PVC (soft)	●	PBT	●	PAN	
PVC (rigid)	●	PA	○		
Rubber	●	POM	○		

● Recommended to use, ○ Conditional use, × Not recommended to use.

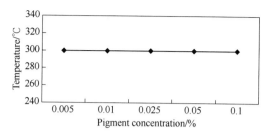

Figure 4.55　Heat resistance of Pigment Red 122 in HDPE (full shade)

Variety characteristics　The chemical structure of the Pigment Red 122 is 2, 9 dimethyl quinacridone. Therefore, compared with un-substituted quinacridone pigment puple19 (γ-modification), the light fastness and weather resistance is better. Pigment Red 122 has excellent comprehensive fastness, and it is suitable for coloring general polyolefin and engineering plastics. However, when used at low concentration, it often occur color difference and light fastness decrease, for micro dissolution of Pigment Red 122 in polymer

Pigment Red 122 is suitable for the coloring of polypropylene before spinning and during spinning. It is cautious recommended when used in polyester and nylon 6. Pigment Red 122 can cause moderate warpage when used in partially crystalline HDPE plastics. What's more, it can become the standard color in bluish red zone.

(4) **C.I. Pigment Red 202**　C.I. structure number: 73907. Formula: $C_{20}H_{10}N_2O_2$. CAS registry number: [3089-17-6].

Structure formula

Color characterization　Pigment Red 202 is binary mixtures, the main ingredient is 2, 9 dichloro quinacridone. And the color is more blue and darker than Pigment Red 122. Pigment Red 202 has a good tinting strength, only need 0.20% pigment to prepare 1/3 SD of flexible PVC with 1% titanium dioxide, and 0.26% pigment to prepare 1/3 SD of HDPE with 1% titanium dioxide.

Main properties　Shown in Table 4.184～Table 4.186 and Figure 4.56.

Table 4.184　Application properties of Pigment Red 202 in PVC

Project		Pigment	Titanium dioxide	Light fastness degree	Weather resistance degree (3000h)	Migration fastness degree
PVC	Full shade	0.1%		8	5	5
	Reduction	0.2%	2%	8	4～5	5

Table 4.185 Application properties of Pigment Red 202 in HDPE

Project		Pigment	Titanium dioxide	Light fastness degree	Weather resistance degree (3000h)
HDPE	Full shade	0.1 %		8	4
	Reduction	0.1%	1%	7~8	3

Table 4.186 Application range of Pigment Red 202

General plastics		Engineering plastics		Spinning	
LL/LDPE	●	PS/SAN	●	PP	●
HDPE	●	ABS	●	PET	●
PP	●	PC	×	PA6	●
PVC (soft)	●	PET	●	PAN	
PVC (rigid)	●	PA	○		
Rubber	●	POM	○		

● Recommended to use, ○ Conditional use, × Not recommended to use.

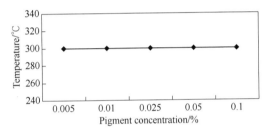

Figure 4.56 Heat resistance of Pigment Red 202 in HDPE (full shade)

Variety characteristics Compared with Pigment Red 122, Pigment Red 202 has excellent comprehensive fastness, and it is suitable for coloring of polyolefins and general engineering plastics. It is also suitable for coloring of PP, polyester and nylon 6. What's more, it can cause slight warpage when used in partially crystalline HDPE plastics.

4.7.3 Perylene and Pyrene Ketone

The name of perylene pigments which treats tetracarboxydiimide as parent structure comes from *N*, *N*-disubstituted 3,4,9,10-tertracarboxylic acid diimide derivatives or 3,4,9,10-perylenetetracarboxylic dianhydride. The pigments had been found early in 1931, mainly used as dyestuff. Based on the evident planarity and symmetry of molecule structure, some species can form intermolecular hydrogen bonds. This kind of pigment has excellent heat resistance and weather resistance. Therefore, perylene pigments are a class of important species, widely used in plastics and synthetic liquid coloring.

Perylene pigments have wide color regions. The color regions of perylene pigments range wide from orange to bluish red spectrum with high covering power and good transparency.

General structure formula of perylene pigments:

The main varieties of perylene pigments applied in plastics are shown in Table 4.187.

Table 4.187 The main varieties of Perylene pigments applied in plastics

Variety	C.I. structure No.	CAS No.	X	Shade
Pigment Red149	71137	[4948-15-6]	N with two CH$_3$ groups	Red
Pigment Red 178	71155	[3049-71-6]	N—N=N—	Red
Pigment Red 179	71130	[5521-31-3]	N—CH$_3$	Red to Reddish Purple
Pigment Purple 29	71129	[81-33-4]	HN	Red to Purplish Red

When perylene pigments are used in polyolefin with hindered amine light stabilizer, problems may occur. With the concentration of perylene pigments changed, hindered amine light stabilizers will lose their effect. As a result, it can't meet the requirements of plastics. If it must be used in a recipe, it is recommended to assess the evaluation before selection.

(1) **C.I. Pigment Red 149** C.I. structure number: 71137. Formula: 71137. CAS registry number: [4948-15-6].

Structure formula

Color characterization Pigment Red 149 is a bright positive red pigment. It has good tinting strength, only need 0.12% pigment to prepare 1/3 SD of flexible PVC with 1% titanium dioxide, and 0.15% pigment to prepare 1/3 SD of HDPE with 1% titanium dioxide.

Main properties Shown in Table 4.188~Table 4.190 and Figure 4.57.

Table 4.188 Application properties of Pigment Red 149 in PVC

	Project	Pigment	Titanium dioxide	Light fastness degree	Weather resistance degree (5000h)	Migration fastness degree
PVC	Full shade	0.1%		8	3~4	5
	Reduction	0.1%	1%	7~8		

Chapter 4 Main Types and Properties of Organic Pigments

Table 4.189 Application properties of Pigment Red 149 in HDPE

Project		Pigment	Titanium dioxide	Light fastness degree	Weather resistance degree (3000h)
HDPE	Full shade	0.1%		8	4
	Reduction	0.1%	1%	8	3

Table 4.190 Application range of Pigment Red149

General plastics		Engineering plastics		Spinning	
LL/LDPE	●	PS/SAN	●	PP	●
HDPE	●	ABS	○	PET	○
PP	●	PC	○	PA6	●
PVC (soft)	●	PET	○	PAN	
PVC (rigid)	●	PA6	○		
Rubber	●	PMMA	●		

● Recommended to use, ○ Conditional use.

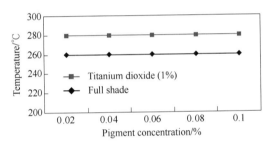

Figure 4.57 Heat resistance of Pigment Red 149 in HDPE

Variety characteristics Because the melting point is higher than 450℃, Pigment Red 149 has particularly good heat resistance which is independent to the concentration over a wide range. After being diluted with titanium dioxide, the heat resistance is greater than quinacridone pigment red. Pigment Red 149 has excellent light fastness and weather resistance. But the weather resistance will destroy light stabilizers of hindered amine series in plastics. Pigment Red 149 has excellent comprehensive fastness, and it is suitable for coloring general polyolefin and engineering plastics.

Pigment Red 149 is suitable for the coloring of polypropylene and polyester. It should be noted that Pigment Red 149 will be dissolved into polyester to change the color from red to orange when spinning in low concentration. Therefore, it is recommended for nylon 6. Pigment Red 149 can cause warpage of HDPE plastic product with injection molding, and the influence descends with temperature increasing.

(2) **C.I. Pigment Red 178** C.I. structure number: 71155. Formula: $C_{48}H_{26}N_6O_4$. CAS registry number: [3049-71-6].

Structure formula

[Structure: phenyl-N=N-phenyl-N(C=O)(C=O)-perylene-N(C=O)(C=O)-phenyl-N=N-phenyl]

Color characterization Pigment Red 178 is slightly dark yellow, it has general tinting strength, and needs 0.18% pigment to prepare 1/3 SD of flexible PVC with 1% titanium dioxide, and 0.26% pigment to prepare 1/3 SD of HDPE with 1% titanium dioxide.

Main properties Shown in Table 4.191~Table 4.193 and Figure 4.58.

Table 4.191　Application properties of Pigment Red 178 in PVC

	Project	Pigment	Titanium dioxide	Light fastness degree	Migration fastness degree
PVC	Full shade	0.1%		8	5
	Reduction	0.1%	1%	7	

Table 4.192　Application properties of Pigment Red 178 in HDPE

	Project	Pigment	Titanium dioxide	Light fastness degree	Weather resistance degree (3000h)	Migration fastness degree
HDPE	Full shade	0.1%		8	3~4	5
	Reduction	0.1%	1%	7		

Table 4.193　Application range of Pigment Red 178

General plastics		Engineering plastics		Spinning	
LL/LLDPE	●	PS/SAN	●	PP	●
HDPE	●	ABS	○	PET	○
PP	●	PC	○	PA	○
PVC (soft)	●	PET	●	PAN	
PVC (rigid)	●	PA6	○		
Rubber	●	PMMA	●		

● Recommended to use, ○ Conditional use.

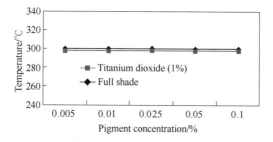

Figure 4.58　Heat resistance of Pigment Red 178 in HDPE

Variety characteristics The light fastness and weather resistance of Pigment Red

178 is worse than Pigment Red 149, but equal to or slightly higher than the Pigment Red 122. The degree of the light fastness is up to 8 in the 0.1% full shade of HDPE. Pigment Red 178 has excellent comprehensive fastness, and it is suitable for coloring general polyolefin and engineering plastics polypropylene before spinning and during spinning. When using in partially crystalline plastics, it causes warpage most serious in the perylene pigments.

(3) **C.I. Pigment Red 179** C.I. structure number: 71130. Formula: $C_{26}H_{14}N_2O_4$. CAS registry number: [5521-31-3].

Structure formula

Color characterization Pigment Red 179 has a variety of different shade, its early variety is jujube red, and its recent variety is bright reddish. Pigment Red 179 has good tinting strength, only need 0.13% pigment to prepare 1/3 SD of flexible PVC with 1% titanium dioxide, and 0.17% pigment to prepare 1/3 SD of HDPE with 1% titanium dioxide.

Main properties Shown in Table 4.194~Table 4.196 and Figure 4.59.

Table 4.194 Application properties of Pigment Red 179 in PVC

Project		Pigment	Titanium dioxide	Light fastness degree	Weather resistance degree (5000h)	Migration fastness degree
PVC	Full shade	0.1%		8	5	5
	Reduction	0.1%	1%	8	4~5	

Table 4.195 Application properties of Pigment Red 179 in HDPE

Project		Pigment	Titanium dioxide	Light fastness degree	Weather resistance degree (3000h)	Migration fastness degree
HDPE	Full shade	0.1%		8	5	5
	Reduction	0.1%	1%	8	4	

Table 4.196 Application range of Pigment Red 179

General plastics		Engineering plastics		Spinning	
LL/LDPE	●	PS/SAN	●	PP	●
HDPE	●	ABS	○	PET	○
PP	●	PC	○	PA	●
PVC (soft)	●	PBT	○	PAN	
PVC (rigid)	●	PA6	○		
Rubber	●	PMMA	○		

● Recommended to use, ○ Conditional use.

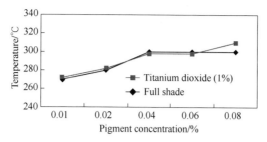

Figure 4.59　Heat resistance of Pigment Red 179 in HDPE

Variety characteristics　Pigment Red 179 has excellent light fastness and weather resistance, equal to or higher than Pigment Red 122. The heat-resistance of Pigment Red 179 is high, while is less than other perylene pigments.

Pigment Red 179 has excellent comprehensive fastness, and it is suitable for coloring general polyolefin and engineering plastics. It is cautious recommended for nylon 6. It will make chromatic aberration due to the reduction of nylon melt. What's more, Pigment Red 179 can cause moderate warpage when it is used in partially crystalline plastics.

(4) **C.I. Pigment Purple 29**　C.I. structure number: 71129. Formula: $C_{24}H_{10}Cl_2O_2$. CAS registry number：[81-33-4].

Structure formula

Color characterization　Pigment Purple 29 is dark reddish purple, and the saturation is not high. The tinting strength is high, only need 0.18% pigment to prepare 1/3 SD of HDPE with 1% titanium dioxide, higher than the carbazole two Oxazine Purple 23 of 0.09%, lower than the Quinacridone Purple 19 of 0.23%.

Main properties　As shown in Table 4.197~Table 4.199 and Figure 4.60.

Table 4.197　Application properties of Pigment Purple 29 in PVC

Project		Pigment	Titanium dioxide	Light fastness degree	Weather resistance degree (3000h)	Migration fastness degree
PVC	Full shade	0.1%		8	5	5
	Reduction	0.1%	1%	8	3~4	

Table 4.198　Application properties of Pigment Purple 179 in HDPE

Project		Pigment	Titanium dioxide	Light fastness degree	Weather resistance degree (3000h)
HDPE	Full shade	0.2%		8	4
	Reduction	0.2%	2%	8	3

Table 4.199 Application range of Pigment Purple 29

General plastics		Engineering plastics		Spinning	
LL/LDPE	●	PS/SAN	●	PP	●
HDPE	●	ABS	○	PET	●
PP	●	PC	○	PA6	●
PVC (soft)	●	PBT	○	PAN	
PVC (rigid)	●	PA	○		
Rubber	●	PMMA	●		

● Recommended to use, ○ Conditional use.

Figure 4.60 Heat resistance of Pigment purple 29 in HDPE

Variety characteristics Pigment Purple 29 has excellent comprehensive fastness, and it is suitable for coloring general polyolefin and engineering plastics in outdoor, and suitable for coloring of polypropylene, polyester and nylon 6. When it is used for coloring the polyester condensation, Pigment Purple 29 can resistant the temperature of 290℃ for 5～6 hours.

(5) **C.I. Pigment orange 43**   C.I. structure number: 71105. Formula: $C_{26}H_{12}N_4O_2$. CAS registry number: [4424-06-07].

Structure formula

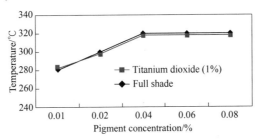

Color characterization The chemical structure of Pigment Orange 43 is trans-form of pi ketone pigment, showing bright reddish orange. Pigment Orange 43 has good tinting strength, only need 0.9% pigment to prepare 1/3 SD of flexible PVC with 5% titanium dioxide, 0.25% with 1% titanium dioxide.

Main properties As shown in Table 4.200～Table 4.202 and Figure 4.61.

Table 4.200 Application properties of Pigment orange 43 in PVC

Project		Pigment	Titanium dioxide	Light fastness degree	Weather resistance degree (2000h)	Migration fastness degree
PVC	Full shade	0.1%		8	4	4～5
	Reduction	0.1%	0.5%	7～8		

Table 4.201　Application properties of Pigment Orange 43 in HDPE

Project		Pigment	Titanium dioxide	Light fastness degree	Weather resistance degree (3000h)
HDPE	Full shade	0.2%		8	5
	1/3 SD	0.2%	1%	8	

Table 4.202　Application range of Pigment Orange 43

General plastics		Engineering plastics		Spinning	
LL/LDPE	●	PS/SAN	●	PP	●
HDPE	●	ABS	●	PET	×
PP	●	PC	●	PA6	×
PVC (soft)	●	PBT	●	PAN	●
PVC (rigid)	●	PA	○		
Rubber	●	POM	×		

● Recommended to use, ○ Conditional use, × Not recommended to use.

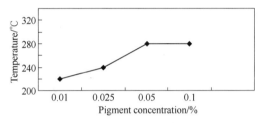

Figure 4.61　Heat resistance of Pigment Orange 43 in HDPE (full shade)

Variety characteristics　Pigment Orange 43 has excellent light fastness which can be up to degree eight even diluted to a low concentration. It has good heat resistance when used for polyolefin, the heat-resistance temperature is about 280 ℃ in PE and PET. However the heat resistance significantly decreases when the concentration is below 0.1% in PE. It is also dissolve and its color change to yellow with low concentration in PET. Pigment Orange 43 is suitable for coloring general polyolefin and engineering plastics in outdoor. For PVC, The permeability resistance is great. But the bleeding phenomenon occurs in the pigment with low concentration and plasticizer with high concentration. Pigment Orange 43 is suitable for coloring of polypropylene. Pigment orange 43 can cause serious warpage of crystal plastic products.

4.7.4　Anthraquinone and Anthraquinone Ketone Pigments

Anthraquinones and anthraquinone ketone pigments are vat dyes with high molecular weight and complex structure. After specific surface treatment, the pigments are converted into valuable organic pigments. This kind of pigments have high transparency, medium to high tinting strength, color area covering yellow, bluish red and reddish blue. The pigments also have excellent comprehensive performance.

(1) **C.I. Pigment Yellow 199** C.I. structure number: 65320. Formula: $C_{52}H_{60}N_2O_6$.

Structure formula

Color characterization excellent transparency, reddish yellow varieties.

Main properties As shown in Table 4.203～Table 4.205 and Figure 4.62.

Table 4.203 Application properties of Pigment Yellow 199 in PVC

Project	Pigment	Titanium dioxide	Light fastness degree	Weather resistance degree (2000h)	Migration fastness degree	Heat resistance	
						180℃ /30min	200℃ /10min
PVC	Not suitable for PVC because of poor dispersion						

Table 4.204 Application properties of Pigment Yellow 199 in HDPE

Project		Pigment	Titanium dioxide	Light fastness degree	Weather resistance degree (3000h)	Migration fastness degree
HDPE	Full shade	0.1%		7~8	4~5	4.2
	Reduction	0.1%	1%	7~8	3~4	4.2

Table 4.205 Application range of Pigment Yellow 199

General plastics		Engineering plastics		Spinning	
LL/LDPE	●	PS/SAN	●	PP	●
HDPE	●	ABS	○	PET	○
PP	●	PC	○	PA6	×
PVC (soft)	×	PBT	○	PAN	
PVC (rigid)	×	PA6	×		
Rubber	●	POM	●		

● Recommended to use, ○ Conditional use, × Not recommended to use.

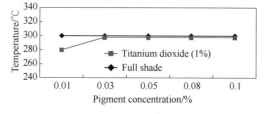

Figure 4.62 Heat resistance of Pigment Yellow 199 in HDPE

Variety characteristics Pigment Yellow 199 shows excellent light fastness and meets the requirements for long-term outdoor exposure. Owing to the good heat resistance, in addition to the coloring of general polyolefins, it can also be used in the coloring of polystyrene based engineering plastics. Pigment Yellow 199 is suitable for the coloring of polypropylene before spinning and during spinning, and plastics that contact with foodstuff. When used for crystal HDPE plastics, it will not cause warpage.

(2) **C.I. Pigment Red 177**　C.I. structure number: 65300. Formula: $C_{28}H_{16}N_2O_4$. CAS registry number: [4051-63-2].

Structure formula

Color characterization　Pigment Red 177 is blue light red with good transparency. When matched with molybdate red pigment, the color will be bright deep red, and present excellent light fastness and weather resistance. Tinting strength of Pigment Red 177 is not good, need 0.3% pigment to prepare 1/3 SD of HDPE with 1% titanium dioxide.

Main properties　As shown in Table 4.206~Table 4.208 and Figure 4.63.

Table 4.206　Application properties of Pigment Red 177 in PVC

Project		Pigment	Titanium dioxide	Light fastness degree	Weather resistance degree (2000h)	Migration fastness degree	Heat resistance	
							180℃/30min	200℃/10min
PVC	Full shade	0.1%		8	4~5	5	5	4~5
	Reduction	0.2%	2%	7~8	2	5	5	5

Table 4.207　Application properties of Pigment Red 177 in HDPE

Project		Pigment	Titanium dioxide	Light fastness degree	Weather resistance degree (3000h)	Migration fastness degree
HDPE	Full shade	0.1%		7~8	3	5
	Reduction	0.1%	1%	7~8	1	5

Table 4.208　Application range of Pigment Red 177

General plastics		Engineering plastics		Spinning	
LL/LDPE	●	PS/SAN	○	PP	●
HDPE	●	ABS	○	PET	×
PP	●	PC	○	PA6	×
PVC (soft)	●	PBT	○	PAN	
PVC (rigid)	●	PA6	◐		
Rubber	●	PMMA	●		

● Recommended to use, ○ Conditional use, × Not recommended to use.

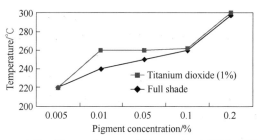

Figure 4.63　Heat resistance of Pigment Red 177 in HDPE

Variety characteristics　Pigment Red 177 has excellent heat resistance, its heat-resistance temperature is 270~280℃ in PE. The heat resistance significantly decreases when the concentration is below 0.1%. Pigment Red 177 is suitable for coloring general polyolefin plastics, and the permeability resistance is poor for flexible PVC. Besides, it is suitable for the coloring of polypropylene and nylon 6. Owing to no nucleation, Pigment Red 177 will not cause warpage when used for crystal plastics such as HDPE.

(3) **C.I. Pigment blue 60**　C.I. structure number: 69800. Formula: $C_{28}H_{14}N_2O_4$. CAS registry number: [81-77-6].

Structure formula

Color characterization　Pigment Blue 60 is pure reddish blue, and its shade is more reddish than α-phthalocyanine blue. Pigment Blue 60 has good tinting strength, needs 0.15% pigment to prepare 1/3 SD of HDPE with 1% titanium dioxide, its concentration is higher than phthalocyanine blue of 0.08%.

Main properties　As shown in Table 4.209~Table 4.211 and Figure 4.64.

Table 4.209　Application properties of Pigment Blue 60 in PVC

Project		Pigment	Titanium dioxide	Light fastness degree	Weather resistance degree (2000h)	Migration fastness degree	Heat resistance	
							180℃ /30min	200℃ /10min
PVC	Full shade	0.1%		7~8	4~5	5	5	5
	Reduction	0.2%	2%	6~7	3~4	5	5	5

Table 4.210　Application properties of Pigment Blue 60 in HDPE

Project		Pigment	Titanium dioxide	Light fastness degree	Weather resistance degree (3000h)	Migration fastness degree
HDPE	Full shade	0.1%		7~8	4~5	5
	Reduction	0.1%	1%	8	4	5

Table 4.211 Application range of Pigment Blue 60

General plastics		Engineering plastics		Spinning	
LL/LDPE	●	PS/SAN	●	PP	●
HDPE	●	ABS	○	PET	●
PP	●	PC	○	PA6	●
PVC (soft)	●	PBT	●	PAN	
PVC (rigid)	●	PA6	○		
Rubber	●	PMMA	●		

● Recommended to use, ○ Conditional use.

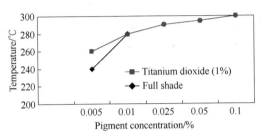

Figure 4.64 Heat resistance of Pigment Blue 60 in HDPE

Variety characteristics Pigment Blue 60 has excellent comprehensive fastness, and it is suitable for coloring general polyolefin and engineering plastics, polypropylene, polyester and nylon 6 before spinning. It will not cause warpage when used for crystal plastics such as HDPE.

4.7.5 Isoindolinone and Isoindoline Pigments

The molecule structure of isoindolinone pigments is composed of two parts: two four-chlorosubstituted isoindoline rings and two diamine substituted derivatives. The Pigment Yellow 109 and 110 can be obtained through changing the structure of substituted derivatives of diamine. Isoindolinone and isoindoline pigments were advanced organic pigments which developed in the middle 1960s after quinacridone and dioxazine pigments. The pigments have excellent heat resistance, migration and light fastness, especially in low concentration and the reduction of titanium dioxide

The general structure formula of isoindolinone pigments:

The main varieties of isoindolinone and isoindoline pigments applied in plastics are shown in Table 4.212.

Table 4.212 Main varieties of isoindolinone and isoindoline pigments applied in plastics

Variety	C.I.number	CAS	R	Shade
Pigment Yellow 109	56284	[5045-40-9]	CH$_3$-phenyl	Greenish yellow
Pigment Yellow 110	56280	[5590-18-1]	phenyl	Reddish yellow
Pigment Orange 61	11265	[40716-47-0]	CH$_3$-phenyl-N=N-phenyl	Orange

Isoindolinone pigments are another class of practical value organic pigments whose molecular structure has isoindolinone ring and tautomeric form. The general structure formula as follow:

$$R_1, R_2, R_3, R_4 \text{ substituted isoindoline structure}$$

Based on the substituents of R_1 and R_2, the pigments can be divided into structure symmetric pigments (such as Pigment Yellow 139 and Pigment Brown 138) and structure asymmetric pigments. Owing to the carbonyl and imine groups of molecule to form hydrogen bonds, isoindoline pigments have excellent light fastness and weather resistance.

(1) Isoindolinone pigments

① **C.I. Pigment Yellow 109**   C.I. structure number: 56284. Formula: $C_{23}H_8Cl_8N_4O_2$. CAS registry number: [5045-40-9].

Structure formula

Color characterization   Pigment Yellow 109 is bright greenish yellow, and the tinting strength is low, needs 0.4% pigment to prepare 1/3 SD of HDPE with 1% titanium dioxide.

Main properties   As shown in Table 4.213～Table 4.215 and Figure 4.65.

Table 4.213 Application properties of Pigment Yellow 109 in PVC

Project		Pigment	Titanium dioxide	Light fastness degree	Weather resistance degree (2000h)	Migration fastness	Heat resistance	
							180℃/30min	200℃/10min
PVC	Full shade	0.1%		8	5	5	5	4～5
	Reduction	0.2%	2%	8	4～5	5	5	5

Table 4.214 Application properties of Pigment Yellow 109 in HDPE

Project		Pigment	Titanium dioxide	Light fastness degree	Weather resistance degree (3000h)	Migration fastness degree
HDPE	Full shade	0.1%		7~8	4	5
	Reduction	0.1%	1%	7~8	2~3	5

Table 4.215 Application range of Pigment Yellow 109

General plastics		Engineering plastics		Spinning	
LL/LDPE	●	PS/SAN	●	PP	●
HDPE	●	ABS	○	PET	×
PP	●	PC	○	PA6	×
PVC (soft)	○	PET	○	PAN	×
PVC (rigid)	○	PA6	○		
Rubber	●	POM			

● Recommended to use, ○ Conditional use, × Not recommended to use.

Figure 4.65 Heat resistance of Pigment Yellow 109 in HDPE

Variety characteristics Pigment Yellow 109 has excellent light fastness, but it can't bear exposure outdoor for a long time when used in rigid PVC, this phenomenon is particularly evident under high content of titanium dioxide. Pigment Yellow 109 also has excellent heat resistance. When used for coloring of polyolefin, the heat-resistance temperature of 1/3 SD sample with 1% titanium dioxide is up to 300℃, but the 1/25 SD sample is reduced to 250℃. In high temperature, the shade will be darken. Pigment Yellow 109 is suitable for coloring general polyolefin and engineering plastics, polypropylene. It causes warpage when used for HDPE.

② **C.I. Pigment Yellow 110** C.I. structure number: 56280. Formula: $C_{22}H_6C_{18}N_4O_2$. CAS registry number: [5590-18-1].

Structure formula

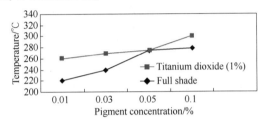

Color characterization Pigment Yellow 110 is strong reddish yellow, and The tinting strength is medium, needs 1.4%~1.9% pigment to prepare 1/3 SD of PVC with 5% titanium dioxide.

Main properties As shown in Table 4.216~Table 4.218 and Figure 4.66.

Table 4.216 Application properties of Pigment Yellow 110 in PVC

Project		Pigment	Titanium dioxide	Light fastness degree	Weather resistance degree (2000h)	Migration fastness degree	Heat resistance	
							180℃ /30min	200℃ /10min
PVC	Full shade	0.1%		8	4~5	5	5	5
	Reduction	0.2%	2%	8	4~5	5	5	5

Table 4.217 Application properties of Pigment Yellow 110 in HDPE

Project		Pigment	Titanium dioxide	Light fastness degree	Weather resistance degree (3000h)	Migration fastness degree
HDPE	Full shade	0.1%		7~8	4~5	5
	Reduction	0.1%	1%	8	4	5

Table 4.218 Application range of Pigment Yellow 110

General plastics		Engineering plastics		Spinning	
LL/LDPE	●	PS/SAN	●	PP	●
HDPE	●	ABS	●	PET	×
PP	●	PC	×	PA6	×
PVC (soft)	●	PBT	×	PAN	×
PVC (rigid)	●	PA6	×		
Rubber	●	POM	●		

● Recommended to use, × Not recommended to use.

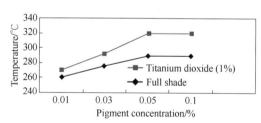

Figure 4.66 Heat resistance of Pigment Yellow 110 in HDPE

Variety characteristics In terms of the comprehensive fastness, the Pigment Yellow 110 isconsidered to be the best one in the reddish yellow organic pigments. And in terms of the light fastness and weather resistancethe, it is one of the best varieties in the yellow organic pigments. When used in the rigid PVC, it can resist outdoor exposure for long time. Therefore, it is widely used. Pigment Yellow 110 also has excellent heat resistance. When used for coloring of polyolefin, the heat-resistance temperature of 1/3 SD HDPE is up to 290℃, and the 1/25 SD HDPE is 270℃ without shade changes. The heat resistance of low concentration should be noticed.

Besides for coloring general polyolefin, Pigment Yellow 110 is suitable for polystyrene-based engineering plastics, such as PS and polypropylene before spinning and during spinning. It causes serious warpage when used for crystal plastics such as HDPE. As the acicular crystal, Pigment Yellow 110 is very sensitive to shear stress in process.

③ **C.I. Pigment Orange 61**   C.I. structure number: 11265. Formula: $C_{29}H_{12}Cl_8N_6O_2$. CAS registry number: [40716-47-0].

Structure formula

Color characterization   Pigment Orange 61 is yellowish orange and the color saturation is not high. The tinting strength is low, needs 1.4% pigment to prepare 1/3 SD of flexible PVC with 5% titanium dioxide. The shade is similar to Pigment Orange 13, but the tinting strength is just 1/2 of Orange Pigment 13.

Pigment Orange 61 has excellent comprehensive fastness, The light fastness can reach to degree of 7~8 in 1/25 SD PVC with 5% titanium dioxide.

Variety Characteristics   Pigment Orange 61 has excellent heat resistance. Besides for coloring general polyolefin, it is suitable for phenethylene-based engineering plastics, polyurethane and unsaturated polyester plastics. Pigment Orange 61 is suitable for the coloring of polypropylene and acrylic fiber before spinning and during spinning. Besides, it causes warpage when used for crystal plastics such as HDPE.

Main properties   As shown in Table 4.219~Table 4.221 and Figure 4.67.

Table 4.219   Application properties of Pigment Orange 61 in PVC

	Project	Pigment	Titanium dioxide	Light fastness degree	Weather resistance degree (2000h)	Migration fastness degree	Heat resistance degree 180℃ /30min	Heat resistance degree 200℃ /10min
PVC	Full shade	0.1%		8	4~5	4.9	5	5
	Reduction	0.2%	2%	7~8	4~5	4.8	5	5

Table 4.220   Application properties of Pigment Orange 61 in HDPE

	Project	Pigment	Titanium dioxide	Light fastness degree	Weather resistance degree (3000h)	Migration fastness degree
HDPE	Full shade	0.1%		7~8	4~5	5
	Reduction	0.1%	1%	7~8	3	5

Table 4.221 Application range of Pigment Orange 61

General plastics		Engineering plastics		Spinning	
LL/LDPE	●	PS/SAN	●	PP	●
HDPE	●	ABS	○	PET	×
PP	●	PC	×	PA6	×
PVC (soft)	●	PET	×	PAN	
PVC (rigid)	●	PA6	×		
Rubber	●	POM	●		

● Recommended to use, ○ Conditional use, × Not recommended to use.

Figure 4.67 Heat resistance of Pigment Orange 61 in HDPE

(2) Isoindolinne pigments

C.I. Pigment Yellow 139   C.I. Structure Number: 56298. Molecular formula: $C_{16}H_9N_5O_6$. CAS registry number: [36888-99-0].

Structure formula

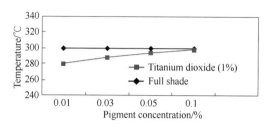

Color characterization   Pigment Yellow 139 is reddish yellow. Pigment Yellow 139 has high tinting strength, and it needs 1% pigment to formulate flexible PVC of 1/3 standard depth with 5% titanium dioxide, as well as 0.2% pigment to formulate HDPE of 1/3 standard depth with 1% titanium dioxide.

Main properties   As shown in Table 4.222～Table 4.224 and Figure 4.68.

Table 4.222 Application performance of Pigment Yellow 139 in PVC

Project		Pigment	Titanium dioxide	Light fastness degree	Weather resistance degree (2000h)	Migration resistance degree	Heat resistance degree	
							180℃/30min	200℃/10min
PVC	Full shade	0.1%		7～8	4	5	4～5	4～5
	Reduction	0.2%	2%	7	3	5	4～5	4～5

Table 4.223 Application performance of Pigment Yellow 139 in HDPE

Project		Pigment	Titanium dioxide	Light fastness degree	Weather resistance degree (3000h)	Migration resistance degree
HDPE	Full shade	0.1%		7～8	3	5
	Reduction	0.1%	1%	7	2	5

Table 4.224 Applications of Pigment Yellow 139

General plastics		Engineering plastics		Spinning	
LL/LDPE	●	PS/SAN	○	PP	●
HDPE	○	ABS	×	PET	×
PP	○	PC	×	PA6	×
PVC(soft)	●	PET	×	PAN	
PVC(rigid)	●	PA6	×		
rubber	○	POM			

● Recommended to use, ○ Conditional use, × Not recommended to use.

Variety characteristic Pigment Yellow 139 has moderate fastness properties and high performance-price ratio. Pigment Yellow 139 has good light fastness, moderate heat resistance, and heat resistance temperature is up to 250℃ for HDPE of 1/3 standard depth. At higher temperatures pigment will decompose, which will darken the shade. Pigment yellow 139 can be used for general polyolefin plastics coloring, it has good migration resistance in flexible PVC. Pigment Yellow 139 is suitable for polypropylene spun dyeing and spinning. It will have an effect on warping deformation of HDPE when Pigment Yellow 139 is used.

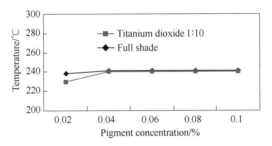

Figure 4.68 Heat resistance of Pigment Yellow 139 in HDPE

4.7.6 Diketopyrrolo-pyrrolo Pigment

1,4-diketopyrrolopyrrole pigment (referred to as DPP pigment) is a class of high-performance organic pigments which has new structures. The advent of the series of varieties is known as a new milestone in the history of organic pigments. The molecular structure of DPP pigment has good symmetry, molecules are arranged in a plane, and hydrogen bonding is formed between molecules. Although DPP pigments has low molecular weight, it has excellent light fastness resistance, heat resistance, solvent resistance, good dispersion. It can be comparable with high performance pigments such as perylene red pigment and benzimidazolone. DPP pigments have pure and bright hue, high tinting strength, and high hiding power, color area covers orange, yellowish red, bluish red, including transparent and opaque varieties.

General structure of DPP pigments is as follows.

Main varieties of DPP pigments applied in plastics are shown in Table 4.225.

Table 4.225 Main varieties of DPP pigments applied in plastics

Pigment	C.I. Structure No.	CAS RN	X	Y	Shade
Orange71	561200	[84632-50-8]	H	CN	Brilliant orange
Orange73	56117	[84632-59-7]	$C(CH_3)_3$	H	Brilliant orange
Red254	56110	[84632-65-5]	Cl	H	Bright red
Red264	561300	[88949-33-1]	C_6H_5	H	Bluish red
Red272	561150	[30249-32-0]	CH_3	H	Red

(1) C.I. Pigment Orange71 CI structure number: 561200. Molecular formula: $C_{20}H_{10}N_4O_2$. CAS registry number: [71832-85-4].

Structure formula

Color characterization Pigment Yellow 71 is high transparent yellowish orange. Pigment Yellow 71 has high tinting strength, and it needs 1% pigment to formulate flexible PVC of 1/3 standard depth with 0.17% titanium dioxide, as well as 0.2% pigment to formulate HDPE of 1/3 standard depth with 0.23% titanium dioxide.

Main properties As shown in Table 4.226～Table 4.228 and Figure 4.69.

Table 4.226 Application performance of Pigment Orange 71 in PVC

Project		Pigment	Titanium dioxide	Light fastness degree	Weather resistance degree (2000h)	Migration resistance degree	Heat resistance degree	
							180℃ /30min	200℃ /10min
PVC	Full shade	0.1%		7～8	2	5	5	5
	Reduction	0.2%	2%	7～8	1	5	5	5

Table 4.227 Application performance of Pigment Orange 71 in HDPE

Project		Pigment	Titanium dioxide	Light fastness degree	Weather resistance degree (3000h)	Migration resistance degree
HDPE	Full shade	0.1%		7~8	4	5
	Reduction	0.1%	1%	7~8	2	5

Table 4.228 Applications of Pigment Orange 71

General plastics		Engineering plastics		Spinning	
LL/LDPE	●	PS/SAN	●	PP	●
HDPE	●	ABS	○	PET	×
PP	●	PC		PA6	×
PVC(soft)	●	PBT	×	PAN	
PVC(rigid)	○	PA6	×		
Rubber	●	POM	●		

● Recommended to use, ○ Conditional use, × Not recommended to use.

Varieties characteristics Pigment Yellow 71 has excellent fastness properties, and meets the long-term open-outdoor requirements. Pigment Orange 71 can be used for PVC and general plastics coloring besides polystyrene based engineering plastics. Pigment Orange 71 is used in HDPE, and only has little influence on warpage variant.

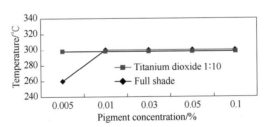

Figure 4.69 Heat resistance of Pigment Orange 71 in HDPE

(2) **C.I. Pigment Orange 73** C.I. structure number: 56117. Molecular formula: $C_{26}H_{28}N_2O_2$. CAS registry number: [71832-85-4].

Structure formula

Color characterization Pigment Orange 73 is bright orange, and has acceptable tinting strength. It needs 0.37% pigment to formulate PVC of 1/3 standard depth with 1% titanium dioxide, as well as 0.35% pigment to formulate HDPE of 1/3 standard depth with 1% titanium dioxide.

Main properties As shown in Table 4.229～Table 4.231 and Figure 4.70.

Table 4.229 Application properties of Pigment Orange 73 in PVC

Project		Pigment	Titanium dioxide	Light fastness degree	Weather resistance degree (2000h)	Migration resistance degree	Heat resistance degree	
							180℃/30min	200℃/10min
PVC	Full shade	0.1%	2%	7～8	4～5	4.6	5	5
	Reduction	0.2%		6～7	3～4	4.7	5	4～5

Table 4.230 Application properties of Pigment Orange 73 in HDPE

Project		Pigment	Titanium dioxide	Light fastness degree	Weather resistance degree (3000h)	Migration resistance degree
HDPE	Full shade	0.1%	1%	8	4	4.9
	Reduction	0.1%		7	2～3	4.9

Table 4.231 Applications of Pigment Orange 73

General plastics		Engineering plastics		Spinning	
LL/LDPE	○	PS/SAN	×	PP	
HDPE	○	ABS	×	PET	×
PP	○	PC	×	PA6	×
PVC(soft)	●	PBT	×	PAN	×
PVC(rigid)	●	PA	×		
Rubber	●	POM			

● Recommended to use, ○ Conditional use, × Not recommended to use.

Varieties characteristics Pigment Orange 73 has excellent overall fastness, and meets the long-term open-outdoor requirements. White reduction with titanium dioxide and changes of pigments concentration will lead the weather resistance degree reducing 1～2. It should pay attention to the heat resistance of low concentration coloring. Mixing Pigment Orange 73 and Pigment Red 254 will produce mixed crystal, causing discoloration. Pigment Orange 73 can be used for PVC and polyolefin plastics coloring besides polystyrene-based engineering plastics.

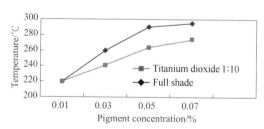

Figure 4.70 Heat resistance of Pigment Orange 73 in HDPE

(3) **C.I. Pigment Red 254** C.I. structure number: 56110. Molecular formula: $C_{18}H_{10}Cl_2N_2O_2$. CAS registry number: [84632-65-5].

Structure formula

$$\text{Cl}-\!\!\!\bigcirc\!\!\!-\underset{\underset{H}{N}}{\overset{O}{\underset{\|}{C}}}\!\!-\!\!\bigcirc\!\!-\text{Cl}$$

Color characterization Pigment Red 254 is the first successfully developed product by Switzerland's Ciba Specialty Chemicals Inc. Pigment Red 254 is positive bright red, and has excellent transparency and saturation. Pigment Red 254 has a high tinting strength, and it needs 0.15% pigment to formulate PVC of 1/3 standard depth with 1% titanium dioxide, as well as 0.16% pigment to formulate HDPE of 1/3 standard depth with 1% titanium dioxide.

Main properties As shown in Table 4.232~Table 4.234 and Figure 4.71.

Table 4.232 Application properties of Pigment Red 254 in PVC

Project		Pigment	Titanium dioxide	Light fastness degree	Weather resistance degree (2000h)	Migration resistance degree	Heat resistance degree	
							180℃/30min	200℃/10min
PVC	Full shade	0.1%		8	5	5	5	5
	Reduction	0.2%	2%	8	3~4	5	5	5

Table 4.233 Application properties of Pigment Red 254 in HDPE

Project		Pigment	Titanium dioxide	Light fastness degree	Weather resistance degree (3000h)	Migration resistance degree
HDPE	Full shade	0.1%		8	4	5
	Reduction	0.1%	1%	8	2	5

Table 4.234 Applications of Pigment Red 254

General plastics		Engineering plastics		Spinning	
LL/LDPE	●	PS/SAN	●	PP	●
HDPE	●	ABS	○	PET	×
PP	●	PC	×	PA6	×
PVC(soft)	●	PBT	×	PAN	
PVC(rigid)	●	PA6	×		
Rubber	●	POM	●		

● Recommended to use, ○ Conditional use, × Not recommended to use.

Varieties characteristics Pigment Red 254 has excellent overall fastness, and is suitable for outdoor applications. Pigment Red 254 can be used for general PVC and polyolefin coloring besides polystyrene-based engineering plastics. However, it will be dissolved to form a bright fluorescent yellow at a temperature higher than 320℃ in PC.

Pigment Red 254 is used in HDPE, and only has little influence on warpage variant. It has no effect on warpage variant of HDPE when Pigment Red 254 is produced through special treatment. Pigment Red 254 has become the standard color of red zone for plastic coloring.

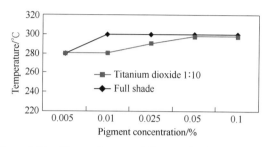

Figure 4.71 Heat resistance of Pigment Red 254 in HDPE

(4) C.I. Pigment Red 264 C.I. structure number: 561300. Molecular formula: $C_{30}H_{20}N_2O_2$. CAS registry number: [177265-40-5].

Structure formula

Color characterization Pigment Red 264 is a bright bluish red pigment with high transparency. Pigment Red 264 has the highest tinting strength in the DPP pigments, and it needs 0.09% pigment to formulate PVC of 1/3 standard depth with 1% titanium dioxide, as well as 0.11% pigment to formulate HDPE of 1/3 standard depth with 1% titanium dioxide (it requires 0.16% Pigment Red 254).

Main properties As shown in Table 4.235～Table 4.237 and Figure 4.72.

Table 4.235 Application properties of Pigment Red 264 in PVC

Project		Pigment	Titanium dioxide	Light fastness degree	Weather resistance degree (2000h)	Migration resistance degree	Heat resistance degree	
							180℃ /30min	200℃ /10min
PVC	Full shade	0.1%		8	5	5	5	5
	Reduction	0.2%	2%	7～8	2～3	5	5	5

Table 4.236 Application properties of Pigment Red 264 in HDPE

Project		Pigment	Titanium dioxide	Light fastness degree	Weather resistance degree (3000h)	Migration resistance degree
HDPE	Full shade	0.1%		8	4	5
	Reduction	0.1%	1%	8	2	5

Table 4.237 Applications of Pigment Red 264

General plastics		Engineering plastics		Spinning	
LL/LDPE	●	PS/SAN	○	PP	●
HDPE	●	ABS	●	PET	×
PP	●	PC	×	PA6	×
PVC(soft)	●	PET	○	PAN	×
PVC(rigid)	●	PA6	○		
Rubber	●	POM	●		

● Recommended to use, ○ Conditional use, × Not recommended to use.

Varieties characteristics Pigment Red 264 has excellent overall fastness, and is suitable for outdoor applications. Pigment Red 264 has excellent heat resistance, but the heat resistance will decrease with the addition of titanium dioxide. Pigment Red 264 can be used for general PVC and polyolefin coloring besides polystyrene-based engineering plastics. Migration resistance of Pigment Red 264 is 5 degree in flexible PVC, and it also can be applied in engineering plastics such as nylon. Pigment Red 264 has no effect on warping variants of HDPE.

Figure 4.72 Heat resistance of Pigment Red 272 in HDPE

(5) **C.I. Pigment Red 272** C.I. structure number: 561150. Molecular formula: $C_{20}H_{16}N_2O_2$.

Structure formula

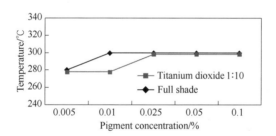

Color characterization Pigment Red 272 is bright positive red, and has high hiding power and saturation. Pigment Red 272 has high tinting strength, and it needs 0.18% pigment to formulate PP of 1/3 standard depth with 1% titanium dioxide.

Main properties As shown in Table 4.238～Table 4.240 and Figure 4.73.

Table 4.238　Application properties of Pigment Red 272 in PVC

Project		Pigment	Titanium dioxide	Light fastness degree	Weather resistance degree (2000h)	Migration resistance degree	Heat resistance degree	
							180℃ /30min	200℃ /10min
PVC	Full shade	0.1%		7～8	4～5	4.9	5	5
	Reduction	0.2%	2%	7	2	5	5	5

Table 4.239　Application properties of Pigment Red 272 in HDPE

Project		Pigment	Titanium dioxide	Light fastness degree	Weather resistance degree (3000h)	Migration resistance degree
HDPE	Full shade	0.1 %		7～8	3～4	5
	Reduction	0.1%	1%	7～8	1～2	5

Table 4.240　Applications of Pigment Red 272

General plastics		Engineering plastics		Spinning	
LL/LDPE	●	PS/SAN	○	PP	○
HDPE	●	ABS	×	PET	×
PP	●	PC	×	PA6	×
PVC(soft)	●	PET	×	PAN	×
PVC(rigid)	●	PA6	×		
Rubber	●	POM			

● Recommended to use, ○ Conditional use, × Not recommended to use.

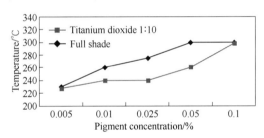

Figure 4.73　Heat resistance of Pigment Red 272 in HDPE

Varieties characteristics　Pigment Red 272 has excellent overall fastness, the full shade is suitable for outdoor applications. Pigment Red 272 can be used for PVC and general polyolefin plastics coloring. The Migration resistance is 5 degree in flexible PVC. Pigment Red 272 is used in HDPE, and only has little influence on warpage variant.

4.7.7　Quinophthalone Pigments

Quinophthalone are a class of ancient compounds which have been synthesized in 1882. However, the chemical structure was determined in 1907. There are a few derivatives of quinophthalone that can be used as pigments, and only yellow species have commercial value. The typical species is Pigment Yellow 138.

C.I. Pigment Yellow 138 C.I. structure number: 56300. Molecular formula: $C_{26}H_6Cl_8N_2O_4$. CAS registry number: [30125-47-4].

Structure formula

Color characterization Pigment Yellow 138 is a bright greenish yellow pigments which has high hiding power and good saturation, and the color is very bright. Pigment Yellow 138 has high tinting strength, and it needs 0.21% pigment to formulate HDPE of 1/3 standard depth with 1% titanium dioxide

Main properties As shown in Table 4.241~Table 4.243 and Figure 4.74.

Table 4.241 Application properties of Pigment Yellow 138 in PVCs

Project		Pigment	Titanium dioxide	Light fastness degree	Migration resistance degree
PVC	Full shade	0.1%		7~8	4~5
	Reduction	0.1%	1%	7	

Table 4.242 Application properties of Pigment Yellow 138 in HDPE

Project		Pigment	Titanium dioxide	Light fastness degree	Migration resistance degree
HDPE	Full shade	0.1%		8	4~5
	Reduction	0.1%	1%	7	

Table 4.243 Applications of Pigment Yellow 138

General plastics		Engineering plastics		Spinning	
LL/LDPE	●	PS/SAN	●	PP	●
HDPE	●	ABS	○	PET	○
PP	●	PC	●	PA6	×
PVC(soft)	●	PBT	○	PAN	
PVC(rigid)	●	PA6	×		
Rubber	●	PMMA	○		

● Recommended to use, ○ Conditional use, × Not recommended to use.

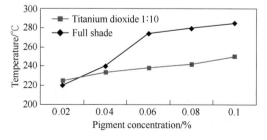

Figure 4.74 Heat resistance of Pigment Yellow 138 in HDPE

Variety characteristic Pigment Yellow 138 has excellent overall fastness, and weather resistance of natural color (deep color) is also very excellent. When it is long-term exposure to the outdoors the color is still very bright. The weather resistance will decrease sharply after white reduction with titanium dioxide. It should pay attention to the heat resistance of low-concentration coloring. Pigment Yellow 138 can be used for PVC and general polyolefin plastics coloring besides polystyrene-based engineering plastics. Pigment Yellow 138 can boost the nucleation of plastics, and when used in HDPE plastic it has negative effect on warpage variant. However, with an increase of processing temperature the effect will decrease. Most varieties of Pigment Yellow 138 have large particle size, small specific surface area and excellent hiding power. Some also have specific surface area, good transparency and high tinting strength.

4.7.8 Metal Complex-based Pigments

Such pigments are complexes of transition metals, azo compounds and methine compounds. Before complexation with the metal, the colors of azo compounds and methine compounds are very vivid. After complexation with the metal, the resulting color shade becomes darker. The advantage of complexation is obtaining high heat resistance, light fastness and weather resistance for the pigment.

(1) **C.I. Pigment Yellow 150**　C.I. structure number: 12764. Molecular formula: $C_8H_6N_6O_6$. CAS registry number: [68511-62-6].

Structure formula

Color characterization　Pigment Yellow 150 is dark yellow, its full shade is acceptable, and reduction has low color saturation, which has very high heat resistance and good light fastness.

Main properties　As shown in Table 4.244～Table 4.246 and Figure 4.75.

Table 4.244　Application properties of Pigment Yellow 150 in PVC

Project		Pigment	Titanium dioxide	Light fastness degree	Migration resistance degree
PVC	Full shade	0.1%		8	5
	Reduction	0.1%	1%	8	

Table 4.245　Application properties of Pigment Yellow 150 in HDPE

Project		Pigment	Titanium dioxide	Light fastness degree
HDPE	Full shade	0.1%		8
	Reduction	0.1%	1%	8

Table 4.246 Applications of Pigment Yellow 150

General plastics		Engineering plastics		Spinning	
LL/LDPE	●	PS/SAN	●	PP	●
HDPE	●	ABS	○	PET	●
PP	●	PC	●	PA	●
PVC(soft)	●	PBT	●	PAN	
PVC(rigid)	●	PA	●		
Rubber	●	POM	●		

● Recommended, ○ Conditional use.

Figure 4.75 Heat resistance of Pigment Yellow 150 in HDPE

Main properties Light fastness of pigment yellow 150 is 8 degree with high heat resistance. It's especially recommended for nylon 6, nylon 66 coloring.

(2) **C.I. Pigment Orange 68** C.I. structure number: 48615. Molecular formula: $C_{29}H_{18}N_4O_3Ni$. CAS registry number: [42844-93-9].

Structure formula

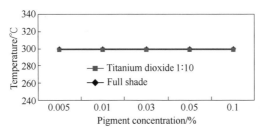

Color characterization Pigment Orange 68 is dark reddish orange, the full shade has poor saturation, and the reduction is acceptable. Pigment Orange 68 contains two varieties, one is coarse particles with great hiding power, another is smaller particles with high transparency and tinting strength.

Pigment Orange 68 has good tinting strength, and it needs 0.15% pigment to formulate HDPE of 1/3 standard depth with 1% titanium dioxide.

Main properties As shown in Table 4.247~Table 4.249 and Figure 4.76.

Table 4.247 Application properties of Pigment Orange 68 in PVC

Project		Pigment	Titanium dioxide	Light fastness degree	Weather resistance degree (2000h)	Migration resistance degree
PVC	Full shade	0.1%		7~8	4	5
	1/3 SD	0.1%	0.5%	7~8		

Table 4.248 Application properties of Pigment Orange 68 in HDPE

Project		Pigment	Titanium dioxide	Light fastness degree	Weather resistance degree (3000h)
HDPE	Full shade	0.2 %		8	3
	Reduction	0.2 %	2%	7	

Table 4.249 Applications of Pigment Orange 68

General plastics		Engineering plastics		Spinning	
LL/LDPE	●	PS/SAN	●	PP	○
HDPE	●	ABS	●	PET	×
PP	●	PC	●	PA	●
PVC(soft)	●	PBT	●	PAN	×
PVC(rigid)	●	PA	●		
Rubber	●	POM	●		

● Recommended to use, ○ Conditional use, × Not recommended to use.

Figure 4.76 Heat resistance of Pigment Orange 68 in HDPE (full shade)

Main properties Pigment Orange 68 has excellent overall fastness and heat resistance. It can be used for coloring of general PVC, polyolefin plastic, general engineering plastics, specialty plastics. Pigment Orange 68 is applied to spun dyeing and spinning of nylon 6.

4.7.9 Thiazide Pigments

C.I. Pigment Red 279

Structure formula

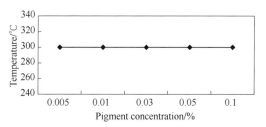

Color characterization Pigment Red 279 is very bright. It is yellowish red, and has high tinting strength. It needs 0.25% pigment to formulate HDPE of 1/3 standard depth with 1% titanium dioxide.

Main properties As shown in Table 4.250～Table 4.252 and Figure 4.77.

Table 4.250 Application properties of Pigment Red 279 in PVC

Project		Pigment	Titanium dioxide	Light fastness degree	Weather resistance degree (3000h)	Migration resistance degree
PVC	Full shade	0.1%		6～7	3	5
	Reduction	0.1%	0.5%	7		

Table 4.251 Application properties of Pigment Red 279 in HDPE

Project		Pigment	Titanium dioxide	Light fastness degree	Weather resistance degree (3000h, 0.2%)
HDPE	Full shade	0.1 %		7～8	3
	Reduction	0.1 %	1 %	7	

Table 4.252 Applications of Pigment Red 279

General plastics		Engineering plastics		Spinning	
LL/LDPE	●	PS/SAN	●	PP	●
HDPE	●	ABS	●	PET	×
PP	●	PC	×	PA	×
PVC(soft)	●	PBT	×	PAN	○
PVC(rigid)	●	PA	×		
Rubber	○	POM	●		

● Recommended to use, ○ Conditional use, × Not recommended to use.

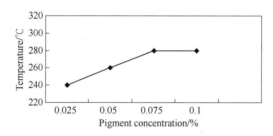

Figure 4.77 Heat resistance of Pigment Red 279 in HDPE (full shade)

Variety characteristics Pigment Red 279 has excellent heat resistance, light fastness and weather resistance. Except for the general polyolefins coloring it can also be used for polystyrene-based engineering plastics coloring. Pigment Red 279 is used in crystallinic plastics such as HDPE, and only has little influence on warpage variant.

4.7.10 Pteridine Pigments

Pigment Yellow 215

Color characterization Basic tone presents micro greenish yellow, and the pigment is bright. In polyolefins it needs 0.18% pigment to formulate 1/3 standard depth with 1% titanium dioxide.

Main properties As shown in Table 4.253~Table 4.255 and Figure 4.78.

Table 4.253 Application properties of Pigment Yellow 215 in HDPE

Project		Pigment	Titanium dioxide	Light fastness degree
HDPE	Full shade	0.1 %		7
	Reduction	0.1 %	1 %	7

Table 4.254 Application properties of Pigment Yellow 215 in PA6

Project		Pigment	Titanium dioxide	Light fastness degree	Migration resistance degree	Heat resistance /℃
PA6	Full shade	0.1 %				
	1/3SD	0.1 %	0.5%	6~7	5	320

Table 4.255 Applications of Pigment Yellow 215

General plastics		Engineering plastics		Spinning	
LL/LDPE	●	PS/SAN	●	PET	×
HDPE	●	ABS	●	PA	×
PP	●	PC	×	PAN	×
PVC(soft)	○	PBT	×		
PVC(rigid)	○	PA	●		
Rubber	○	POM	×		

● Recommended to use, ○ Conditional use, × Not recommended to use.

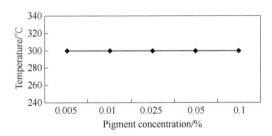

Figure 4.78 Heat resistance of Pigment Yellow 215 in HDPE (full shade)

Variety characteristic Pigment Yellow 215 has excellent heat resistance, bright colors, and it is a kind of important yellow pigment for polyamide coloring. The heat-resistant temperature is up to 320℃, and it has good anti-reducing properties.

Tips: This section has introduced the chemical structures of several major organic pigments in detail which are suitable for plastics coloring, as well as the main application performance in varieties of plastics. For plastic coloring, it is very important to understand the chemistry of the colorants and the indicators, so that it is possible to directly evaluate the quality of the colored plastic. According to the requirements of customer colorists make full use of each colorant's data and meet customer's needs through experiments and tests. Thus, the author collected and analyzed data published by well-known paint suppliers.

It should be emphasized that some data are only used as a guide rather than accurate data of plastics coloring products. Heat resistance, light fastness, weather resistance are obtained by international standard test methods. In fact, the real compositions of colored resin and the final products are very complex. And it needs to add a lot of materials to meet the requirements of customer. In addition, each colorant in the market with different suppliers has different tones and performances mainly due to post-treatment process which result in different pigment particle sizes and distribution, so the performance and other aspects of the data are different.

Chapter 5
Main Varieties and Properties of Solvent Dyestuffs

Solvent dyestuffs are initially named after the dissolution in various organic solvents. Oil-soluble dyestuffs refer to the dyes that can be dissolved in nonpolar or low polarity solvents (aliphatic hydrocarbons, toluene, xylene, fuel oil, paraffin, etc.), and alcohol-soluble dyestuffs mean that the dyes can be dissolved in polar solvents (ethanol, acetone). Currently the main solvent dyestuffs used in the coloring of plastics are oil-soluble dyestuffs that solvent dyestuffs are conventionally called oil-soluble dyestuffs.

The application of solvent dyestuffs is usually limited in the coloring of amorphous polymers, such as polystyrene, polycarbonate, polyester, non-plasticized polyvinyl chloride, the spin-dyeing of polyester and polyamid fibers. And the consumption is great in plastics. It should be noted that the solvent dyestuffs would migrate when they are applied to certain thermoplastic plastics, especially the polyolefin and plasticized PVC.

5.1 Development History of Solvent Dyestuffs

Solvent dyestuffs are a smaller category in dyes, some structures were found as early as 150 years ago. For example, Solvent Yellow 1 (aniline yellow) was discovered in 1861, Solvent Black 5 and 7 were discovered in 1867. In fact many varieties of solvent dyestuffs have different names with the same chemical structure, especially prominent in disperse, acid, basic, direct and vat dyestuffs. Many disperse dyestuffs are applied not only to the coloring of polyester but also to other polymers. Because of containing dispersants the disperse dyestuffs would seriously affect the transparency and luminance of products in the coloring of plastics, thus a certain pure dyes of disperse dyestuffs are regarded as solvent dyestuffs. Previously the traditional area of application of dyes was the coloring of textiles, in recent years other areas were discovered. According to the intended area of application, the color index differentiates between solvent and disperse dyestuffs.

In addition, the chemical structure was not recognized for the reason of competition with the dyes trended to the market, which also results in some chemical structure identical colorants are classified both as solvent and disperse dyestuffs. Therefore the same structure dyes tend to have two index numbers (Table 5.1).

Table 5.1 Different color index of solvent and disperse dyes with the same chemical structure

Shade	Chemical structure	Solvent dyestuff color index	Disperse dyestuff color index
Yellow	Quinoline	Disperse Yellow 54	Solvent Yellow 114
	Quinoline	Disperse Yellow 64	Solvent Yellow 176
	Methylene	Disperse Yellow 201	Solvent Yellow 179
Orange	Multi-methylene	Disperse Orange 47	Solvent Orange 107
Red	Anthraquinone	Disperse Red 9	Solvent Red 111
	Benzopyran	Disperse Red 277	Solvent Red 197
	Anthraquinone	Disperse Red 60	Solvent Red 146
Violet	Anthraquinone	Disperse Violet 26	Solvent Violet 59
		Disperse Violet 31	Solvent Violet 59
		Disperse Violet 28	Solvent Violet 31

5.2 Types, Properties and Varieties of Solvent Dyestuffs

The main types of solvent dyestuffs are azo (monoazo and disazo), azo metal complex, triarylmethane free base, anthraquinone, phthalocyanine (copper phthalocyanine, aluminum phthalocyanine), and heterocyclic, etc., which almost cover the various structural types of dyes (Table 5.2). In common with organic pigments, the coloring property of solvent dyestuffs depends on the chemical structure and application.

In engineering plastics and the spin-dyeing of chemical fibers, not all of the solvent dyes can be selected, for example, some necessary properties and requirements of colored products should be considered, such as heat stability, lightfastness, chemical resistance, migration resistance, etc.. Currently the types and varieties of solvent dyestuffs for the coloring of plastics are as follows.

Table 5.2 Types and varieties of solvent dyestuffs

Type	Azo			Anthraq uinone	Triarylm ethane	Methy lene	Methan imine	Phthaloc yanine	Acid/ Alkali dye complex	Hetero cyclic	Further
	Monoazo	Disazo	Metal complex								
Yellow	17	6	14	5	1	4	1	0	1	23	6
Orange	9	1	14	2	0	2	0	0	0	7	3
Red	15	16	18	20	1	0	1	0	6	21	4
Violet	0	0	5	14	2	0	1	0	1	4	0
Blue	3	0	1	27	6	0	0	7	1	3	4
Green	0	0	0	5	1	0	0	0	0	3	2
Brown	6	2	4	2	0	0	3	0	0	1	2
Black	2	1	12	0	0	0	0	0	9	4	3
Total	52	26	68	76	11	6	6	7	9	66	24

Anthraquinones are the most important solvent dyestuffs which including blue, green, violet, yellow, orange and red. Their properties are very stable, although the solubility of anthraquinones is worse than azo solvent dyestuffs, the light fastness and heat resistance are better. So they play an important role in the coloring of plastics.

Anthraquinone solvent dyestuffs refer to introducing varieties substituents to anthraquinone ring (Figure 5.1), the 1-substituted derivative of anthraquinone ring can get the red and violet solvent dyestuffs, especially the hydrogen of amino, alkylamino and arylamino substituents on 1-anthraquinone can form intramolecular hydrogen bond with anthraquinone carbonyl, which contributes to improve the properties. The main important varieties have Solvent Red 111 and so forth.

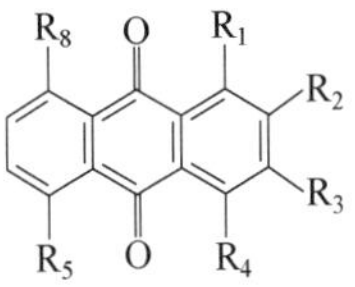

Figure 5.1 Anthraquinone solvent dyestuffs formula

The variety of anthraquinone 1,4-disubstituted solvent dyestuffs is numerous, the color spectrum coverorange, red, blue, green (Table 5.3).

Table 5.3 Varieties of anthraquinone 1,4-disubstituted solvent dyestuffs

Dye	R_1	R_4
Solvent Orange 86	—OH	—OH
Solvent Violet 13	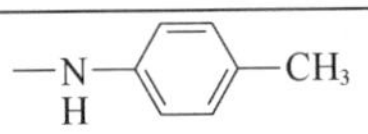 —NH—C$_6$H$_4$—CH$_3$	—OH
Disperse Violet 57	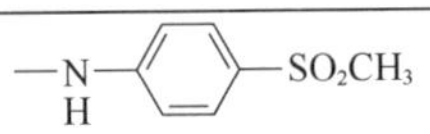 —NH—C$_6$H$_4$—SO$_2$CH$_3$	—OH
Solvent Blue 35	—NH(CH$_2$)$_3$CH$_3$	—NH(CH$_2$)$_3$CH$_3$
Solvent Blue 104	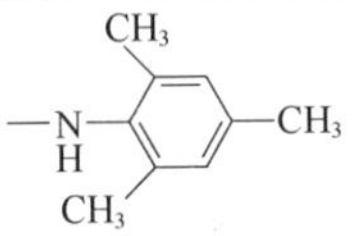 —NH—C$_6$H$_3$(CH$_3$)$_2$—CH$_3$	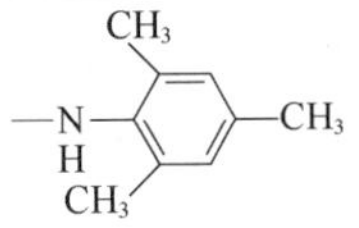 —NH—C$_6$H$_3$(CH$_3$)$_2$—CH$_3$

continued

Dye	R$_1$	R$_4$
Solvent Blue 122	—NH—C$_6$H$_4$—NHCOCH$_3$	—OH
Solvent Blue 97	—NH—C$_6$H$_3$(C$_2$H$_5$)$_2$—CH$_3$	—NH—C$_6$H$_3$(C$_2$H$_5$)$_2$—CH$_3$
Solvent Green 3	—NH—C$_6$H$_4$—CH$_3$	—NH—C$_6$H$_4$—CH$_3$

Anthraquinone solvent dyestuffs also have 1,4-diamino anthraquinone-2,3-disubstituted derivatives, and the main variety is Solvent Violet 59. Anthraquinone 1,5-disubstituted derivatives and 1,8-disubstituted derivatives, and the main variety is Solvent Yellow 163.

The variety of heterocyclic solvent dyestuffs is numerous next to anthraquinones. Heterocyclic solvent dyestuffs show bright shade, many varieties have strong fluorescence and good fastness. They are widely used in the coloring of spinning of engineering plastics and synthetic fibers.

Aminoketone solvent dyestuffs refer that the dyes are composed through 1,8-naphthalene and derivatives or phthalic anhydride and derivatives with aromatic diamine compounds to closed loop. The color spectrum includes yellow, orange and red. These dyes have excellent light fastness, thermal resistance, and can be used in the spin-dyeing of polyester or polyamide fibers. The main varieties are Solvent Orange 60, Solvent Red 135, Solvent Red 179, and Pigment Yellow 192.

Quinophthalone solvent dyestuffs are only limited to yellow and show good light fastness. To improve the thermal resistance the quinophthalone molecule is introduced with the substituted alkyl or sulfonyl compounds. They are suitable for the spin-dyeing of polyester, and the main varieties are Solvent Yellow 114, Solvent Yellow 33, and Solvent Yellow 157.

The main variety of perylene solvent dyestuffs is Solvent Green 5. Coumarin solvent dyestuffs have strong fluorescence, bright shade and show average light fastness and heat resistance. The color spectrum of coumarin solvent dyestuffs is only limited to yellow. In addition to the application in the coloring of plastics, coumarin solvent dyestuffs are regard as the raw materials of fluorescent pigments. The main varieties are Solvent Yellow 160∶1 and Solvent Yellow 145. Thioindigo solvent dyestuffs have Vat red 41 and Pigment Red 181.

Fused ring solvent dyestuffs show good thermal resistance and anti-reducibility and are suitable for the pre-coloring of spinning of PA-6 and PA-66 synthetic fibers. The main varieties are Solvent Red 52, Solvent Orange 63.

The main varieties of methine solvent dyestuffs are Solvent Yellow 93, Solvent Yellow 133, Solvent Yellow 179, Solvent Orange 80 and Solvent Orange 107.

The main varieties of azomethine solvent dyestuffs are Solvent Violet 49 and Solvent Orange 53.

Azo solvent dyestuffs include yellow, orange, red varieties and show high tinting strength, the main varieties are Solvent Red 195, Solvent Orange 116. Azo metal complex solvent dyestuffs refer the dye molecules containing coordinated metal atoms, and prepared by complexing certain dyes (direct, acid, neutral dyes) with metal ions (copper, cobalt, chromium, nickel ions). Most complex dyes are $1:2$ metal complex dyes and a few are $1:1$. Their main shades are yellow, orange, red, and so forth. The main varieties are Solvent Yellow 21, Solvent Red 225, and can be used in the pre-coloring of spinning of nylon fibers.

Phthalocyanine solvent dyestuffs are only limited to blue, it's the ammonium or organic amine salt of sulfonated or chlorosulfonated phthalocyanine. Phthalocyanine solvent dyestuffs show good solubility and light fastness, and the main variety used in plastics is Solvent Blue 67.

5.3 Application Characteristics of Solvent Dyestuffs in Coloring of Plastics

The application of solvent dyestuffs is limited to amorphous polymers (such as PS, ABS engineering plastics, etc.), because these polymers have high glass transition temperature (Table 5.4). Under normal conditions of use, such as room temperature (far below the glass transition temperature), dyes do not migrate from the amorphous polymers. Because dyes are dissolved in the amorphous polymers in the coloring of polymers, the molecular motion of dye molecules is completely restricted in the range of polymer chains, there is no recrystallization or moving to the surface of the polymers. Experiments show that the polymer products colored with a dye are placed for a long time above the glass transition temperature. At this temperature, the polymer chain and the dye molecules are no longer restricted in their molecular motion, and the migration of the dye is observed. If the dyes are applied to partially crystalline polymers such as polyolefin the dyes migrated immediately because their glass transition temperatures are far below room temperature.

Table 5.4 Glass transition temperature of several polymers

Polymer	Glass transition temperature/°C	Polymer	Glass transition temperature/°C
Glass	500~700	PVC (rigid)	80
PS	98~100	PA6	60~70
SB / ABS / SAN	80~105	PE (amorphous)	-80
PMMA	105	PP (isotactic)	+3
PC	143~150	PP (random)	-5

5.3.1 Solubility

The common characteristic of solvent dyestuffs applied to the coloring of plastics is that the solubility is excellent in the process of many plastics and they form a stable solution in the plastic melt. Concrete data about the solubility of solvent dyestuffs in the polymer melt are not possible, however, an accurate assessment is possible because their solubilities are constant in methacrylate and styrene, and additionally the solubilities vary greatly in the polar solvents such as ethanol (Table 5.5).

Table 5.5 Solubility of solvent dyestuffs

Dye	Solubility/(g/L)		
	Methyl methacrylate	Styrene	Ethanol
Solvent Yellow 160:1	2.4	4.7	0.3
Disperse Yellow 201	110	400	1.3
Disperse Yellow 54	1.8	3.1	0.7
Solvent Yellow 130	0.1	0.2	<0.1
Solvent Orange 86	8.5	13	1
Disperse Orange 47	4	6	21
Solvent Red 111	7.5	13	0.7
Solvent Red 179	1.6	4.5	0.1
Disperse Violet 31	35	25	1
Solvent Blue 97	18	55	<0.1
Solvent Green 3	4	11	<0.1
Solvent Green 28	10	25	<0.1

Because solvent dyestuffs dissolve in the polymer melt, there are no dispersion problems in the coloring of plastics. The complete dissolution and uniform distribution of solvent dyestuffs in the polymer melt are essential to avoid defects in the final products. In the coloring of plastics, it could cause color streaks because of the inhomogeneous distribution in the polymer melt which must be avoided.

The application of solvent dyestuffs for the pre-coloring of spinning of synthetic fibers is masterbatch. Owing to the relatively high concentration in masterbatch, generally solvent dyestuffs can't be completely dissolved in the resins. If the mixing shear dispersion is not well in the process of making masterbatch, it may affect the spinnability, filtration and coloring properties of masterbatch.

5.3.2 Sublimation

The solvent dyestuffs dissolved in polymers cause a physical phenomenon-sublimation. Sublimation refers to a substance changes from the solid state directly into the gaseous substances without the liquid state with the increasing temperature. Solvent dyestuffs with different structures have different sublimation fastness. Solvent Red 111 is a very typical example, which sublimes at the normal processing temperatures of amorphous polymers. Different solvent dyestuffs have different sublimation temperatures (Figure 5.2).

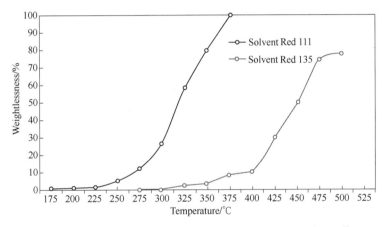

Figure 5.2 Sublimation temperatures in different solvent dyestuffs

The sublimation phenomenon of Solvent Red 111 is used in pyrotechnics for military and civil purposes.

Though the sublimation does not affect the coloring of plastics, it can influence the operational performance of solvent dyestuffs. Under the high processing temperatures a portion of dyes dissolved in the polymer melt turn into the gaseous state when the colored polymer melt fills the mold, and deposited on the relatively cold surface of the mold, slowly forming an increasing deposit. If the deposit is not removed in time, it can cause defects on the surface of the injection molded plastic parts. In theory sublimation can be avoided by reducing the processing temperature, but in practice this is impossible, because each molding process of plastics requires for its processing temperature. The only possibility to avoid deposits is to exchange the solvent dyestuff of better sublimation fastness.

Sublimation occurs not only in the injection molding process of plastics, but also in the drying of a resin and the production process of masterbatch which may pollute equipments. for example, it is essential to dry for the pre-coloring of spinning of PA or PET.

At present the concern on sublimation fastness of solvent dyestuffs in the coloring of plastics is growing, because the demand increases significantly in the coloring of species dark varieties, and the sublimation surface defects is more obvious. In order to increase production, the molding temperature of plastics is improved significantly (sublimation has great influence on fast cycles, pin gate, hot runner) and the degree of automation is also improved (it is no longer possible to clean mold because of the construction of the mold and economic reasons).

5.3.3 Melting Point

The application of solvent dyestuffs in the coloring of plastics is processed into masterbatch, and requires a relatively high processing temperature to accelerate the dissolving, and a sufficient mixing shear profile to reach a homogeneous distribution of the molten or dissolved dyes in the polymer melt. Rapid distribution of the molten dyes in the polymer melt is very important, not only because of the large differences in viscosity between the low-viscous dye melt and the high-viscous polymer melt but also to avoid local oversaturation, which would negatively affect the speed of dissolution. The melting point of several dyes is listed in Table 5.6.

Table 5.6 Melting points of several dyes

Dye	Melting point/°C	Dye	Melting point/°C
Solvent Yellow 160:1	209	Solvent Red 135	318
Disperse Yellow 201	115	Solvent Red 52	280
Solvent Yellow 93	181	Disperse Violet 31	186
Disperse Yellow 54	264	Solvent Violet 13	189
Solvent Yellow 130	300	Solvent Violet 36	213
Solvent Orange 60	230	Solvent Blue 97	200
Solvent Orange 86	180	Solvent Green 3	213
Solvent Red 179	255	Solvent Green 28	245

5.4 Market and Produce of Solvent Dyestuffs

As a part of the dyes, solvent dyestuffs have both the properties of dyes and their own characteristics.

All major foreign countries and companies almost have their own production of solvent dyestuffs. At present a total foreign production of 351 different structures of solvent dyestuffs are shown in Table 5.7.

Table 5.7 Situation of solvent dyestuffs production

Project	Yellow	Orange	Red	Violet
C.I. Dye index number	191	113	248	61
Current products	78	39	102	27
Project	Blue	Green	Brown	Black
C.I. Dye index number	143	49	63	134
Current products	52	11	20	22

Foreign production of solvent dyestuffs is concentrated in Western Europe and the United States. In recent years, foreign production is affected by the products of developing countries, the companies of Western Europe change the business strategy, which is to differentiate and recombine marketing agencies, and the products tend to high-grade. At the same time, imports crude or cake of dyes from China or India to process, then exports dyes.

Major foreign manufacturers are Germany LANXESS Macrolex®; Clariant Sandoplast®, Polynthren®, Hostasol®, Fat®; BASF Oracet®; Sumitomo Sumiplast®; US Keystone Keyplast®.

The production of solvent dyestuffs starts from manufacturers which produce disperse dyes, acid dyes and direct dyes in China, and then developed into the professional manufacturer of solvent dyestuffs.

Major domestic manufacturers are Ningbo Longxin Fine Chemical Co., Ltd., Jiangsu Daobo Chemical Co., Ltd. (YABANG), Haining Modern Chemical Co., Ltd., Tongling Qinghua Technology Co., Ltd., Kunshan Organic Chemical Company, and so forth.

5.5 Main Varieties and Properties of Solvent Dyestuffs

In this section the chemical structure of the main varieties of solvent dyestuffs in the coloring of plastics and the application property in different plastics is detailed introduced. For the plastic colorists, it is very important to understand the chemistry and the performance data of colorants so that the quality of the colored products would be directly evaluated in the color matching. And according to the demands of customers, they make full use of the data of colorants and through the experiments and tests to meet. For this reason, the author provides a summary of the published data of well-known suppliers to readers. It should be emphasized that some data can be used only as a guide but not accurate data in the coloring of plastics. Because of the various technologies of production and post-processing in every company, the application data would be difference and only as a reference for readers in the formulation of the coloring of plastics.

Although there are significant differences between dyes and pigments, given a precise definition is impossible. Few organic pigments can be dissolved in certain polymers, behaves like a dye, and the main application is the pre-coloring of spinning of chemical fibers and the coloring of engineering plastics. Then organic pigments are included in this section.

5.5.1 Anthraquinone Solvent Dyestuffs

Anthraquinone solvent dyestuffs are bright shade, good fastness, and high melting point (250~300℃), so it has excellent thermal resistance and occupies an important position in solvent dyestuffs. For the structure the types of foreign production anthraquinone solvent dyestuffs are 76, ranking first in the structure dyes. Anthraquinone solvent dyestuffs play an important role in violet, blue and green spectrum.

Anthraquinone solvent dyestuffs can be introduced with various substituents on the anthraquinone ring (Figure 5.3).

Anthraquinone solvent dyestuffs include anthraquinone 1-substituted derivatives (C.I. Solvent Red 111, C.I. Solvent Red 168, C.I. Solvent Red 169, C.I. Solvent Yellow 167, etc.); anthraquinone 1,4-disubstituted derivatives (C.I. Solvent Violet 13, C.I. Solvent Blue 97, C.I. Solvent Blue 104, C.I. Solvent Green 3, etc.); 1,5-and 1,8-disubstituted derivatives (C.I. Solvent Yellow 163, C.I. Solvent Yellow 189, C.I. Solvent Red 207, C.I. Solvent Violet 36, etc.).

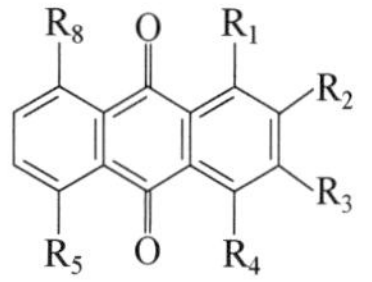

Figure 5.3 Anthraquinone solvent dyestuffs formula

5.5.1.1 Anthraquinone 1-Substituted Derivatives

Anthraquinone 1-substituted derivatives are not much, the most important variety is Solvent Red 111.

C.I. Solvent Red 111　C.I.: 60505. Formula: $C_{15}H_{11}NO_2$. CAS [82-38-2]. Yellowish red, melting point 170℃.

Structural formula

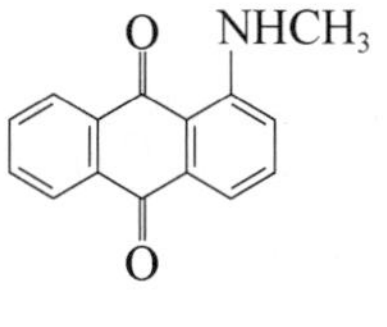

Main properties Shown in Table 5.8.

Table 5.8 Main properties of C.I. Solvent Red 111

Project		PS	ABS	PC	
Tinting strength (1/3 SD)	Dye / %	0.47	0.94	0.245	PET unsuitable
	Titanium dioxide / %	2	4	1	
Thermal resistance /°C	Full shade 0.05%	300	280		
	White reduction 1:20	280	280		
Light fastness degree	Full shade 0.05%	6~7		7	
	1/3 SD	4		4	

Application range Shown in Table 5.9.

Table 5.9 Application range of C.I. Solvent Red 111

PS	●	SB	●	ABS	●
SAN	●	PMMA	●	PC	○
PVC-(U)	●	PPO	●	PET	×
POM	○	PA6 / PA66	×	PBT	×
PES fiber	×				

● Recommended to use, ○ Conditional use, × Not recommended to use.

Variety characteristics Solvent Red 111 is an economical variety in the coloring of engineering plastics, but there is sublimation phenomenon in processing applications.

5.5.1.2 Anthraquinone 1,4-Disubstituted Derivatives

Anthraquinone 1,4-disubstituted derivatives have a wide variety, and the color spectrum contains orange, red, blue, green, but the main color is blue.

① **C.I. Solvent Violet 13** C.I.: 60725. Formula: $C_{21}H_{15}NO_3$. CAS [81-48-1]. Bluish violet, melting point 189 ℃.

Structural formula

Main properties Shown in Table 5.10.

Table 5.10 Main properties of C.I. Solvent Violet 13

Project		PS	ABS	PC	PEPT
Tinting strength (1/3 SD)	Dye / %	0.085	0.097	0.085	0.065
	Titanium dioxide / %	1.0	1.0	1.0	1.0
Light fastness degree	1/3 SD white reduction	6	5	7~8	7~8
	1/25 SD transparent	7~8	6	8	8
Thermal resistance (1/3 SD) / (℃/5min)		300	290	310	290

Application range Shown in Table 5.11.

Table 5.11 Application range of C.I. Solvent Violet 13

PS	●	SB	●	ABS	●
SAN	●	PMMA	●	PC	●
PVC-(U)	●	PPO	●	PET	●
POM	○	PA6 / PA66	×	PBT	●
PES fiber	×				

● Recommended to use, ○ Conditional use, × Not recommended to use.

Variety characteristics Solvent Violet 13 is high tinting strength, excellent light fastness and thermal resistance, and can basically meet the requirements for long-term exposure. It's an economic variety with well performance-price ratio, and could be used in the coloring of engineering plastics such as polycarbonate. Solvent Violet 13 is also used for pre-coloring of spinning of PET.

② **C.I. Disperse Violet 57** Formula: $C_{21}H_{15}NO_6S$. CAS [1594-08-7].

Reddish violet, high transparency in the full shade coloring of HIPS and ABS.

Structural formula

Main properties Shown in Table 5.12.

Table 5.12 Main properties of C.I. Disperse Violet 57

Project	PS	ABS	PC	PET
Dye / %	0.05	0.1	0.05	0.02
Titanium dioxide / %	1.0	1.0		
Light fastness degree	4~5	4	6~7	6~7
Thermal resistance /℃	280	280	300	290
Weather resistance degree (3000h)			4~5	

Application range Shown in Table 5.13.

Table 5.13 Application range of C.I. Disperse Violet 57

PS	●	SB	●	ABS	○
SAN	●	PMMA	●	PC	○
PVC-(U)	×	PA6 / PA66	×	PET	●
POM	●			PBT	●
PES fiber					

● Recommended to use, ○ Conditional use, × Not recommended to use.

Variety characteristics Disperse Violet 57 has good light fastness, excellent thermal resistance, and can be used in the coloring of engineering plastics. Because of its good compatibility with polyester, it is suitable for pre-coloring of spinning of PET and also for the toning of carbon black and phthalocyanine blue.

③ **C.I. Solvent Violet 36**   CAS [61951-89-1].

Reddish violet, melting point 213 °C.

Main properties   Shown in Table 5.14.

Table 5.14   Main properties of C.I. Solvent Violet 36

Project		PS	ABS	PC	PET
Tinting strength (1/3 SD)	Dye / %	0.22	0.44	0.125	
	Titanium dioxide / %	2	4	1	
Thermal resistance /°C	Pure tone 0.05%	300	260~280	350	290
	White reduction 1:20	300	260~280	350	
Light fastness degree	Pure tone 0.05%	7~8		7~8	
	(1/3 SD)	5~6		6~7	

Application range   Shown in Table 5.15.

Table 5.15   Application range of C.I. Solvent Violet 36

PS	●	SB	●	ABS	●
SAN	●	PMMA	●	PC	●
PVC-(U)	●	PPO	●	PET	●
POM	○	PA6 / PA66	×	PBT	○
PES fiber	×				

● Recommended to use, ○ Conditional use, × Not recommended to use.

Variety characteristics   Solvent Violet 36 has good light fastness, excellent thermal resistance, and can be used in the coloring of engineering plastics. It is also applied to the pre-coloring of spinning of PET.

④ **C.I. Solvent Blue 35**   C.I.: 61554. Formula: $C_{22}H_{26}N_2O_2$. CAS [17354-14-2].

Greenish blue, melting point 127 °C.

Structural formula

Main properties   Shown in Table 5.16.

Table 5.16   Main properties of C.I. Solvent Blue 35

Project		PS	ABS	PC	PMMA
1/3 SD (2% titanium dioxide)	Thermal resistance/°C	280	260	300	300
	Light fastness degree	7	6	7	7
	Weather resistance degree	4~5	2	3	4~5

Application range   Shown in Table 5.17.

Table 5.17   Application range of C.I. Solvent Blue 35

PS	•	SB	•	ABS	•
SAN	•	PMMA	•	PC	•
PVC-(U)	•	PA6 / PA66	×		
POM	•				
PES fiber					

• Recommended to use, × Not recommended to use.

Variety characteristics   Solvent Blue 35 has high tinting strength, excellent thermal resistance and light fastness. The light fastness decreases with the increasing degree of white reduction. It is an economical variety and can be used in the coloring of PS and ABS.

⑤ **C.I. Solvent Blue 45**: CAS [23552-74-1].

Reddish blue, melting point 200 ℃.

Main properties   Shown in Table 5.18.

Table 5.18   Main properties of C.I. Solvent Blue 45

Project		PS	ABS	PC	PEPT
Tinting strength (1/3 SD)	Dye / %	0.18	0.19	0.16	0.12
	Titanium dioxide / %	1.0	1.0	1.0	1.0
Light fastness degree	1/3 SD white reduction	6~7	5~6	6	6
	1/25 SD transparent	7~8	5~6	7	7
Thermal resistance (1/3 SD) / (℃/5min)		300	300	330	310

Application range   Shown in Table 5.19.

Table 5.19   Application range of C.I. Solvent Blue 45

PS	•	PMMA	•	ABS	•
SAN	•	PA6 / PA66	×	PC	•
				PET	•
				PBT	•

• Recommended to use, × Not recommended to use.

Variety characteristics   Solvent Blue 45 has excellent light fastness and thermal resistance, and can be applied to the coloring of engineering plastics. It is particularly suitable for the pre-coloring of spinning of PET, and the fabric is excellent in light fastness, wet processing, and friction fastness. It is also suitable for the blow molding of polyester bottles.

⑥ **C.I. Solvent Blue 97**   C.I.: 615290. Formula: $C_{36}H_{38}N_2O_2$. CAS [61969-44-6].

Reddish blue, melting point 200 ℃.

Structural formula

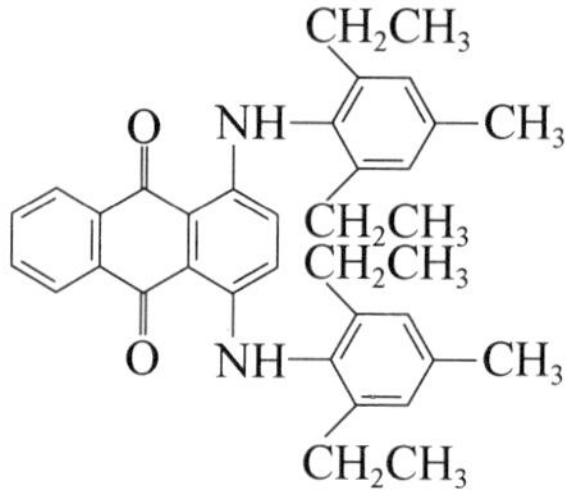

$$\text{anthraquinone core with } -NH-\text{aryl substituents}$$

(1,4-bis[(2,6-diethyl-4-methylphenyl)amino]anthracene-9,10-dione)

Main properties   Shown in Table 5.20.

Table 5.20   Main properties of C.I. Solvent Blue 97

Project		PS	ABS	PC
Tinting strength (1/3 SD)	Dye / % Titanium dioxide / %	0.23 2	0.46 4	0.126 1
Thermal resistance/℃	Full shade 0.05% White reduction 1:20	300 300	260~280 260~280	340 340
Light fastness degree	Full shade 0.05% 1/3 SD	7 6		

Application range   Shown in Table 5.21.

Table 5.21   Application range of C.I. Solvent Blue 97

PS	●	SB	●	ABS	●
SAN	●	PMMA	●	PC	●
PVC-(U)	●	PPO	·	PET	●
POM	○	PA6 / PA66	●	PBT	○
PES fiber	○				

● Recommended to use, ○ Conditional use.

Variety characteristics   Solvent Blue 97 has high tinting strength, good light fastness, excellent thermal resistance, and can be used in the coloring of engineering plastics. It is applied to the pre-coloring of spinning of PET. The thermal resistance of Solvent Blue 97 is up to 300℃ in PA6 and 280℃ in PA66, which is suitable for pre-coloring of spinning of PA.

⑦ **C.I. Solvent Blue 104**   C.I.: 61568. Formula: $C_{32}H_{30}N_2O_2$. CAS [116-75-6]. Reddish blue, melting point 240℃.

Structural formula

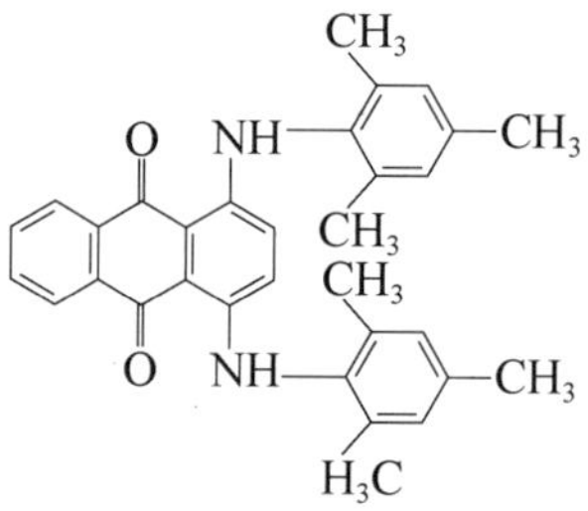

(1,4-bis[(2,4,6-trimethylphenyl)amino]anthracene-9,10-dione)

Main properties   Shown in Table 5.22.

Table 5.22   Main properties of C.I. Solvent Blue 104

Project		PS	ABS	PC	PEPT
Tinting strength (1/3 SD)	Dye / %	0.1	0.114	0.096	0.067
	Titanium dioxide / %	1.0	1.0	1.0	1.0
Light fastness degree	1/3 SD white reduction	6	4	6	5~6
	1/25 SD transparent	7~8	5	7~8	7
Thermal resistance (1/3 SD) / (℃/5min)		300	300	340	320

Application range   Shown in Table 5.23.

Table 5.23   Application range of C.I. Solvent Blue 104

PS	●	PMMA	●	ABS	●
PVC-(U)	●	PPO	●	PC	●
		PA6 / PA66	●	PET	●
				PBT	●

● Recommended to use.

Variety characteristics   Solvent Blue 104 has excellent thermal resistance, light fastness, and weather resistance. It is an important variety of solvent dyestuffs and extensively applied to the coloring of engineering plastics such as PET, PC, PA, and so forth. It is also suitable for pre-coloring of spinning of PET, PA6 and PA66.

⑧ **C.I. Solvent Blue 122**   C.I.: 60744. Formula: $C_{22}H_{16}N_2O_4$. CAS [67905-17-3]. Reddish blue, melting point 239 ℃.

Structural formula

Main properties   Shown in Table 5.24.

Table 5.24   Main properties of C.I. Solvent Blue 122

Project		PS	ABS	PC	PEPT	PA
Tinting strength (1/3 SD)	Dye / %	0.090	0.097	0.088	0.063	Not recommended
	Titanium dioxide / %	1.0	1.0	1.0	1.0	
Light fastness degree	1/3 SD white reduction	6-7	5	7-8	7	
	1/25 SD transparent	7	6	8	8	
Thermal resistance (1/3 SD) / (℃/5min)		300	300	300	290	

Application range   Shown in Table 5.25.

Table 5.25 Application range of C.I. Solvent Blue 122

PS	●	PMMA	●	ABS	●
SAN	●	PPO	●	PC	●
PVC-(U)	×	PA6 / PA66	×	PET	●
				PBT	●

● Recommended to use, × Not recommended to use.

Variety characteristics Solvent Blue 122 has high tinting strength, good light fastness, excellent thermal resistance, and can be used in the coloring of engineering plastics. It is particularly suitable for the pre-coloring of spinning of PET, and the fabric is excellent in light fastness, wet processing, and friction fastness. It is also suitable for the blow molding of polyester bottles.

⑨ **C.I. Solvent Blue 132** Bright reddish blue.

Main properties Shown in Table 5.26.

Table 5.26 Main properties of C.I. Solvent Blue 132

Project	PA6 fiber (110 dtex/24 f-round cross-section)	
Dye	0.1%	1.0%
Light fastness degree	5~6	7

Variety characteristics Solvent Blue 132 has high tinting strength, excellent light fastness and thermal resistance, and it is particularly suitable for PA6. Because of its good wet processing, Solvent Blue 132 is mainly used in carpet fibers and interior textiles.

⑩ **C.I. Solvent Green 3** C.I.: 61565. Formula: $C_{28}H_{22}N_2O_2$. CAS [128-80-3]. Bluish green, melting point 215℃.

Structural formula

Main properties Shown in Table 5.27.

Table 5.27 Main properties of C.I. Solvent Green 3

Project		PS	ABS	PC	PEPT
Tinting strength (1/3 SD)	Dye / %	0.096	0.117	0.102	0.084
	Titanium dioxide / %	1.0	1.0	1.0	1.0
Light fastness degree	1/3 SD white reduction	4~5	4	6	5~6
	1/25 SD transparent	7	6	7~8	7
Thermal resistance (1/3 SD) / (℃/5min)		300	300	340	300

Application range Shown in Table 5.28.

Table 5.28 Application range of C.I. Solvent Green 3

PS	●	SB	●	ABS	●		
SAN	●	PMMA	●	PC	●		
PVC-(U)	●	PPO	●	PET	●		
POM	○	PA6 / PA66	×	PBT	○		
PES fiber	●						

● Recommended to use, ○ Conditional use, × Not recommended to use.

Variety characteristics Solvent Green 3 has high tinting strength, excellent light fastness and thermal resistance, and it is applied to the coloring of styrenic engineering plastics. It is also suitable for the pre-coloring of spinning of PET.

5.5.1.3 Anthraquinone 1,5- and 1,8-Disubstituted Derivatives

In anthraquinone dyes the varieties of these structures are few and the structural formula is as follows:

R_1, R_2 are the same or different arylamino group, cyclohexylamino group or phenylthio group.

① **C.I. Solvent Yellow 163** C.I.: 58840. Formula: $C_{26}H_{16}O_2S_2$. CAS [13676-91-0]. Positive reddish medium yellow, melting point 193 ℃.

Structural formula

Main properties Shown in Table 5.29.

Table 5.29 Main properties of C.I. Solvent Yellow 163

Project	PS	ABS	PC	PET
Dye / %	0.05	0.1	0.05	0.02
Titanium dioxide / %	1.0	1.0		
Light fastness degree	7	7	8	8
Thermal resistance/℃	300	300	360	300
Weather resistance degree (3000h)			4~5	5

Application range Shown in Table 5.30.

Table 5.30   Application range of C.I. Solvent Yellow 163

PS	•	SB	•	ABS	•
SAN	•	PMMA	•	PC	•
PVC-(U)	•	PPO	•	PET	•
POM	•	PA6 / PA66	×	PBT	•

• Recommended to use, × Not recommended to use.

Variety characteristics   Solvent Yellow 163 has excellent light fastness, thermal resistance and weather resistance, and migration fastness. It can be used in the coloring of engineering plastics, especially in outdoor (automobile decoration). It is also suitable for the pre-coloring of spinning of PET.

② **C.I. Solvent Red 207**   C.I.: 617001. Formula: $C_{26}H_{30}N_2O_2$. CAS [15958-68-6]. Reddish blue, melting point 243 ℃.

Structural formula

Main properties   Shown in Table 5.31.

Table 5.31   Main properties of C.I. Solvent Red 207

	Project	PS	Project	PS
Tinting strength	Dye / %	0.05	Thermal resistance/℃	300
	Titanium dioxide / %	1.0	Light fastness degree	7~8

Application range   Shown in Table 5.32.

Table 5.32   Application range of C.I. Solvent Red 207

PS	•	PMMA	•	ABS	•
SAN	•			PC	•
PVC-(U)	•			PET	○
				PBT	•

• Recommended to use, ○ Conditional use.

Variety characteristics   Solvent Red 207 has good light fastness and excellent thermal resistance which is up to 300℃. suitable used in the coloring of engineering plastics.

5.5.1.4   Other Substituted Derivatives of Anthraquinone

① **C.I. Solvent Red 146 (Disperse Red 60)**   C.I.: 60756. Formula: $C_{20}H_{13}NO_4$. CAS [12223-37-9].

Bright bluish red, melting point 213 ℃.
Structural formula

(structure: 1-amino-2-phenoxy-4-hydroxyanthraquinone)

Main properties Shown in Table 5.33.

Table 5.33 Main properties of C.I. Solvent Red 146

Project	PS	ABS	PC	PET
Dye / %	0.05	0.1	0.05	0.02
Titanium dioxide / %	1.0	1.0		
Light fastness degree	5	4～5	8	7～8
Thermal resistance/℃	300	280	360	300
Weather resistance degree (3000h)			4～5	

Application range Shown in Table 5.34.

Table 5.34 Application range of C.I. Solvent Red 146

PS	●	PMMA	●	ABS	●
SAN	●			PC	●
PVC-(U)	×			PET	●
POM	●			PBT	×

● Recommended to use, × Not recommended to use.

Variety characteristics Solvent Red 146 has excellent thermal resistance and general light fastness, and particularly suitable for the light shade and the pink modulated with titanium dioxide. It is a common product with high performance-price ratio and applied to the coloring of POM.

② **C.I. Solvent Violet 31(Disperse Violet 28)** C.I.: 61102. Formula: $C_{14}H_8Cl_2N_2O_2$. CAS [70956-27-3].

Bright violet, melting point 245 ℃.
Structural formula

(structure: 1,4-diamino-2,3-dichloroanthraquinone)

Main properties Shown in Table 5.35.

Table 5.35 Main properties of C.I. Solvent Violet 31

	Project	PS	Project	PS
Tinting strength	Dye / %	0.05	Light fastness degree	6~7
	Titanium dioxide / %	1.0	Thermal resistance / ℃	300

Application range Shown in Table 5.36.

Table 5.36 Application range of C.I. Solvent Violet 31

PS	●	SB	●	ABS	●
SAN	●	PMMA	●	PC	●
PVC-(U)	●	PPO	●	PET	●
		PA6 / PA66	○	PBT	○

● Recommended to use, ○ Conditional use.

Variety characteristics Solvent Violet 31 has good light fastness and thermal resistance, and can be used in the coloring of engineering plastics. It is also suitable for the pre-coloring of spinning of PET.

③ **C.I. Solvent Violet 37** Formula: $C_{37}H_{40}N_2O_3$. CAS [71701-33-2].

Bright bluish violet.

Main properties Shown in Table 5.37.

Table 5.37 Main properties of C.I. Solvent Violet 37

	Project	PS	ABS	PC	PEPT	PA
Tinting strength (1/3 SD)	Dye / %	0.168	0.184			
	Titanium dioxide / %	1.0	1.0			
Light fastness degree	1/3 SD white reduction	6~7	4	Not recommended		
	1/25 SD transparent	7	6			
Thermal resistance (1/3 SD) / (℃/5min)		300	300			

Application range Shown in Table 5.38.

Table 5.38 Application range of C.I. Solvent Violet 37

PS	●	PMMA	●	ABS	●
SAN	●	PA6 / PA66	×	PC	×
PVC (U)	●			PET	×
				PBT	×

● Recommended to use, × Not recommended to use.

Variety characteristics Solvent Violet 37 is excellent in thermal resistance and general in light fastness, and can be used in the coloring of engineering plastics.

④ **C.I. Solvent Violet 59 (Disperse Violet 26, Disperse Violet 31)** C.I.: 62025. Formula: $C_{26}H_{18}N_2O_4$. CAS [6408-72-6].

Bright reddish violet, melting point 186℃.
Structural formula

Main properties Shown in Table 5.39.

Table 5.39 Main properties of C.I. Solvent Violet 59

Project		PS	ABS	PC	PEPT
Tinting strength (1/3 SD)	Dye / %	0.093	0.1	0.094	0.084
	Titanium dioxide / %	1.0	1.0	1.0	1.0
Light fastness degree	1/3 SD white reduction	6	5	6~7	6
	1/25 SD transparent	7~8	6	8	7
Thermal resistance (1/3 SD) / (℃/5min)		300	300	330	280

Application range Shown in Table 5.40.

Table 5.40 Application range of C.I. Solvent Violet 59

PS	●	SB	●	ABS	●
SAN	●	PMMA	○	PC	●
PVC-(U)	●	PPO	●	PET	●
POM	○	PA6 / PA66	×	PBT	●

● Recommended to use, ○ Conditional use, × Not recommended to use.

Variety characteristics Solvent Violet 59 has high tinting strength, excellent thermal resistance and light fastness, and can be used in the coloring of engineering plastics. It is also used for the pre-coloring of spinning of PET.

⑤ **C.I. Solvent Green 28** C.I.: 625580. Formula: $C_{34}H_{34}N_2O_4$. CAS. [4851-50-7]. Yellowish green, melting point 245 ℃.

Structural formula

Main properties Shown in Table 5.41.

Table 5.41 Main properties of C.I. Solvent Green 28

Project		PS	ABS	PC	PET
Tinting strength (1/3 SD)	Dye / %	0.15	0.17	0.163	0.12
	Titanium dioxide / %	1.0	1.0	1.0	1.0
Light fastness degree	1/3 SD white reduction	7~8	5	7~8	7
	1/25 SD transparent	8	6	8	7~8
Thermal resistance (1/3 SD) / (℃/5min)		300	300	310	300

Application range Shown in Table 5.42.

Table 5.42 Application range of C.I. Solvent Green 28

PS	●	SB	●	ABS	●
SAN	●	PMMA	●	PC	●
PVC-(U)	●	PPO	●	PET	●
		PA6 / PA66	×	PBT	○

● Recommended to use, ○ Conditional use, × Not recommended to use.

Variety characteristics Solvent Green 28 has good light fastness and excellent thermal resistance, and can be used in the coloring of engineering plastics. It is also suitable for the pre-coloring of spinning of PET.

⑥ **C.I. Pigment Yellow 147** C.I.: 60645. Formula: $C_{37}H_{21}N_5O_4$. CAS [4118-16-5]. Reddish yellow, melting point above 245℃.

Structural formula

Main properties Shown in Table 5.43.

Table 5.43 Main properties of C.I. Pigment Yellow 147

Project	PS	ABS	PC	PET
Pigment / %	0.05	0.1	0.05	0.02
Titanium dioxide / %	1.0	1.0		
Light fastness degree	6-7	6	8	8
Thermal resistance/℃	300	280	340	300
Weather resistance degree (3000h)			4	5

Application range Shown in Table 5.44.

Table 5.44 Application range of C.I. Pigment Yellow 147

PS	○	PMMA	○	ABS	●
SAN	○	PA6	○	PC	●
PVC-(U)	○	PA66	×	PET	●
POM	●			PBT	●

● Recommended to use, ○ Conditional use, × Not recommended to use.

Variety characteristics Pigment Yellow 147 is excellent in thermal resistance, sublimation resistance, and light fastness. It is good compatibility with polyester, particularly suitable for the pre-coloring of spinning of polyester and polyether sulfone fibers, and can be used in automotive decorations fibers, clothes, interior textiles.

5.5.2 Heterocyclic Solvent Dyestuffs

The variety of heterocyclic solvent dyestuffs is wide, which is second to anthraquinone and azo metal complex dyes. Its color spectrum is complete but the predominant are yellow and red. Heterocyclic solvent dyestuffs show bright shade, good fastness, and many varieties have a strong fluorescence. It is widely used in the coloring of plastics and the pre-coloring of spinning of synthetic fibers.

5.5.2.1 Aminoketones

Aminoketone solvent dyestuffs refer to those dyes which are compound through 1,8-naphthalene anhydride or phthalic anhydride or their derivatives with aromatic diamine compounds through removing two molecules of water. The color spectrum includes yellow, orange, red, violet and blue. These dyes have excellent light fastness and thermal resistance, and do not migrate and sublimate in the polar resins.

① **C.I. Solvent Orange 60** C.I.: 564100. Formula: $C_{18}H_{10}N_2O$. CAS [61969-47-9]. Yellowish orange, melting point 230℃.

Structural formula

Main properties Shown in Table 5.45.

Table 5.45 Main properties of C.I. Solvent Orange 60

	Project	PS	ABS	PC	PET
Tinting strength (1/3 SD)	Dye / %	0.28	0.56	0.155	0.119
	Titanium dioxide / %	2	4	1	1

continued

Project		PS	ABS	PC	PET
Thermal resistance/℃	Pure tone 0.05%	300	280	350	
	White reduction 1:20	300	280	350	290
Light fastness degree	Pure tone 0.05%	8		8	
	1/3 SD	6		7	7~8

Application range　Shown in Table 5.46.

Table 5.46　Application range of C.I. Solvent Orange 60

PS	●	SB	●	ABS	●
SAN	●	PMMA	●	PC	●
PVC-(U)	○	PPO	●	PET	●
POM	○	PA6 / PA66	○	PBT	○
PES fiber	×				

● Recommended to use,　○ Conditional use,　× Not recommended to use.

Variety characteristics　Solvent Orange 60 has high tinting strength and excellent thermal resistance, light fastness and weather resistance, which can be used in the coloring of engineering plastics such as polyamide. It is also suitable for the pre-coloring of spinning of PET and should be careful for PA.

② **C.I. Solvent Red 135**　C.I.: 564120. Formula: $C_{18}H_6Cl_4N_2O$. CAS [71902-17-5]. Yellowish red, standard color, bluer than Solvent Red 179, melting point 318℃.

Structural formula

Main properties　Shown in Table 5.47.

Table 5.47　Main properties of C.I. Solvent Red 135

Project		PS	ABS	PC	PEPT
Tinting strength (1/3 SD)	Dye / %	0.23	0.27	0.23	0.17
	Titanium dioxide / %	1.0	1.0	1.0	1.0
Light fastness degree	1/3 SD white reduction	7	7	7~8	7
	1/25 SD transparent	8	7	8	8
Thermal resistance (1/3 SD) / (℃/5min)		300	290	340	320

Application range　Shown in Table 5.48.

Table 5.48 Application range of C.I. Solvent Red 135

PS	●	SB	●	ABS	●
SAN	●	PMMA	●	PC	●
PVC-(U)	×	PPO	●	PET	●
POM	○	PA6 / PA66	○	PBT	○
PES fiber	●				

● Recommended to use, ○ Conditional use, × Not recommended to use.

Variety characteristics Solvent Red 135 not only has good thermal resistance, light fastness, and migration fastness but also has excellent weather resistance. It tends to sublimate at the temperature above the processing temperatures of amorphous polymer. Solvent Red 135 is applied to the high-performance products. It is suitable for the pre-coloring of spinning of PET, and the fabric is excellent in light fastness, wet processing, and friction fastness. Solvent Red 135 is also used in the blow molding of polyester bottles and should be careful in PA6.

③ **C.I. Solvent Red 179** C.I.: 564150. Formula: $C_{22}H_{12}N_2O$. CAS [89106-94-5]. Yellowish red, melting point 255℃.

Structural formula

Main properties Shown in Table 5.49.

Table 5.49 Main properties of C.I. Solvent Red 179

Project		PS	ABS	PC	PEPT
Tinting strength (1/3 SD)	Dye / %	0.16	0.18	0.155	0.113
	Titanium dioxide / %	1.0	1.0	1.0	1.0
Light fastness degree	1/3 SD white reduction	6	5	5~6	3~4
	1/25 SD transparent	7	7	8	6~7
Thermal resistance (1/3 SD) / (℃/5min)		300	300	340	320

Application range Shown in Table 5.50.

Table 5.50 Application range of C.I. Solvent Red 179

PS	●	SB	●	ABS	●
SAN	●	PMMA	●	PC	●
PVC-(U)	●	PPO	●	PET	●
POM	○	PA6 / PA66	○	PBT	●
PES fiber	×				○

● Recommended to use, ○ Conditional use, × Not recommended to use.

Variety characteristics   Solvent Red 179 has good light fastness and the thermal resistance is up to 300 ℃. It is suitable for the coloring of engineering plastics and pre-coloring of spinning of PET., it is also applied to the pre-coloring of spinning of PA6.

④ **C.I. Pigment Yellow 192**   C.I.: 507300. Formula: $C_{19}H_{10}N_4O_2$. CAS [56279-27-7].

Reddish yellow.

Structural formula

Main properties   Shown in Table 5.51.

Table 5.51   Main properties of C.I. Pigment Yellow 192

Project		PA6	PET
Tinting strength (1/3 SD)	Pigment / %	1.2	0.9
	Titanium dioxide / %	1	1
Thermal resistance/℃	1/3 SD	300	320
Light fastness degree	1/25 SD	7～8	8
	1/3 SD	7	7～8

Application range   Shown in Table 5.52.

Table 5.52   Application range of C.I. Pigment Yellow 192

PS	×	PA6	●	ABS	×
SAN	×	PA66	●	PC	×
PVC-(U)	×			PET	○
POM	○			PBT	×
PES fiber	×				

● Recommended to use, ○ Conditional use, × Not recommended to use.

Variety characteristics   Pigment Yellow 192 has good light fastness, 7～8 (1/25 SD) in PA6, but 7 (1/3 SD prepared with 1% titanium dioxide) in PA6. Its thermal resistance is up to 300℃ in PA6 and could be dedicated to the coloring of PA6 and PA66.

5.5.2.2   Coumarin Solvent Dyestuffs

The color spectrum of coumarin solvent dyestuffs is complete but the predominant are yellow and red. Coumarin solvent dyestuffs show bright shade, and many varieties have a strong fluorescence and good fastness.

① **C.I. Solvent Yellow 160∶1**   CAS [94945-27-4].

Greenish yellow with fluorescence, melting point 209 ℃.

Main properties Shown in Table 5.53.

Table 5.53 Main properties of C.I. Solvent Yellow 160:1

Project		PS	ABS	PC	PET
Tinting strength (1/3 SD)	Dye / %	0.2		0.1	
	Titanium dioxide / %	2	4	1	
Thermal resistance/℃	1/3 SD	300	280	350	280
Light fastness degree	Pure tone 0.05%	6		6~7	
		3~4		5	

Application range Shown in Table 5.54.

Table 5.54 Application range of C.I. Solvent Yellow 160:1

PS	●	SB	●	ABS	●
SAN	●	PMMA	●	PC	●
PVC-(U)	●	PPO	●	PET	●
POM	○	PA6 / PA66	○	PBT	○
PES fiber	×				

● Recommended to use, ○ Conditional use, × Not recommended to use.

Variety characteristics The light fastness of Solvent Yellow 160:1 is bad and decreases with increasing titanium dioxide. But it is excellent in thermal resistance and can be used in the coloring of engineering plastics. It is suitable for the pre-coloring of spinning of PET and should be careful for PA6.

② **C.I. Solvent Yellow 145** Greenish yellow with fluorescence, melting point 240℃.

Main properties Shown in Table 5.55.

Table 5.55 Main properties of C.I. Solvent Yellow 145

Project	PS	ABS	PC	PET
Dye / %	0.05	0.1	0.05	0.02
Titanium dioxide / %	1.0	1.0		
Light fastness degree	5	4~5	7~8	7~8
Thermal resistance/℃	300	280	360	300
Weather resistance degree (3000h)			3~4	

Application range Shown in Table 5.56.

Table 5.56 Application range of C.I. Solvent Yellow 145

PS	●	PMMA	●	ABS	●
SAN	●	PA6	○	PC	●
PVC-(U)	○	PA66	×	PET	●
POM	●			PBT	●

● Recommended to use, ○ Conditional use, × Not recommended to use.

Variety characteristics Solvent Yellow 145 has good light fastness, excellent thermal resistance, and bright transparent color. Owing to its various application properties in different polymers, it is limited for the coloring of engineering plastics. It tends to sublimate at the processing temperature above 280 ℃.

5.5.2.3 Quinophthalone Solvent Dyestuffs

Quinophthalone solvent dyestuffs are only limited to yellow and show good light fastness. Solvent Yellow 114 (Disperse Yellow 54) is also an important disperse dyestuff.

① **C.I. Solvent Yellow 33**   C.I.: 47000. Formula: $C_{18}H_{11}NO_2$. CAS [8003-22-3]. Yellow, melting point 240 ℃.

Structural formula

Main properties   Shown in Table 5.57.

Table 5.57   Main properties of C.I. Solvent Yellow 33

Project	PS	ABS	PC	PET
Dye / %	0.05	0.1	0.05	0.02
Titanium dioxide / %	1.0	1.0		
Light fastness degree	6~7	6	8	8
Thermal resistance/℃	300	280	300	300
Weather resistance degree (3000h)			4	5

Application range   Shown in Table 5.58.

Table 5.58   Application range of C.I. Solvent Yellow 33

PS	●	PMMA	●	ABS	●
SAN	●	PA6	×	PC	●
PVC-(U)	●	PA66	×	PET	●
POM	●			PBT	●

● Recommended to use, × Not recommended to use.

Variety characteristics   Solvent Yellow 33 has good light fastness and excellent thermal resistance, which can be used in the coloring of engineering plastics.

② **C.I. Solvent Yellow 114 (Disperse Yellow 54)**   C.I.: 47020. Formula: $C_{18}H_{11}NO_3$. CAS [75216-45-4].

Bright greenish yellow, melting point 264 ℃.

Structural formula

Main properties Shown in Table 5.59.

Table 5.59 Main properties of C.I. Solvent Yellow 114

Project		PS	ABS	PC
Tinting strength (1/3 SD)	Dye / %	0.12	0.24	0.065
	Titanium dioxide / %	2	4	1
Thermal resistance/℃	Pure tone 0.05%	300	280	340
	White reduction 1∶20	300	280	340
Light fastness degree	Pure tone 0.05%	8		8
	1/3 SD	7~8		7~8

Application range Shown in Table 5.60.

Table 5.60 Application range of C.I. Solvent Yellow 114

PS	●	SB	●	ABS	●
SAN	●	PMMA	●	PC	●
PVC-(U)	●	PPO	●	PET	●
POM	○	PA6 / PA66	×	PBT	○
PES fiber	×				

● Recommended to use, ○ Conditional use, × Not recommended to use.

Variety characteristics Solvent Yellow 114 has high purity and excellent light fastness. Its thermal resistance is up to 300 ℃ and can be used in the coloring of engineering plastics (limited to polyether plastics). It is also suitable for the pre-coloring of spinning of PET.

③ **C.I. Solvent Yellow 157** C.I.: 470180. Formula: $C_{18}H_7Cl_4NO_2$. CAS [27908-75-4].

Greenish yellow, melting point 323 ℃.

Structural formula

Main properties Shown in Table 5.61.

Table 5.61 Main properties of C.I. Solvent Yellow 157

Project		PS	Project	PS
Tinting strength	Dye/%	0.05	Light fastness degree	7~8
	Titanium dioxide/%	1.0	Thermal resistance/(℃ / 5min)	300

Application range　Shown in Table 5.62.

Table 5.62 Application range of C.I. Solvent Yellow 157

PS	●	PMMA	●	ABS	●
SAN	●	PA6	×	PC	●
PVC-(U)	●	PA66	×	PET	●
POM	●			PBT	○

● Recommended to use,　○ Conditional use,　× Not recommended to use.

Variety characteristics　Solvent Yellow 157 shows excellent light fastness and thermal resistance, which is suitable for the pre-coloring of spinning of PET.

④ **C.I. Solvent Yellow 176**　C.I.: 47023. Formula: $C_{18}H_{10}BrNO_3$. CAS [10319-14-9].

Reddish yellow, melting point 218℃.

Structural formula

Main properties　Shown in Table 5.63.

Table 5.63 Main properties of C.I. Solvent Yellow 176

Project		PS	Project	PS
Tinting strength	Dye/%	0.05	Light fastness degree	7
	Titanium dioxide/%	1.0	Thermal resistance/(℃ / 5min)	280

Application range　Shown in Table 5.64.

Table 5.64 Application range of C.I. Solvent Yellow 176

PS	●	PMMA	●	ABS	●
SAN	●	PA6	×	PC	●
PVC-(U)	●	PA66	×	PET	●
POM	●			PBT	×

● Recommended to use, × Not recommended to use.

Variety characteristics　Solvent Yellow 176 has good light fastness and thermal resistance. It is suitable for the pre-coloring of spinning of PET.

5.5.2.4 Perylene Solvent Dyestuffs

C.I. Solvent Green 5 C.I.: 59075. Formula: $C_{30}H_{28}O_4$. CAS [79869-59-3]. Bright greenish yellow with fluorescence.

Structural formula

$$(CH_3)_2CHCH_2OOC\text{—}\bigcirc\text{—}COOCH_2CH(CH_3)_2$$

Main properties Shown in Table 5.65.

Table 5.65 Main properties of C.I. Solvent Green 5

	Project	PS	ABS	PC	PMMA
1/3 SD (2% titanium dioxide)	Thermal resistance/℃	260	300	310	300
	Light fastness degree	3	4	3~4	4
	Weather resistance degree	1~2	2	1	1~2

Application range Shown in Table 5.66.

Table 5.66 Application range of C.I. Solvent Green 5

PS	●	SB	●	ABS	●
SAN	●	PMMA	●	PC	●
PVC-(U)	●	PA6 / PA66	×	PET	○
POM	●			PBT	○

● Recommended to use, ○ Conditional use, × Not recommended to use.

Variety characteristics Solvent Green 5 has excellent thermal resistance and light fastness (pure tone), but the light fastness rapidly decreases with the increasing of white reduction. It is applied to the coloring of engineering plastics such as PS, SAN, PMMA and PC, and it is recommended to test for the coloring of SB, ABS, ASA. It is also suitable for the pre-coloring of spinning of PET.

5.5.2.5 Thioindigo Solvent Dyestuffs

① **C.I. Vat Red 41** C.I.: 73300. Formula: $C_{16}H_8O_2S_2$. CAS [522-75-8]. Bright red with fluorescence, brilliant transparent color, melting point 312 ℃.

Structural formula

Main properties Shown in Table 5.67.

Table 5.67 Main properties of C.I. Vat Red 41

Project		PS	ABS	PC
Light fastness degree	1/3 SD white reduction	3	3~4	5~6
	1/25 SD transparent	4	6	6~7
Thermal resistance (1/3 SD) / (℃/5min)		300	250	320

Application range Shown in Table 5.68.

Table 5.68 Application range of C.I. Vat Red 41

PS	●	PMMA	●	ABS	○
SAN	●	PA6 / PA66	×	PC	●
PVC (U)	●			PET	○
				PBT	○

● Recommended to use, ○ Conditional use, × Not recommended to use.

Variety characteristics Vat Red 41 has excellent thermal resistance and good light fastness (pure tone), but the light fastness decreases with the increasing of white reduction. It is applied to the coloring of engineering plastics.

② **C.I. Pigment Red 181 (Vat Red 1)** C.I.: 73360. Formula: $C_{18}H_{10}Cl_2O_2S_2$. CAS [2379-74-0].

Bright red with fluorescence, melting point above 300 ℃.

Structural formula

Main properties Shown in Table 5.69.

Table 5.69 Main properties of C.I. Pigment Red 181

Project	PS	ABS	PC	PET
Pigment / %	0.05	0.1	0.05	0.02
Titanium dioxide / %	1.0	1.0		
Light fastness degree	5	4~5	8	7~8
Thermal resistance / ℃	300	280	360	300
Weather resistance degree (3000h)			4~5	

Application range Shown in Table 5.70.

Table 5.70 Application range of C.I. Pigment Red 181

PS	●	PMMA	●	ABS	○
SAN	●	HIPS	●	PC	●
PVC-(U)	○	PA6	×	PET	●
POM	○	PA66	×	PBT	●

● Recommended to use, ○ Conditional use, × Not recommended to use.

Variety characteristics Pigment Red 181 has good light fastness, and thermal resistance, which can be used in the coloring of engineering plastics. It shows excellent light fastness and weather resistance in the coloring of PET and PC with unique bluish red.

5.5.2.6 Fused Ring Solvent Dyestuffs

① **C.I. Solvent Yellow 98** C.I.: 56238. Formula: $C_{26}H_{45}NO_2S$. CAS [27870-92-4]. Greenish yellow with fluorescence, melting point 115℃.
Structural formula

Main properties Shown in Table 5.71.

Table 5.71 Main properties of C.I. Solvent Yellow 98

Project		PS	ABS	PC	PEPT
Light fastness degree	1/3 SD white reduction	4~5	3	6	7
	1/25 SD transparent	7	5~6	7~8	7
Thermal resistance (1/3 SD) / (℃/5min)		300	300	340	300

Application range Shown in Table 5.72.

Table 5.72 Application range of C.I. Solvent Yellow 98

PS	●	PMMA	●	PC	●
SAN	●	PA6 / PA66	○	PET	●
PVC-(U)	●	ABS	●	PBT	○

● Recommended to use, ○ Conditional use.

Variety characteristics Solvent Yellow 98 has good light fastness which decreases with the increasing of white reduction. It shows excellent thermal resistance in the coloring of engineering plastics with brilliant transparent color. It is suitable for the pre-coloring of spinning of PET, PA6 and should be careful for PA66.

② **C.I. Solvent Red 52** C.I.: 68210. Formula: $C_{24}H_{18}N_2O_2$. CAS [81-39-0]. Bluish red, melting point 280℃.
Structural formula

Main properties   Shown in Table 5.73.

Table 5.73   Main properties of C.I. Solvent Red 52

Project		PS	ABS	PC	PET
Tinting strength (1/3 SD)	Dye / %	0.195	0.39	0.100	
	Titanium dioxide / %	2	4	1	
Light fastness degree	1/3 SD white reduction	3~4		4~5	
	1/25 SD transparent	7		7	
Thermal resistance (1/3 SD) / (°C/5min)		280	280	350	290

Application range   Shown in Table 5.74.

Table 5.74   Application range of C.I. Solvent Red 52

PS	●	SB	●	ABS	●
SAN	●	PMMA	●	PC	●
PVC-(U)	●	PPO	●	PET	●
POM	○	PA6 / PA66	●	PBT	○
PES fiber	○				

● Recommended to use, ○ Conditional use.

Variety characteristics   Solvent Red 52 has good light fastness which decreases with the increasing of white reduction. It shows excellent thermal resistance and can be used in the coloring of engineering plastic such as PA6 and PET. Its thermal resistance is up to 300°C in PA6 and it should be carefully applied to the pre-coloring of spinning of PA6 and PA66.

③ **C.I. Solvent Orange 63**   C.I.: 68550. Formula: $C_{23}H_{12}OS$. CAS [16294-75-0]. Reddish orange with fluorescence, transparent, bright color, melting point 320 °C.

Structural formula

Main properties   Shown in Table 5.75.

Table 5.75   Main properties of C.I. Solvent Orange 63

Project			PS	ABS	PC	PEPT
Pure tone	0.05%	Light fastness degree	7	6~7	7	7
		Thermal resistance (°C/5min)	300	300	340	300
White reduction	0.2% TiO$_2$ 1%	Light fastness degree	4	4	6	4
		Thermal resistance (°C/5min)	7	6~7	7	7

Application range   Shown in Table 5.76.

Table 5.76   Application range of C.I. Solvent Orange 63

PS	●	PMMA	●	ABS	●
SAN	●	PA6 / PA66	○	PC	●
PVC-(U)	●			PET	●
				PBT	●

● Recommended to use,   ○ Conditional use.

Variety characteristics   Solvent Orange 63 shows excellent light fastness and thermal resistance, and the recommend concentration is less than 0.3%, which can be used in the coloring of engineering plastics. It is also applied to the pre-coloring of spinning of PET, PA6 and should be careful for PA66.

5.5.3   Methine Solvent Dyestuffs

The variety of methine solvent dyestuffs is few and limited to the light chromatography. There are six varieties abroad of which the most important are Solvent Yellow 93, Solvent Yellow 133, Solvent Yellow 145, Solvent Yellow 179, Solvent Orange 80 and Solvent Orange 107.

① **C.I. Solvent Yellow 93**   C.I.: 48160. Formula: $C_{21}H_{18}N_4O_2$. CAS [4702-90-3]. Mid shade yellow, melting point 180 ℃.

Structural formula

Main properties   Shown in Table 5.77.

Table 5.77   Main properties of C.I. Solvent Yellow 93

Project		PS	ABS	PC
Tinting strength (1/3 SD)	Dye / %	0.26	—	0.142
	Titanium dioxide / %	2	—	1
Thermal resistance/ (℃/5min)	Pure tone 0.05%	300	Not applicable	340
	White reduction 1:20	300		340
Light fastness degree	Pure tone 0.05%	8		8
	1/3 SD	7~8		7

Application range   Shown in Table 5.78.

Table 5.78 Application range of C.I. Solvent Yellow 93

PS	●	SB	●	ABS	×
SAN	●	PMMA	●	PC	●
PVC (U)	●	PPO	●	PET	●
POM	○	PA6 / PA66	×	PBT	○
PES fiber	×				

● Recommended to use, ○ Conditional use, × Not recommended to use.

Variety characteristics Solvent Orange 93 has good light fastness and excellent thermal resistance, which can be used in the coloring of styrenic engineering plastics but not recommend in ABS. It is the standard color of greenish yellow and has good performance price ratio.

② **C.I. Solvent Yellow133**   C.I.: 48580. Formula: $C_{30}H_{24}N_2O_6$. CAS [51202-86-9]. Bright greenish yellow.

Structural formula

$$H_3CO-\!\!\!\bigcirc\!\!\!-CH=\!\!\!\bigcirc\!\!\!-\!\!\!\bigcirc\!\!\!-OCH_3$$

Main properties Shown in Table 5.79.

Table 5.79 Main properties of C.I. Solvent Yellow 133

Project		PC	PET
Tinting strength (1/3 SD)	Dye / %	0.36	0.32
	Titanium dioxide / %	1	1
Thermal resistance/ ℃	1/3 SD	330	300
Light fastness degree	1/25 SD (pure tone)	5	6
	1/3 SD (white reduction)	5	5

Application range Shown in Table 5.80.

Table 5.80 Application range of C.I. Solvent Yellow 133

PS	×	PMMA	○	PC	●
PVC-(U)	×	PA6 / PA66	×	PET	●
PES fiber	×	ABS	×	PBT	●

● Recommended to use, ○ Conditional use, × Not recommended to use.

Variety characteristics Solvent Yellow 133 has high tinting strength, acceptable light fastness and excellent thermal resistance, which can be used in the coloring of engineering plastics. It is particularly recommended for the pre-coloring of spinning of

PET and shows good washing and abrasion fastness during heat setting process of fibers.

③ **C.I. Solvent Yellow 179 (Disperse Yellow 201)**   CAS [80748-21-6].

Greenish yellow, melting point 115℃.

Main properties   Shown in Table 5.81.

Table 5.81   Main properties of C.I. Solvent Yellow 179

Project		PS	ABS	PC
Tinting strength (1/3 SD)	Dye / %	0.36	0.165	0.070
	Titanium dioxide / %	2	4	1
Thermal resistance/ (℃/5min)	Pure tone 0.05%	300	240~260	350
	White reduction 1:20	300	240~260	350
Light fastness degree	Pure tone 0.05%	8		8
	1/3 SD	7~8		7

Application range   Shown in Table 5.82.

Table 5.82   Application range of C.I. Solvent Yellow 179

PS	●	SB	●	ABS	●
SAN	●	PMMA	●	PC	●
PVC-(U)	●	PPO	●	PET	●
POM	○	PA6 / PA66	×	PBT	○
PES fiber	○				

● Recommended to use, ○ Conditional use, × Not recommended to use.

Variety characteristics   Solvent Yellow 179 has good light fastness and excellent thermal resistance, which can be used in the coloring of engineering plastics. It is particularly recommended for the pre-coloring of spinning of PET.

④ **C.I. Solvent Orange 107 (Disperse Orange 47)**   CAS [185766-20-5].

Reddish orange, melting point 115℃.

Main properties   Shown in Table 5.83.

Table 5.83   Main properties of C.I. Solvent Orange 107

Project		PS	ABS	PC
Tinting strength (1/3 SD)	Dye / %	0.090	0.18	0.045
	Titanium dioxide / %	2	4	1
Thermal resistance/ (℃/5min)	Pure tone 0.05%	300	280	320
	White reduction 1:20	300	280	320
Light fastness degree	Pure tone 0.05%	7~8		8
	1/3 SD	5~6		5

Application range   Shown in Table 5.84.

Table 5.84　Application range of C.I. Solvent Orange 107

PS	●	SB	●	ABS	●
SAN	●	PMMA	●	PC	●
PVC-(U)	●	PPO	●	PET	●
POM	○	PA6 / PA66	×	PBT	●
PES fiber	×				

● Recommended to use, ○ Conditional use, × Not recommended to use.

Variety characteristics　Solvent Orange 107 has excellent thermal resistance, acceptable light fastness which decreases with the increasing of white reduction. It can be used in the coloring of engineering plastics and the spin-dyeing of polyester.

5.5.4　Azo Solvent Dyestuffs

Azo solvent dyestuffs have simple structure and are easily synthesized, the main color spectrums are red, orange and yellow, but the properties are not good. In the application of plastics the variety of azo solvent dyestuffs is few and azo metal complex structure plays an important role in the solvent dyestuffs.

①　**C.I. Disperse Yellow 241**　C.I.: 128450. Formula: $C_{14}H_{10}Cl_2N_4O_2$.

Bright greenish yellow, melting point 254 ℃.

Structural formula

Main properties　Shown in Table 5.85.

Table 5.85　Main properties of C.I. Disperse Yellow 241

Project		PS	PC
Tinting strength (1/3 SD)	Dye / %	0.08	0.042
	Titanium dioxide / %	2	1
Thermal resistance/℃	1/3 SD	280	350
Light fastness degree	Pure tone 0.05%	7	6～7
	1/3 SD	5	6

Application range　Shown in Table 5.86.

Table 5.86　Application range of C.I. Disperse Yellow 241

PS	●	SB	●	ABS	×
SAN	●	PMMA	●	PC	●
PVC-(U)	●	PPO	×	PET	●
POM	○	PA6 / PA66	×	PBT	●
PES fiber	×				

● Recommended to use, ○ Conditional use, × Not recommended to use.

Variety characteristics Disperse Yellow 241 has good light fastness and thermal resistance, which can be used in the coloring of engineering plastics but not recommend in ABS.

② **C.I. Solvent Orange 116** Bright reddish orange.

Main properties Shown in Table 5.87.

Table 5.87 Main properties of C.I. Solvent Orange 116

Project	PS	ABS	PC	PET	PA6
Dye / %	0.05	0.1	0.05	0.02	0.1
Titanium dioxide / %	1	1			0.5
Thermal resistance / ℃	300	300	360	310	300
Light fastness degree	6	6~7	7~8	8	
Weather resistance degree	3~4		4~5	4	

Application range Shown in Table 5.88.

Table 5.88 Application range of C.I. Solvent Orange 116

PS	●	SB	●	ABS	●
SAN	●	PMMA	●	PC	●
PVC-(U)	●	PA6 / PA66	●	PET	●
				PBT	●

● Recommended to use.

Variety characteristics Solvent Orange 116 has high tinting strength and excellent light fastness, thermal resistance, weather resistance and sublimation resistance, which is especially recommended for the coloring of PA6 and PA66.

③ **C.I. Solvent Red 195** Bright bluish red, melting point 214 ℃.

Main properties Shown in Table 5.89.

Table 5.89 Main properties of C.I. Solvent Red 195

Project		PS	ABS	PC	PEPT
Tinting strength (1/3 SD)	Dye / %	0.56	0.58	0.59	0.4
	Titanium dioxide / %	1.0	1.0	1.0	1.0
Light fastness degree	1/3 SD white reduction	5	5~6	7	7
	1/25 SD transparent	8	6~7	8	8
Thermal resistance (1/3 SD) / (℃/5min)		300	280	330	310

Application range Shown in Table 5.90.

Table 5.90 Application range of C.I. Solvent Red 195

PS	●	SB	●	ABS	●
SAN	●	PMMA	●	PC	○
PVC-(U)	●	PPO	×	PET	●
POM	×	PA6 / PA66	×	PBT	○
PES fiber	×				

● Recommended to use, ○ Conditional use, × Not recommended to use.

Variety characteristics Solvent Red 195 has excellent light fastness, thermal resistance, and weather resistance, which can be used in the coloring of engineering plastics. It is recommended for the coloring of spinning of PET.

④ **C.I. Solvent Yellow 21** C.I.: 18690. Formula: $C_{34}H_{24}CrN_8O_6 \cdot H$. CAS [5601-29-6].

Reddish yellow.

Structural formula

Main properties Shown in Table 5.91.

Table 5.91 Main properties of C.I. Solvent Yellow 21

Project	PA6 fiber 110 dtex/24 f-round cross-section	
Dye	0.1%	1.0%
Light fastness degree	6	7

Variety characteristics Solvent Yellow 21 has excellent light fastness, thermal resistance, and weather resistance. It shows good compatibility with PA6 and good color fastness in textiles. It is also recommended for the coloring of carpet fibers.

⑤ **C.I. Solvent Red 225** Yellowish red.

Main properties Shown in Table 5.92.

Table 5.92 Main properties of C.I. Solvent Red 225

Project	PA6 fiber 110 dtex/24 f-round cross-section	
Dye	0.1%	1.0%
Light fastness degree	6~7	7

Variety characteristics Solvent Red 225 has high tinting strength, excellent light fastness and thermal resistance. It shows good compatibility with PA and can be used in the coloring of PA6. It is also suitable for carpet and textile fibers.

5.5.5 Azomethine Solvent Dyestuffs

The variety of azomethine solvent dyestuffs is only six, in addition to Solvent Yellow 79, the rest are metal complexes. Besides traditional nickel complex, there are cobalt and copper complexes. Azomethine solvent dyestuffs have particularly excellent light fastness.

① **C.I. Solvent Violet 49**   C.I.: 48520. Formula: $C_{27}H_{14}N_4Ni_4O_4$. CAS [205057-15-4].

Dark reddish violet, melting point above 300℃.

Structural formula

Main properties   Shown in Table 5.93.

Table 5.93   Main properties of C.I. Solvent Violet 49

Project		ABS	PC	PEPT
Tinting strength (1/3 SD)	Dye / %	0.090	0.064	0.051
	Titanium dioxide / %	1.0	1.0	1.0
Light fastness degree	1/3 SD white reduction	6~7	7~8	7~8
	1/25 SD transparent	6~7	8	8
Thermal resistance (1/3 SD) / (℃/5min)		240	300	320

Application range   Shown in Table 5.94.

Table 5.94   Application range of C.I. Solvent Violet 49

PS	○	PMMA	●	ABS	○
SAN	○	PA6 / PA66	×	PC	●
PVC-(U)	×			PET	●
				PBT	●

● Recommended to use,   ○ Conditional use,   × Not recommended to use.

Variety characteristics   Solvent Violet 49 has excellent light fastness and thermal resistance, high tinting strength, which is recommended for the spin-dyeing of polyester.

② **C.I. Solvent Brown 53**   C.I.: 48525. Formula: $C_{18}H_{10}N_4NiO_2$. CAS [64696-98-6].

Dark reddish brown, melting point above 350 ℃.

Structural formula

Main properties   Shown in Table 5.95.

Table 5.95 Main properties of C.I. Solvent Brown 53

Project		PS	ABS	PC	PEPT
Tinting strength (1/3 SD)	Dye / %	0.12	0.13	0.11	0.082
	Titanium dioxide / %	1.0	1.0	1.0	1.0
Light fastness degree	1/3 SD white reduction	7	7	8	8
	1/25 SD transparent	8	8	8	8
Thermal resistance (1/3 SD) / (°C/5min)		300	300	340	320

Application range Shown in Table 5.96.

Table 5.96 Application range of C.I. Solvent Brown 53

PS	●	PMMA	●	ABS	○
SAN	●	PA6 / PA66	×	PC	●
PVC-(U)	×			PET	●
				PBT	●

● Recommended to use, ○ Conditional use, × Not recommended to use.

Variety characteristics Solvent Brown 53 has excellent light fastness and thermal resistance. It is particularly suitable for the pre-coloring of spinning of PET, and the fabric is excellent in light fastness, wet processing, and friction fastness. It is also suitable for the blow molding of polyester bottles.

5.5.6 Phthalocyanine Solvent Dyestuffs

Phthalocyanine solvent dyestuffs show bright greenish blue, many of which are copper phthalocyanine derivatives. They are insoluble in most of organic solvents, but can be converted into solvent dyestuffs of good solubility in organic solvents by introducing some lipophilic groups to molecules. They also have bright greenish blue varieties which anthraquinone solvent dyestuffs do not character, and show good fastness properties. There are only seven varieties abroad, and the main variety in the application of engineering plastics is Solvent Blue 67.

C.I. Solvent Blue 67 CAS [81457-65-0]. Greenish blue.

Main properties Shown in Table 5.97.

Table 5.97 Main properties of C.I. Solvent Blue 67

Project	PS	ABS	PC	PET
Dye / %	0.05	0.1	0.05	0.02
Titanium dioxide / %	1.0	1.0		
Light fastness degree	5	3~4	3~4	7
Thermal resistance / °C	260	290	290	300
Weather resistance degree (3000h)			3~4	

Application range　Shown in Table 5.98.

Table 5.98　Application range of C.I. Solvent Blue 67

PS	○	PMMA	○	ABS	○
SAN	●	PA6	○	PC	○
PVC-(U)	●	PA66	×	PET	●
POM	×			PBT	●

● Recommended to use, ○ Conditional use, × Not recommended to use.

Variety characteristics　Solvent Blue 67 has good light fastness, thermal resistance and sublimation resistance. It is suitable for the pre-coloring of spinning of PET, and can be used in clothing and indoor decorations.

Chapter 6
Plastic Colorants Inspection Method and Standard

6.1 Inspection and Standard

6.1.1 Importance of Inspection

The inspection of colorant includes the quality assurance and the property. affirmation. For the instruments and equipment applied to the inspection of colorant, there is a series of executive standards that are recognized all over the world. It is the common language for all these colorant industrial chain, including producer, retailer, user, these standards is the common language for judging the quality and application performance of the same product. They are commonly used throughout the entire industrial chain while include the material analyze, the quality control, the quality restriction of finished product and the judgment in application.

The character of colorant product is the same as other industrial products which is a very complicated process. Whether raw material, equipment, environment, personnel operation, or else a lot of factors could influence the production process directly or indirectly, eventually lead to the fluctuation of product's quality. Furthermore, the production of colorants consist of a series procedures, it is impossible for each procedure to always stay the same. In the enterprise, the production is a complex process, people, machines, materials, environments and the other factors will influence the production. Namely each working procedure can not keep absolutely steady, so it is merited that the quality of the product can fluctuate. The role of the test is to keep the fluctuation of the quality of the products in a reasonable and acceptable range, and to ensure the quality of the final product is within the limited range of the standard. Therefore, for the production and application of colorant, it should be stressed to strict the inspection, to standardize the inspecting approach, to form a industry mechanism, and to be put into practice.

The unification and standardization of the inspection is foundamental work of great importance in the colorants inspection. Therefore, when the inspection methods are unified and standarlized the data and results from the inspections get the same cognitive of all sides. And it will be favorable for the communication among all relevant parties, thus ensure that the inspection data can process true practical values.

6.1.2 Adopt International Standards, Improve Enterprise Competitiveness

Standardization is the key to promote trading and technical exchanges. With the rapid development in science and technology and the fast progress of the process of global economic integration, the degree of marketization of the society is higher and higher. As one of the important elements of modernization, standardization, is also committed to improve efficiency and quality, and to reflect the scientific and reasonable of whatever work. Standard is the passport to the enormous international market with the progress of global economic integration.

Through the essence of market economy, standardization is a effective measure to improve the market competitiveness. Adopting international standards has become key to the promotion of trading and technical exchanges, and it helps improve the quality of products, enhance the competitiveness of products and expand exportation.

There are different kinds of standards, including the international standards, the government standards, the industrial standards, the enterprise standards and so on. Right now, the standardization of the colorant for plastic is relatively hysteretic. Practitioner is transforming the test methods from the ones used in printing ink to the ones used in plastic especially, while still a lot of the currently-used test methods can only be applied in the extensive mode of production, unable to reflect the real application properties of pigments in plastic. China formulated 13 pieces of industrial inspection standards, including the dispersibility, heat resistance, immigration and color properties of pigments and so on, which are highly anastomotic to the current international standards, and filled the gaps in the area of pigment standards at home. The successful formulation and implementation of the standards laid a solid foundation for improving; the international development and the competitiveness in the global market of the nation's colorant products. Meanwhile it also provides a platform for communication among the producer and the users of the industry, which contributes to the development of pigment and color masterbatch industries. This section generally introduces the professional test methods of colorant applied in plastic through taking the international standards for examples.

6.1.3 Quality Control Test and Application Performance Test

Quality control test and application performance test of colorant are two different aspects.

The main of quality control test is to control the quality of product to stay stable, and is used to judge whether the product is up to standard or not. With advanced testing methods and strict control of the standards, the quality of our products are ensured to be stable and reliable.

For colorants, acquiring the results of quality control test is not enough. When the colorant is used in plastic, it will be faced with different shaded mediums, all sorts of processing equipments and different processing conditions, which requires the colorant to withstand and get through the harsh test and keep its original features. This requires the colorant to have the features to correspond the requirements of plastic processing, such as the heat-resistant performance, the easy dispersion performance, etc.; Moreover, the colored plastic products may problem when being used. For example, the extravagant exposure under the sun when being used outdoors,the risk of chemical reactions with the product when applied to wrappers, and the safety issue concerning the toxicity when applied to food wrappers or children's toys marketing. All those boils down to application performance index of the colorant. Therefore, we can divide the application performances index into two parts: conforming to the performance of the processing of plastic products and conforming to the performance when being used ultimately. In view of the plastic colorants, all the testing of using performance indexes should completely according to (or simulate) the actual situation of plastic product processing and using conditions.

The final shader quality and color effect of color plastic products are not only based on the color performance of colorant, machinability and application performance, but also closely related to the pitting formula of plastic products, relevant products processing technology, application conditions and the environment, and the shape and size of the products.

Application of colorants in plastics is shown in Figure 6.1.

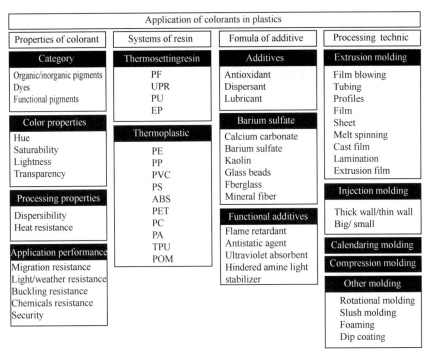

Figure 6.1　Main factors related to color performance of plastics

6.1.4　Standards

6.1.4.1　Grey Scale for Assessing Discoloration (GB/T250—2008 or ISO 105/A02—1993)

(1) Principle

The 5-step, scale for assessing discoloration consists of five pairs of non-glossy grey or white color chips (or swatches of grey or white cloth), which illustrate the perceived color diferences corresponding to fastness ratings 5, 4, 3, 2 and 1.This essential scale may be augmented by the provision of similar chips or swatches illustrating the perceived color differences correspongding to half-step fastness ratings 4-5, 3-4, 2-3 and 1-2, and scales could be termed 9-step scales. The first member of each pair is neutral gray in color and the second member of the pair illustrating fastness rating 5 is identical with the first member. The second members of the remaining pairs are increasingly lighter in color so that each pair illustrates increasing larger color differences by contrast. The perceived color difference of each grade is determined by chromaticity. The perceived color differences are defined colorimetrically. The full colormetric specification is given in Figure 6.2. The colorimetric data is provided by standard GB, ISO, AATCC and DIN, and it could be measured with a spectrophotometer with the specular component included.

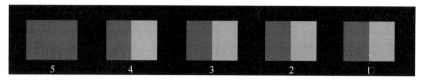

Figure 6.2 Grey scale for assessing discoloration(ISO 105/A02—1993)

(2) Use of the scale

The grey scale for assessing discoloration is used as a criteria in testing the colrants's resistance against weather,acid/alkali or other chemicals when used in plastics.

Place a piece of the unstained, adjacent fabric (the original piece) and the piece which has been part of a composite specimen in a fastness test (the testing specimen) side by side in the same plane. The surrounding should be in a neutral grey color approximately midway between grade 1 and grade 2 of the grey scale for assessing change in color. If necessary to avoid effects of the backing on the appearance of the textiles, use two or more layers of the unstained undyed textile under both original and treated pieces. Illuminate the surfaces with north sky light in the Northern hemisphere, or an equivalent source with an illumination of 600 Lux or more. The light should be incident upon the surfaces at approximately 45°, and the direction of viewing approximately perpendicular to the plane of the surfaces. Compare the visual difference between the original piece and the tested specimen with the differences represented by the grey scale.

If the 5-step grade is used, the degree of fastness of the testing piece is a grade of the grey scale which has a perceived color difference equal to the perceived color difference between the original and the tested pieces. If the latter is judged to be nearer a imaginary grade lying midway between two adjacent pairs than it is to either, the testing piece is given an intermediate grade, like 4-5 or 2-3.

When a number of assessments have been made, it is necessary to compare all the pairs of original and testing specimens which have been given the same grade. This gives a good indication of the consistency of the assessments, and any errors become prominent. Pairs which do not appear to have the same degree of contrast as the remainder of their groups should be re-checked against the grey scale and if it is necessary, the rating should be changed.

6.1.4.2 Grey Scale for Assessing Staining(GB/T 251—2008 or ISO 105/A03—1993)

(1) Principle

The basic, or 5-grade, scale consists of five pairs of non-glossy grey or white color chips (or swatches of grey or white cloth), which illustrate the perceived color differences corresponding to fastness ratings 5, 4, 3, 2 and 1.This basic scale may be augmented by the provision of similar chips or swatches illustrating the perceived color differences correspongding to half fastness ratings 4-5, 3-4, 2-3 and 1-2, and the scales could be termed 9-step scales. The first piece of each pair is white in color and the second piece illustrates the degree of staining. For example, if the second piece indicates a grade 5 fastness, then it will be identical to the first piece. The second pieces of the remaining pairs are increasingly darker in color so that each pair illustrates increasingly larger color differences by contrast, and the perceived color differenceof each grade is determined by chromaticity. The full colormetric specification is given in Figure 6.3.

Figure 6.3　Grey scale for assessing staining (ISO 105/A03—1993)

The color grades of the grey scale are set totally according to the colorimetric data provided by standard GB, ISO, AATCC and DIN, and the colorimetric data of each grade are determined using spectrophotpmeter with high accuracy.

(2) Use of the scale

Grey scale for assessing staining is used as criteria for judging the degree of contamination of the testing object, in tests inspecting the colorant's migration performance, acid/alkaki resistance or other chemical resistance according to the degree of staining of the sample piece.

The Grey scale for assessing staining can be used by referance to the use of the scale in 6.1.4.1.

6.1.4.3　Wool Standards (GB730—1998)

(1) Principle

The blue-colored wool color-fastness to light standards consist of 8 grades, representing 8 color-fastness ratings: 8, 7, 6, 5, 4, 3, 2 and 1. When exposed to sunlight, grade 1 will fade most while grade 8 fade least. If a grade 4 fade to a certain degree after a period of time under certain specific light, then in order to fade to the same degree, a grade 3 will take about half the time, while a grade 5 will take about twice.

The blue wool standards is made of wool fabric dyed with 8 kinds of colorants whose depths are specified (see in Table 6.1).

Table 6.1　Kinds of eight wool reference material

Rating	Reference number	Rating	Reference number
1	C.I. Acid Blue 104	5	C.I. Acid Blue 47
2	C.I. Acid Blue 109	6	C.I. Acid Blue 23
3	C.I. Acid Blue 83	7	C.I. Soludilized Vat Blue 5
4	C.I. Acid Blue 121	8	C.I. Soludilized Vat Blue 8

The American-made samples for their wool standards ranges from L2 to L9. These 8 samples are composed of two kinds of fibers mixed by different percentages, one dyed by C.I. Mordant Blue 1 and the other C.I. Soludilized Vat Blue 8, making the color fastness of a higher grade almost double that of its former grade.

L2 to L9 applies to exposure conditions in America in standards such as GB/T 8427—2008, ISO 105 B02 and AATCC TM 16.

Special attention should be paid that grades 1-8 and grades L2-L9 can't be used together, so as their results.

(2) Use of the scale

The blue wool standards are used for assessing the grades of color-fastness of the test samples after opicical radiation resistance tests.

Both the test samples and the blue wool reference samples are exposed to the same sunlight for the same time. Then compare the degree of fading of the test samples to that of the standard samples to evaluate the grade of color fastness. If the degree of fading of the test samples is similar to one of the standards, then its grade of color-fastness to light is the same as the standard sample. When the degree of fading is between two of the grades, then take it as an intermediate grade, like 4~5,6~7 for example.

6.1.4.4　Standard Depth of Color

In the colorimetric, the absolute depth of pigments can be obtained through the K/S prepared in the standard condition, expressed by the amount necessary to got a certain depth.

The concept of standard depth is well used to evaluate the tinting strength of colorants in palstics. Following is the general method.

Make the plastic colored by the way specified, with certain amount of decolorant agent, TiO_2. The tinting strength is evaluated according to the amount of pigments additive to the plastic through which the plastic can got 1/3 depth of the standard depth.

6.2　Quality Tests and Standards of Color Performance

The test methods in the color performance of colorant(hue, lightness, saturation, tinting strength) abroad as early as the German industrial standards of DIN 53775, but it has been gradually replaced by EN BS14469-1-2004.

Our country has formulated the corresponding industry standards, and the national standard revision of the program are being carried out. The basic method is as follows.

6.2.1 Composition of the Basic Mixtures of PVC

(1) Composition of the basic mixtures of PVC are listed in Table 6.2. Principal properties are listed in Table 6.3.

Table 6.2 Composition of the basic mixtures of PVC

Material	Components/phr	
	Basic mixture A	Basic mixture B
Vinyl chloride homopolymer	65.0	65.0
Plasticizer	33.5	33.5
Epoxidized soybean oil	1.5	1.5
Stabilize	1.3	1.3
Lubricant	0.2	0.2
Titanium dioxide pigment	0	5.34
Total	1015	106.84

Table 6.3 Values of constituents of the composition of the basic mixtures of PVC

Material	Property	Value	Testing Standord
Vinyl chloride homopolymer (PVC)	K value	70±1	EN ISO 1628-2
Diisodecyl phthalate (DIDP)	Refractive index	1.4850~1.4860	EN ISO 489
	Color value	≤200	ISO 6271
Epoxidized soybean oil (ESO)	Refractive index	1.4720~1.4740	EN ISO 489
	Iodine color value ICV/(mg/100ml)	≤3	DIN 6162
	Content of epoxide oxygen/%	6.2~6.8	ASTM D 1652
Liquid Ba/Zn stabilizer system	Barium content/%	10.2~12.2	Determination by atomic absorption spectrometry
	Zinc content/%	1.95~2.35	
Stearic acid	Refractive index	1.4970~1.5010	EN ISO 489
	Density (80 ℃)/(g/ml)	0.840~0.850	EN ISO 12185
	Acid value/(mg/KOH/g)	206~211	EN ISO 2114
Titanium dioxide pigment (TiO2)	Titanium dioxide (TiO_2) content/%	92~98	EN ISO 591-1
	Rutile content of the titanium dioxide part inthe pigment/%	≥98	
	Silicon dioxide (SiO_2) content and/ oraluminium oxide (Al_2O_3) content/%	2~8	

The composition of the basic mixtures of PVC is adapted by test of heat resistance, dispersibility, and migration resistance in this chapter. The technical specifications of every materials in the composition is made clear.

(2) Preparation of basic mixtures

PVC, stabilizer and lubricant are premixed in a high-speed mixer until the mixing temperature is 70 ℃. In the case of basic mixture B, the titanium dioxide pigment is added and mixed for about 2 min (if the mixing time is longer there is a risk of greying due to metal abrasion). The premixed amounts of plasticizer and epoxidized soybean oil are added as a fine stream while the mixer is running. It becomes homogeneous through high - speed agitation, the temperature of which has risen to about 100 ℃ as a result of the mixing procedure, is cooled to room temperature with continuing agitation.

Storage of basic mixtures

The storage time of the basic mixtures should not exceed 2 years, and must be hermetically sealed and protected from exposure to light.

6.2.2 Test of Color Performance

6.2.2.1 Apparatus

Two-roll mill The mill can be heated, and the roll spacing can be adjusted. Roll diameters are between 80 mm and 200 mm, ratio of the speeds of rotation of the two rollers are between 1 : 1.1 and 1 : 1.2. The rolls should have a chromed surface.

Plate press It can be heated and cooled.

Mold Two specular and stainless molds, with thickness of 1.0mm to 1.2mm; cupreous die frame, with thickness of 1.0mm to 1.2mm.

6.2.2.2 Preparation of the Test Specimens

(1) Composition and premixing

The quantity of coloring material specified for the particular test sample is mixed for example for 5 min in a shaking mixer with the PVC basic mixture on which the test sample is based. Alternatively, in the particular case of coloring materials in paste form, it is recommended that premixing be carried out by hand using a spatula in a polyethylene or polypropylene beaker, until the mixture appears homogeneous. Composition of the the test specimens is listed in Table 6.4.

Table 6.4 Composition of the test specimens

Projects		Full shade	Reduced shade
Mixtures of PVC		Basic mixture A	Basic mixture B
Part by mass of the pigments	Organic pigments	0.10%	Pigments:TiO_2=1 : 10
	Inorganic pigments	0.50%	Pigments:TiO_2=1 : 4

(2) Two-roll milling and drawn off

The mix is placed on the two-roll mill, the roll surfaces of which have been adjusted to a temperature of $(160 \pm 5)\,°C$. A temperature difference between the rolls is permissible if it is within the limits. The amount of mix is judged so that once a milled sheet has formed there is always a rotating bank of molten material in the gap between the rolls. The gap between the rolls is adjusted so that the milled sheet has a uniform thickness of 0.4 mm to 0.5 mm. The sheet is formed in such a way that the whole of the material forms a continuous sheet on the front roll. During the rolling process, the milled sheet shall be continuously turned over or removed and reapplied, so as to give thorough mixing. After the mix has been applied it should be worked for 200 roll revolutions. The rolling time shall be at least 5 min and not exceed 10 min.

The milled sheet is then drawn off. To this end, it is permissible to change the gap between the rolls, the speed of rotation and friction.

(3) Pressing of test specimens

It may be necessary to press the milled sheet in order to improve the surface quality, e.g. to achieve high gloss or to create a sufficiently thick sample for photometric measurement. The sections of the sheet and a spacer frame of the desired thickness, are placed between high-gloss chromed plates in a plate press heated to a temperature between 165°C and 170°C, and pressed for not more than 2 min. The pressed sheet is then cooled rapidly until 50°C.

6.2.3 Evaluation of Color Performance

(1) Evaluation of tinting strength

Tinting strength is one of the most important index for all the colorant. It embodies the use value of pigment directly. It means that the inspection of tinting strength must be compared with reference. The comparison are usually between the production and the standard or the same production only.

The reduced color is used to determine the tinting strength of the colorant, for the reduced color will magnify the slight difference of tinting strength.

The testing pigment, PT and reference pigment, PR in same weight are respective put into PVC in same weight. The PVC is reduced by titanium dioxide in same weight. The tinting strength is obtained according to the formula following.

$$\text{tinting strength} = \frac{(K_{PT}/S)}{K_{PR}/S} \times 100\%$$

Where, K indicates Light absorption coefficient of the opacities in the wavelength of λ; S indicates light-scattering coefficient of the opacities in the wavelength of λ; K_{PT}/S

indicates K/S of simple pigment R_∞ or ρ_∞; K_{PR}/S indicates K/S of reference pigment R_∞ or ρ_∞.

At present, with the development and popularity of the testing equipment, K/S data can be accessd by professional color instrument and equipment directly instead of heavy calculation.

If no automatic detection instruments can be obtained, visual for judging will be accepted.

(2) Evaluation of transparency or hiding power

Index of transparency and hiding power is corresponding to each other, for example, the higher the transparency, the lower the covering power, and vice versa.

Colorful plastic samples are made to determination the transparency and hiding power of pigments. Make the samples according to 6.2.2.2, then put the samples on paper Black & White Chequered, and observe the deference in the part Black & White Chequered between the under layer through the specimen to judge the level of transparency.

Measure and compare the color difference with the help of spectrophotometert between the white board and black board. The bigger the color difference, the higher the transparency.

(3) Evaluation of color difference

Differences in color, lightness and saturation between testing pigments and reference pigments can be obtained by chromaticity coordinates in CIE 1976 ($L^*a^*b^*$). According to GB 11186.2, the chromaticity coordinates of test specimen, L_T^* a_T^* b_T^* and chromaticity coordinates of reference samples, L_R^* a_R^* b_R^* will be tested under a specified conditions.

Total color difference, ΔE_{ab}^* is the distance between two location in the CIE 1976($L^*a^*b^*$) of two kind of color. ΔE_{ab}^* can be obtained through the formula following.

$$\Delta E_{ab}^* = [(\Delta L^*)^2 + (\Delta a^*)^2 + (\Delta b^*)^2]^{1/2}$$

Where:

$\Delta L^* = L_T^* - L_R^*$
$\Delta a^* = a_T^* - a_R^*$
$\Delta b^* = b_T^* - b_R^*$

At present, with the development and popularity of the testing equipment, data above can be accessd by professional color instrument and equipment directly instead of heavy calculation.

It is noteworthy that some coloring materials can change color reversibly after moulding and the mouldings should be maintained at room temperature for at least 16h before assessment.

6.3 Evaluation of Color Stability to Heat During Processing of Coloring Materials in Plastics

Heat resistance is one of the most important properties for the pigments to apply in plastics. There are three test methods.

6.3.1 Evaluation by Injection Moulding (HG/4767.2—2014, EN BS 12877-2)

(1) Apparatus

Screw-injection machine with a screw diameter of 18mm to 30mm and an effective screw length of at least 15 times the screw diameter.

(2) Test specimens testing in the specified plastic material

Such as standard material HDPE, PP, ABS or other materials provided by customer.

(3) Concentration of coloring material

For testing in full shade, a concentration of 0.1%(preferred for organic pigments and dye stuffs), or 2%(preferred for inorganic pigments) should be taken.

For testing in reduced shade,1% of titanium dioxide pigment should be added to the plastics material. The concentration of the coloring material in the test medium should correspond to 1/3 standard depth of shade and 1/25 standard depth of shade in accordance with EN 12877-1:1999, Annes A.

(4) Preparation of the test pigment sample

Pre-mix the colored pigment and until well distributed. Using the high-speed mixer to mix the powders and using the low-speed mixer to mix the pellet. With suitable extruder extnded matends into uniform colored pellets.

(5) Preparation of the test sample and evaluation

Process the test material at 200℃. Use these paltes as reference specimens. Discard sufficient mouldings until injection-moulding cycle is then prolonged so that the specified dwell time of 5 min in the barrel is obtained. The resulting plates shall be used as test specimens for this prolonged dwell time. Repeat the same procedure for all further test temperatures. Make allowance for test temperatures in intervals of 20℃. Evaluate the color difference between the test specimen prepared at 200℃ and those prepared at the higher test temperatures.

It is noteworthy that coloring materials can change color reversibly after moulding and the mouldings should be maintained at room temperature for at least 16h before assessment. In the case of more temperature-sensitive coloring materials or test media such as C.I. Pigment Red 48, 53 and 57, intervals of 10℃ may be used.

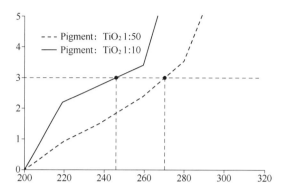

Figure 6.4　The curve of heat resistance of Pigment 138

(6) Evaluation of color difference

Value ΔE 3.0 is used as the measure of color stability to heat in accordance with GB 11186.3. For visual comparison, the industry also uses the grey scale 4 as the indication of heat stability. Express the test results in one of the following ways.

As a graph, giving the color difference as a function of the test temperature as shown in Figure 6.4.

As the temperature at which the color difference on the grey scale does not exceed a defined value.

ΔE 3.0 and grey scale 4 are not representative of the same degree of color difference. Where temperatures are obtained by extrapolation the result shall be rounded to the nearest 10℃.

6.3.2　Evaluation by Oven Test (HG/T 4767.3—2014, EN BS 12877-3)

(1) Principle

Specimens prepared from the coloring material to be tested and the palstics material are subjected to increase temperatures in an air oven for a specified period. The change in the color of these specimens, determined by comparison with the color of untreated specimens, is taken as a measure of the color stability of the coloring material.

(2) Usage scale

The method is mainly used to determination of color stability to heat in PVC or thermosetting plastic.

(3) Apparatus

Air circulating oven, capable of maintaining temperatures up to 250℃ preferably to within ±1℃, but not exeeding ±3℃. The oven shall be capable of regaining the test temperatures witin 30 s after the specimen has been palced in it. It shall be fitted with forced-ventilation equipment.

Aluminium sheets, uncoated and degreased, of thickness 0.2~0.4mm.

(4) Concentration of coloring material.

Table 6.5 Concentration of coloring material

Projects		Full shade	Reduced shade
Part by mass of the pigments	Organic pigments and dyestuff	0.10%	1/3 standard depth of shade and 1/25 standard depth of shade
	Inorganic pigment	2.00%	1% of titanium dioxide pigment

(5) Sampling

Color measurements in accordance with 6.2.2.2. Unless otherwise specifide or agreed, test specimens of demensions 50mm×50mm×1mm, suitable for colorimetry, shall be used.

(6) Procedure

Place the test specimens on aluminium sheets at room temperature. Then place the aluminium sheets with the specimens in the oven, preheated to the specified or agreed test temperature and leave for the specified or agreed period of time. For standard test conditions, see Table 6.6.

Table 6.6 Test conditions of the oven

Material	Temperature/℃	Time/min
PVC	180	30
	200	10

When the specified or agreed period of time is completed, remove the aluminium sheets with the test specimens from the oven and cool down to room temperature. In the case of polyvinyl chloride, press the treated test specimens at 170℃ for 1 min, using the plate press and sufficient pressure to ensure a smooth surface of the test specimen.

(7) Expression of results

Compare the color of the upper surface of the treated test specimens with that of unterated test specimens either colorimetrically or visually. If color measurement is specified, proceed in accordance with GB/T 11186.3. For visual comparison, use the grey scale specified in GB/T 250.

6.3.3 Evaluation by Two-roll Milling

(1) Principle

A milled sheet is prepared from the plastics material and the coloring material to be tested, ensuring that the coloring material is well dispersed. This sheet is then milled at a specified temperature. At appropriate time intervals samples are taken from which test specimens are prepared by heat pressing. The color differences between the test specimens obtained for the different milling times are taken as a measure of color stability of the coloring material.

(2) Usage scale

The method is mainly used to determination of color stability to heat in PVC.

(3) Apparatus

Laboratory two roll-mill, and the specifications is same as above.

(4) Concentration of coloring material

In accordance with Table 6.5.

(5) Sampling

Color measurements in accordance with 6.2.2.2. Take a sample from one of the sheets and prepare a test specimen from it by heat pressing at 170℃ for 1 min, for use as a reference specimen. Test specimens of demensions 50mm×50mm×1mm, suitable for colorimetry, shall be used.

(6) Procedure

Set the roll surface temperatures for PVC (soft) to between 180℃ and for PVC (rigid) to between 190℃ and 195℃. Process a sufficient number of sheets so that the recommended minimum bead diameter is achieved, whilst setting the roller gap to produce a sheet of uniform thickness of about 1 mm. Take samples after milling for 10 min and 20 min. After a further period of 10 min(total testing period 30 min) draw off the mill sheet and take a further sample. Press test specimens from each sample by heat pressing at 170℃ for 1 min, using sufficient pressure to achieve a smooth surface.

(7) Expression of results

Compare the color of the upper surface of the treated test specimens with that of unterated test specimens either colorimetrically or visually. If color measurement is specified, proceed in accordance with GB/T 11186.3. For visual comparison, use the grey scale specified in GB/T 250.

When evaluating pigments in polyvinyl chloride, the test medium shall be adequately stabilized to heat. The color stability of the test medium when subjected to heat shall therefore be tested with and without titanium dioxide pigment, using the same procedure. If there are changes, these shall be taken into account when expressing test result.

6.4 Standards and Methods of Assessment of Dispersibility in Plastics

Dispersibility in plastics is the most important properties for the pigments to be applied in plastics, and it concerns the tinting properties of pigments in processing. The heat resistance performance partly reflects the product positioning and is an important basis for users to make choice according to the product demands of themselves.

There are three methods following for devaluation, including devaluation by two-roll milling(EN BS 13900-2), devaluation by filter pressure value test(EN BS 13900-5), and devaluation by film test(EN BS 13900-6). And the standard of film methods aboard are making up. We can choose the most appropriate methods and pigments according to the demands to dispersibility of different plastics, and adapt the most reasonable process for color mast batch, to get the best benefit of quality/cost.

6.4.1　Evaluation by Two-roll Milling

(1) Principle

Using a two-roll mill, the pigment under test is dispersed at different temperature in the basic compound. The milled sheet obtained in this way is then subjected to the higher shearing forces resulting from two-roll milling at different temperature. The results increase in color strength is a measurement of the complexity of dispersion DH_{PVC-P}.

(2) Apparatus

Two-roll mill, the specifications is same as above.

(3) Formula and sampling

The formula is in accordance with the reduced samples above, and color measurements in accordance with 6.2.2.2. Take a sample from one of the sheets and prepare a test specimen from it by heat pressing at 170℃ for 1 min, for use as a reference specimen. A test specimen having dimensions of 1 mm by at least 50 mm × 50 mm shall be produced from a portion of the milled sheet.

(4) Procedure

The roller temperatures shall be maintained at 130℃ ± 5℃. The milled sheet shall first be passed through the roller gap unfolded, then folded once and passed without delay through the gap again. This procedure (i.e. folded once) shall be repeated ten times. A test specimen having dimensions of 1 mm by at least 50 mm × 50 mm shall be prepared by pressing as specified above.

(5) Evaluation

The color strength of the test specimens prepared shall be measured. These values shall be used to evaluate the color strength by the formula following for the purposes of the calculation of DH_{PVC-P}.

$$DH_{PVC-P}=100\times(F_1/F_2-1)$$

where, F_1 is the color strength value of the test specimen, carried out for a total of

200 rotations of the rolls on 160℃; F_2 is the color strength value of the test specimen, carried out for a further 10 times of the rolls on 130℃.

Evaluation of dispersibility

DH values	Evaluation of dispersibility
<5	Extremely easy to disperse
5~10	Easy to disperse
10~20	General to disperse
>20	Difficult to disperse

6.4.2 Evaluation by Filter Pressure Value Test (HG/T 4768.5—2014, EN BS 13900-5)

(1) Principle

The test mixture, consisting of a color concentration and a basic test polymer, is passed through an extruder fitted with melt pump and screen pack with breaker plate. In front of the screen pack is a melt pressure transducer. The pressure difference between the start pressure and the maximum pressure is used to calculate the filter pressure value [FPV].

(2) Material

① Homogeneous preparation of a colorant in an appropriate thermoplastic polymer.

② Basic test polymer Thermoplastic polymer, polypropylene in general

③ Mixture 1 A test mixture of 200 g (100 %), including 5.0 g colorant (2.5 %) is used. It Usually used for color pigments.

④ Mixture 2 A test mixture of 1000 g (100 %), including 80.0 g colorant (8 %) is used. It Usually used for white and black pigments.

(3) Apparatus

A single screw extruder with non-grooved barrel and a screw without dispersing elements shall be used. A screw with a diameter between 19 mm and 30 mm and with a length of 20 *L/D* to 30 *L/D* (length/diameter) is recommended. Figure 6.5 illustrates the principle construction of the apparatus.

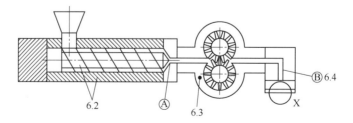

Figure 6.5 Schematic diagram of pressure rise equipment

① Melt pump The melt pump, preferably a metering pump, shall provide a constant throughput of 50 cm³/min to 60 cm³/min.

② Melt pressure transducer The pressure range shall be preferably between 0 bar and 100 bar. The resolution of the pressure measurement should be 0,1 bar.

③ Breaker plate.

④ Screen-pack Screen-packs are used as filter media. The filter media is that part of the system which influences the differential pressure used as the basic data for determining the results of the test.

There are three kinds of Screen-packs depending on the application of it in plastic, film or spinning for examples (in Table 6.7).

Table 6.7 Three kinds of screen-parks

Screen-pack		Kinds of weave	Warp/weft per 25.4 mm	Wire diameter /mm	Weft diameter /mm
Screen-pack 1 Two-layer construction	First layer support mesh	A plain dutch weave A square mesh plain weave	615/108	0.042 0.40	0.14 0.40
Screen-pack 2 Two-layer construction	First layer support mesh	A plain dutch weave A square mesh plain weave	615/132	0.042 0.40	0.13 0.40
Screen-pack 3 Three-layer construction	First layer Support mesh 1 Support mesh 2	A twilled dutch weave A square mesh plain weave A square mesh plain weave	165/1400	0.071 0.16 0.40	0.04 0.16 0.40

(4) Procedure

The complete apparatus should be pre-heated to the processing temperature appropriate for the basic test polymer. The equipment should be cleaned or adequately purged with the basic test polymer before each test is started. Mount a new screen-pack in front of the breaker plate. Allow sufficient time for the screen-pack and the breaker plate to reach the temperature of the equipment. This time will depend on the equipment used. The basic test polymer is plasticized in the extruder and passed through the screen-pack with a defined melt volume throughput until the melting temperature and pressure remain constant. Measure the starting pressure P_s developed by the basic test polymer directly in front of the screen-pack. The start pressure P_s should be constant. When the hopper is empty and the extruder screw is just visible, add the test mixture. After feeding of the test mixture is completed, 100 g basic test polymer are added just as the extruder screw becomes visible again. The test is finished as soon as the extruder screw once again becomes visible. Use the recorded data to evaluate the maximum pressure Pmax and to calculate thefilter pressure value.

(5) Evaluation

The filter pressure value [FPV], defined as the increase of pressure per gram

colorant, is calculated by using the following equation:
$$FPV=(P_{max}-P_s)/M_c$$
where:

FPV——filter pressure value, in bar per gram [bar/g];

P_s——start pressure, in bar;

P_{max}——maximum pressure, in bar;

M_c——colorant quantity used in the test, in gram.

It is recommended to express the filter pressure value accurate to one decimal place, see Figure 6.6.

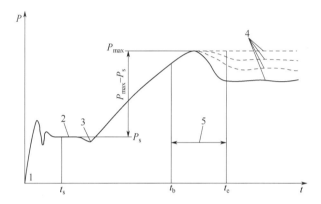

Figure 6.6 Illustrates an example of a typical pressure curve.

1—Start period; 2—Start pressure P_s; 3—Possible pressure drop because of rheological properties; 4—Different pressure development depending on the test mixture; 5—Purging with 100 g of basic test polymer; P—Pressure; P_s—Start pressure; P_{max}—Maximum pressure, t—Time; t_s—Measuring of Ps and filling of the hopper with basic test mixture; t_b—Feeding of the test mixture is completed; t_e—End of monitoring of pressure and determination of the maximum pressure P_{max}.

6.4.3 Evaluation by Film Test (EN BS 13900-6)

(1) Principle

The resin dispersiving with pigments will be tested for the scale and numbers of pigment particles visible or measurable in a particular area after film blowing, to evaluation the dispersibility of the pigment.

(2) Apparatus

Mini film blowing machine for laboratory(in Figure 6.7); detection window equipped with back reflection light source on line or off line; photoelectrical detecting system

(3) Formula and procedure

① Mix the color concentration and LDPE, with 1% pigment in weight, then extrude the mixture and blown film.

② Test the numbers of pigment particles visible or measurable by detection window equipped with back reflection light source on line or off line. Labour counting by visual is available.

(4) Evaluation

This method is traditional to determine the dispersibility of the pigments or fillers. Evaluation of the dipersibility is more and more advanced and the particles visible is diminish gradually. The methods from visual inspection off line by cutting film initially, to photoanalysis and even fine visual system test by machine, will get rid of the stale and bring forth the fresh.

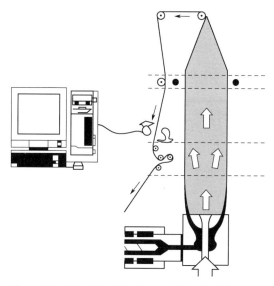

Figure 6.7   Mini film blowing machine for laboratory

6.5 The Test Methods and Standard of Migration (HG/T 4769.4—2014, EN BS 14469-4)

Migration is an important property of plastics and it gets attention of users especially that of sensitive plastic products such as packaging materials of food and health care products, children's toys of plastic. So, both the producers and users must pay high attention to the migration.

(1) Principle

Speed up the bleeding of coloring specimen in higher temperature, meanwhile place the specimen between two white contact sheets ensuring close contact between specimen and sheets for certain time. Then characterize the degree of migration with the staining rating of the white contact sheet.

(2) Apparatus and test specimen

① Test specimen   Test specimens of 50mm×50mm produced preferably by pressing should be used, as described in 6.2.2.2.

② White contact sheets   Prepare in accordance with chapter 2.5.1, minimum size 75mm×75mm×0.5mm

③ Plate glass sheets   With flat surfaces, of a size not less than the contact sheets.

④ Load   With a mass of not less than 500g.

⑤ Drying oven   With forced air circulation and capable of maintaining a temperature of 80~85℃.

⑥ Roller   Tool of roller shape of surface finish, with length of 200mm and diameter of 300~350mm.

(3) Procedure

Place the specimen between two white contact sheets ensuring close contact between specimen and sheets by squeezing out the air with a roller. Then enclose the sheets and specimen between two plates and load with at least 500g. If required, several specimen sheet sets may be placed one on top of the other, as long as they are separated by suitable, non-colored and non-optically whitened paper, e.g. filter pater. Take care to ensure that the specimens lie exactly above one other. Keep the sheets in the drying oven for 24h at 80~85℃. Then separate the white contact sheets from the specimen and carry out the evaluation as soon as they have cooled to room temperature. If the specimen is equally smooth on both sides, either contact sheet may be used. Otherwise, select the white contact sheet which was in contact with the smoother surface.

(4) Evaluation

Make a visual evaluation of any staining of the white contact sheet with the aid of the grey scale for assessing staining in accordance with GB/T251—2008. The reference used for visual evaluation is the white edge of the white contact sheet which was not in contact with the specimen. Degree 5 means no migration, and degree 1 means migration serious.

6.6 The Test Methods and Standards of Light Fastness and Weather Resistance

Light fastness and weather resistance of the colorant are important judgment index which are directly related to whether the colorant can adapt to outdoors for a long time or not.

There are two ways to test light fastness and weather resistance.

One way is the sun exposure or sunlight radiation and aging tests in the specific experimental area outdoor. The feature of this way is very close to the actual application performance, and it has very strong guiding significance to the application selection. In addition, the testing cost is lower. However, the shortage of this test method is time consuming. For regular light fastness or weather resistance test, the time of natural exposure test is equal to that of the service life of the products. In fact, it is unable to spend such long time waiting for a result in the common tests. In addition, the reproducibility of the test result is not ideal under the nature environment. The climate conditions of each year will not be completely consistent, and different seasons have different climate conditions and radiation, and there is also different in the test site and the climate of using area, and all of these account for the reproducibility of the results.

The other way is adopting the accelerated test by the artificial simulation of optical radiation and natural aging process in the laboratory. Because of the enhanced test condition of artificial, the obvious advantages of laboratory test are high speed of processing. In the meantime, the experimental result has a very high reproducibility in this way for the controllability of the artificial conditions. Certainly, there are differences between the artificial simulation condition and the physioclimate, so there are significant differences between the results of artificial simulation condition and nature aging. It is pregnant when used to compare with the test results coming from the same test method.

6.6.1 Sunlight Fastness and Weather Resistance Test

In a general way, a suitable test scheme should be in accordance with the characteristic, usage, designed service environment and service life of the product itself, and exposure or weather resistance aging test should be arranged with test site and equipment specifically. Meanwhile to confirm reasonable sampling time is a very important step as well. In a general way, the initial observation interval can be longer, and it should be shorten immediately when the sample has happened to change. The bigger change, the higher the sampling frequencies, in case that miss critical point of observation.

6.6.1.1 Sunlight Fastness of Natural Light Resistance

(1) Technical requirements of the exposure box of sunlight fastness

Sunlight fastness test is conducted in the exposure box. The exposure box contains a side of transparent glass suffering sunlight and other sides and bottom consist of metal, timber or any other materials. The box should contains effective air-breather; and a single piece of transparent glass should be adopted in the glass lid, with thickness of 2~

3mm and no air bubbles or other defects, and the transmittance of the glass should not less than 90% at 360 nm and the entire visible spectrum area. Transmittance drops to less than 1% under 300 nm and a shorter wavelength. In order to keep these features the glass should be cleaned regular, and the service cycle of the glass will be no more than two years.

The bottom of exposure box need to put on the stand, and the glass plate and sample should face towards the equator. The horizontal plane angle is approximately equal to the exposure latitude, and other angles, 45℃ for example, are useful as well.

(2) Methods

The prepared palette and the card of blue wool standards, are covered half by black thick lining paper, put onto the shelf in exposure box. The sample and blue wool standard are both exposing to sunlight 24 hours each day. Stop the exposure until the blue scale (stage 7) changes the color to gray scale (stage 4). Take out the rest of the sample and the blue scale and put them in a dark place, and evaluate it after half an hour.

(3) Evaluation

Observe the discolor degree of the sample under the scattered beam, and compare the sample with the discolor degree of blue wool standard. If the sample equal to one degree of the blue scale, the degree is the light fastness rating. If the discolor degree is between two levels, the light fastness rating is among the two level, such as 3~4 level.

The light fastness consists of 8 levels. The level 8 is the best to light fastness, while the level 1 is the worst.

6.6.1.2 Weather Resistance Test Under Natural Environment

(1) Exposure site

Exposure test is better to be conducted in professional field, and you can set exposure place nearby by yourself to conduct test. If exposure test is conducted in self-set exposure place and equipment, the place and equipment should conform to normative require as far as possible. That is to say the exposure frame should be placed in the non-residential and non-industrial areas, with low content of dust in the air, and far from being pollution of automobile exhaust gas; make sure that the exposure frame is placed in the sunshine, and the shadow of objects around should not fall onto the sample exposed; exposure frame with a structure rigid fixing the test sample, and the air of back side of the sample should circulate freely.

Exposure frame is laid to the south, with an angle equal to the latitude of exposure place with the horizontal plane.

When exposing, it is better to arrange the instrument for environment monitoring data near the exposure frame, in order to record the environment temperature(the top and bottom every day), the relative humidity(the top and bottom every day), the rainfall time, the total dewing time(rain and dew), the total radiation and ultraviolet radiation power(both broad band and narrow band) and the relative humidity of sample on the same exposure angle(the top and bottom every day). In this way the environment conditions around the exposure frame can be well check, and the recorded data can be used as a part of the test results so that to make the test data more complete.

(2) The angle of the sample

When the simple are exposed outdoor, the sun radiation energy is decided by the angle of the template, which also influence the temperature and the wetting time of damp of the sample. The angle of the sample is based on the place the plastic used. See details by Figure 6.8.

 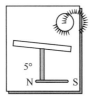

(a) Face to the south, with none angle   (b) Face to the south, with a angle of 5°

Mainly being used of 3 d objects, such as the irradiation of roof, and hardly being used of planar objects, for the hardness to get rid of water with none angle.

Mainly being used of objects such as the car, and the angle of 5° helpful to get rid of water.

(c) Face to the south, with a angle of 45°   (d) Face to the south, with a angle of 90°

The most common angle to irradiate, and being used in most industries.

Mainly being used of objects such as coatings, and the angle of 90° helpful to reduce the sun radiation as well as the temperature and water relatively.

Figure 6.8 The choose of angle to exposure for different simple in application

(3) Test and evaluation

Put the sample on the exposure frame, and expose them 24 hours every day. The exposure cycle is decided by the service life designed for the test sample. Pay attention to the change of sample. When the change of sample get to the minimum permissible degree or the test time set is achieved, finish the test and begin the evaluation phase.

Assess the change in color according to the grey scale(GB25), adopt the 5 levels, level 5 is the best, and level 1 the worst.

6.6.1.3 The Accelerated Outdoor Test Method

Conduct the aging test under the natural climate, and there is another way to shorten the test time through accelerating exposure, which installs the tested sample onto the sample frame rotating as the sun rise and down. In order to ensure the sample to be faced to the direction of the sun and get the maximum natural radiation in the duration of sunshine; the condensing mirror will reflect and concentrated the sunlight to the sample. This instrument track the sun's rays to gather and concentrate automatically, and make the test sample get ceiling exposure. The strength of exposure by this way can afford equivalent eight times of acceleration effect. Some of the instruments is configure with rainmaker. Shown as Figure 6.9.

Figure 6.9   Accelerated testing frame tracking the sun automatically

6.6.2 Light Fastness to Artificial Light and Weather Resistance Tests

The core of the artificial accelerated exposure and aging test adopts artificial light instead of sunlight radiation, supplied by necessary temperature, humidity, and the control of artificial rainfall. The light fastness and weather resistance aging tests are conducted by the accelerating speed in the laboratory.

Exposure or weather aging tests accelerated in laboratory should choose the reasonable test equipment according to the position of the product itself, and the needs of actual use, and decide the test scheme, in order to conform the requirement of the product furthest.

6.6.2.1 Artificial Light Source

At present there are three kinds of artificial light in the world: Carbon arc, Xenon arc, and UV-linght. Due to the difference of the light source characteristics, the overall equipment the three types of laboratory light sources are also different. Studies testify that the simulation and acceleration ratio of the three kinds of artificial light are also variant because of the differences in properties of light sources.

Artificial xenon lamp light source copies all of the solar spectrum, including UV-linght, visible light and infrared light, with the purpose of simulating sunlight furthest; while UV-light aging test is not aimed to copy solar ray, but to copy the execution of sunlight. Carbon arc owns radiation spectrum more different to the sunlight spectrum, and because of the long developing history of this technology, the standard can still be found in earlier Japanese standard, and it is not widespread at present.

(1) Carbon arc lamp

Carbon arc lamp is an ancient technology. At first the carbon arc instrument was used to evaluate the light fastness of dyed textiles by German synthetic dye chemist. Carbon arc lamp includes closed and open carbon arc lamp, whichever owns spectra difference to sunlight obviously. This technology has a long history and was used to original artificial simulation light aging test, and it could be found in the earlier standards.

（2）Xenon arc lamp

Xenon arc radiation test is considered to be the best one to simulate the full solar spectrum, because it can produce ultraviolet radiation, visible light and infrared radiation(Figure 6.10). It is the most widespread method used at home and abroad. But this method also has limitations, that is Xenon arc stability of light source and thereout the complexity of the test system. Xenon arc source must be filtered to reduce the radiation unnecessary. To get the different irradiance distribution, the light produced is usually filtered by a variety of filter glass light. How to choose the glasses depends on the type and end use of the materials tested. Usually there are three types of filter: sunlight, windowpane and expanding UV type filter glass.

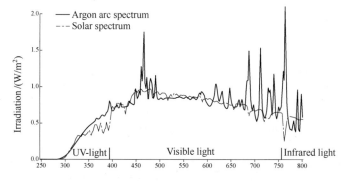

Figure 6.10 Spectrum of Xenon light and natural light

(3) UV lamp

Ultraviolet radiation aging test is to test the destructive effect on materials by using fluorescent UV lamps to simulate the sun's ultraviolet radiation. The electrical principle of fluorescent UV lamps and general lighting fluorescent tubes with cold sunlight are similar, but fluorescent UV lamps generate a stronger ultraviolet radiation instead of visible or infrared radiation. Many artificial light will be aged to weaken while used, distinctively, fluorescent UV lamps's spectral energy distribution will not weaken over time. A trial showed that there is no obvious difference in output power between a lamp use for 2h and a lamp used for 5600h in aging test system equipped with irradiance control, and the light intensity can be maintained constantly by the irradiance control device. In addition, the spectral distribution of energy is not changed, is largely different from the xenon arc lamp. Due to its inherent stability of spectral, the fluorescent UV lamp makes irradiance control easy. This feature improves the reproducibility of the test results, which is a big advantage of ultraviolet light source. Spectrum of UV-light and natural light are in Figure 6.11.

In general, the shorter wave length ultraviolet radiation, the stronger damage effect to the resin material. According to the classification of ultraviolet radiation wave, UVA wavelength range of 320~400nm, UVB is 275~320nm. Therefore, UVB greater impact on the material.

When the sample is exposed to different testing requirements, you can choose lamps with different spectral range of every model, as shown in Table 6.8.

Table 6.8 QUV lamp models, features, and proposal applicability

Lamp type	Radiation wave property	Suggestion application scope
UVA-340	Closest to the sun in the ultraviolet spectral characteristics of radiation. Good correlation with the results of outdoor exposure	Comparative tests. Most of the plastic products, textiles, paints, UV absorber performance test
UVB-351	Fits with the transparent glass filtered sunlight ultraviolet spectrum characteristics.	The light fastness test of Car interior, interior products, home textiles, printing products, etc.
QFS-40 (F40 UVB)	Part of short wavelength UV radiation is much higher than the sun's energy, accelerated aging effect is obvious	Automotive exteriors and coating test.
UVB-313EL	Spectral characteristics similar to QSFS - 40, UV radiation is higher. Accelerated aging function as the most, but the correlation with the results of outdoor exposure.	Need long time outdoor use products aging test. Such as: roof and outdoor decorative materials, outdoor coating, etc.

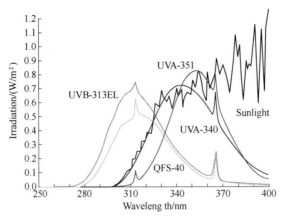

Figure 6.11　Spectrum of UV-light and natural light

6.6.2.2　Artificial Sun Light Fastness Test

In order to unify and specify the control process of exposure test and make the results comparable, it is necessary to make the operation and evaluation methods of tests standard. Hereby, the global federal natural standards institute released and published the standard file ISO 4892 (Plastic laboratory light exposure method), and German institute also released the standard numbered DIN 53378, and similar standards have been promulgated in other countries. At first glance, countries or the society release standards independently and each does things in his own way. However, it is not difficult to find the essence of highly consistent, with some details varied slightly.

(1) Test chamber

Exposure experiment box consist of different designs, but some of the basic requirements and principles must be guaranteed. Housing and accessory materials should be composed of corrosion resistant material. Test chamber contains artificial light sources and fixed sample mounts, and their installation must ensure that all the test sample surface irradiance uniformity.

Test cases should be equipped with temperature detection and control system, to ensure that the temperature in setting range. At the site of the sample frame fixed configuration blackboard thermometer, temperature sensing meter control system, it can effectively test and control the maximum temperature in the process of exposure could reach surface of the specimen exposed to radiant energy source. Make sure that the blackboard thermometer and model have the same subjects exposure intensity of illumination.

(2) Testing and evaluation

According to the section 6.6.1.1

6.6.2.3 Artificial Climate Resistance Test

(1) Equipment and light source

Now the common artificial accelerated simulation natural exposure equipment with UV lamp aging test chamber and xenon lamp aging test chamber, in terms of both compare, UV lamp aging test chamber equipment is simple, cheap, test condition of stable operation, but compared with the natural weathering aging test results, the correlation of xenon lamp test box is better. By the former section (6.6.2.1. Table 6.11), the different QUV tubes have differences with natural daylight, especially QUV - B radiation and sunshine, the difference is bigger, deviation of the spectrum the results increase. Therefore, some special products will prefer to xenon lamp exposure test. Of course, Xenon lamp test takes longer, the price of the equipment is high, operating control demand is high, therefore, the test cost is more expensive.

Equipment cooling and humidity control configuration.

The cooling device of weather resistance test equipment are mainly divided into two categories: a type of air cooled and a type of water cooled. The parameter settings do not have equivalence between them, therefore, there are big differences for parameter setting of different brand and form.

The environmental simulation forms of xenon lamp test box water spray effect of artificial rain or humidity control system, high-level experiment box have both of them. This is largely agree with the climate conditions of products. Artificial rain water spray can lead to exposure temperature drops on the surface of the sample, which to a certain extent, reducing the speed of the material aging, but also is effective to strengthen the thermal shock and mechanical erosion.

The moist environment simulation black light bulb aging test chamber on test depending on the evaporation - condensation cycle. Its principle is that the cooling water flow to the bottom of the box and is heated evaporation, guarantee 100% relative humidity in the box and the relatively high test temperature. Sample for test in the test surface is faced to the chamber, temperature is relatively high, and the back is directly connected to the laboratory environment outside the box, temperature is low, the temperature difference will make the water vapor condensation on surface of the subjects, condensate natural was once again evaporation falls back to the bottom of the box. The cycle system shows many advantages: without additional equipment cooling, the pattern on the surface of condensing water is highly pure water, without pretreatment for water used in test.

(2) Procedure

Using xenon lamp aging tester for proceeding regular weather resistance test, the method stipulated in the recommended according to DIN 53387 / A, set A, B two stages of the cycle, in which A stage lighting 102 minutes every time, B stage light and spray 18 minutes every time, 120 minutes each cycle. Other parameters such as Table 6.9.

Table 6.9 Xenon lamp corrosion resistance test parameters

Light source	6500 W xenon lamp
Irradiation intensity	$0.35 W/m^2$
Wavelength	340nm
The blackboard Temperature	A stage 58~62℃; B stage 18~24℃
Relative humidity	A stage 28%~32%; B stage 95%

(3) Test and evaluation

The test sample according to color with gray CARDS (GB250) and 5 level system: grade 5 is the best, grade 1 is the worst.

Xenon lamp artificial accelerated aging test chamber is the imitation of natural insolate ageing, how can the artificial aging data used to guide practical application? At present general method is to make both (manual testing and the use of natural radiation) in 340~700nm wavelength, 0.35 w/m² under the condition of the radiation intensity in the amount of radiation on the surface of the sample (kJ/m^2) as a basis for conversion. Table 6.10 shows the natural insolate aging (Florida) and convert data between artificial aging test.

Table 6.10 Conversion between natural exposure and artificial sunlight

Exposure	Insolate xenon lamp tester	Florida exposure field
Wavelength 340~700nm, intensity of 0.35 w/m² of radiation	DIN 53-387/A (hour)	5° south (month)
$130 kJ/m^2$	1000	6
$260 kJ/m^2$	2000	12
$390 kJ/m^2$	3000	18
$520 kJ/m^2$	4000	24

Such conversion is just for reference, not as peer to peer! Because the natural climate conditions of artificial aging experiment simulation varies whether in the radiation spectrum, intensity, and other conditions. Strengthen the short-term damage caused by the radiation energy and natural conditions of long-term aging effect also has considerable differences. In addition, the products used for the effects of the atmospheric pollution to the environment can not be simulated by the human exposure device. That is the important reason that there are many tests must spend long time to proceed natural exposure test.

In artificial accelerated exposure test, also note that UV damage to the degradation of resin, most of the resin in a long period of time after exposure test, the surface will produce pulverization yellowing, cracking, and any other phenomenon, which affect the accurate judgment of the color change in colorant itself. Therefore, in order to avoid or reduce these effects, at the beginning of the test specimen preparation, you should consider to add the necessary light stabilizer and ultraviolet absorbent, minimize the influence of resin degradation of aging as much as possible. After completion of the exposure test, sample surface should be cleaned, you can pipe the surface of sample if it is necessary, to remove the fouling or stain on the surface,it is conducive to the accuracy of the evaluation.

6.7 Evaluation of Chemical Stability

Plastic products in the application process are usually exposed to some chemicals, including all kinds of acid, alkali, detergents, solvents and other chemical agents, etc. In the period of this kind of products, of course, it need colorants with the property of resistance to these chemicals, to ensure the stability of color stain and various performance continues to reflect.

Because of the vast majority of dyes in solvent and other soluble in liquid chemicals with the limitations of acid-proof alkaline, here we are talking about chemical resistance mainly for pigments and colorant products of plastic.

6.7.1 Test for acid resistance and alkali resistance

(1) Equipment, material and sample preparation

① Test with polypropylene (PP) coloring injection model Using injection molding machine (EN 12877) injection system board, pigment content, according to 1/3 standard color depth (with 1% titanium dioxide dilute), thickness of PP sample board is 1 mm.

② Standard reagents and concentration 5% hydrochloric acid (HCl) solution, 5% sodium hydroxide (NaOH) solution, 10% (H_2SO_4) sulfate solution, 1% soap solution, 1% detergent (hydrogen peroxide) solution. Etc. Other chemical reagent, it can be used according to the reagent on the premise of ensuring security.

(2) Procedure

Select test reagent, according to the need, and configured to normality.

Reagent infiltration model sample infiltration method has two kinds. Sample directly is put into a container containing reagent, sample part is put into and the rest did not enter the reagent. Use filter paper in the part of the sample surface, wet filter paper with reagent and maintain long time wetting and joint with the sample surface.

Maintain laboratory at room temperature in 23~25℃, guarantee the sample for 24 hours.

After the infiltration, take out the sample, washed with deionization water sample surface, in preparation for evaluation.

(3) Evaluation

Use color gray card(GB/T250—2008 or ISO 105/A02—1993) in sample infiltration and infiltration of off color, according to the color difference degree of qualitative corresponding chemical resistance to evaluate samples color changes such as Table 6.11

Table 6.11 Chemical resistance evaluation a color change

Gray card	Change in color	Corresponding chemical resistance
Level 1	Maximum	Very bad
Level 2	Great	Unsteady
Level 3	Medium	Ordinary
Level 4	Mild	Fine
Level 4-5	Very small	Stable
Level 5	No change	Very stable

6.7.2 Solvent Resistance Test

(1) Definition

After contact with the solvent it may lead to the solvent color or the paint itself fade, it indicates that the pigment on the effect of solvent is very sensitive. The significance of pigment solvent resistance is the characterization of pigment resist the solvent without being dissolved or change their own performance ability.

(2) Materials and equipment

Balance: sensitive quality is 1 mg.

Test tube: capacity is 25 ml, belt grinding plug.

Electric generator: oscillation frequency (280 ± 5) times/min, oscillation amplitude (40 ± 2)mm.

Crucible glass filter holders: grade 5, capacity of 30mL.

Smoke filter: capacity of 125mL.

Color dish: thickness of 0.5cm, 6.4cm.

White color frame: White Background, there are two parallel observation hole in the front, just to insert two color dish.

Color gray scale: GB251-84.

Solvent for testing: the commonly used solvents are: ethanol, ethyl acetate, butanone, xylene, solvent naphtha, dibutyl phthalate, linseed oil, etc. According to the actual application need to choose other solvent for testing.

(3) Procedure

Accurately weigh pigment samples 0.5g (accurate to 1 mg), put in a test tube, measuring 20mL solution and put in the test tube, plug the ground plug. Paints and solvents will be loaded in the horizontal test tube fixed on the electric generator, open oscillation 1 min, remove the test tube after shocks, put the suspension liquid into the glass filter after shocks, vacuum suction filter until get clear filtrate. Collect the filtrate to judge.

(4) Evaluation

Fill up the cuvettes with the solvent and filtrate prepared respectively, and put the two cuvettes onto the cuvettes shelf parallel. Faceing south to the sunligh, the observed objects and incident light are at a angle of 45 ℃, and observation direction is perpendicular to the observed surfaces. Assess the degree of filtrate by visual according to the gray scale. The solvent resistance is characterized by bleeding intensity of pigments in the solvent (in Table 6.12).

Table 6.12 Gray card level

Gray card	Pigment dissolving	Gray card	Pigment dissolving
Level 1	Strong bleeding	Level 4	Very mild bleeding
Level 2	Medium bleeding	Level 4-5	Trace bleeding
Level 3	Mild bleeding	Level 5	No bleeding

In order to ensure the accuracy of the detection, general advice to do two parallel test samples at the same time, the horizontal comparison between two samples, if there are any significant differences, you should test again.

6.8 Evaluation of Warpage and Deformability

6.8.1 Evaluation of Degeneration or Shrinkage

Use the differences of the actual size between plastic injection mould and model to evaluate specific plastic resin injection molding shrinkage. Shrinkage comparison between colored and uncolored pigments injection model to assess the impact of the injection-shaped degeneration. This test method for container shaped products have very good guiding significance such as plastic box, box hollow toys.

(1) Test materials and equipment

The test material: low density polyethylene(HDPE), or polypropylene (PP).

Testing equipment: injection molding machine (according to the regulations of the EN 12877).

Injection mould: cavity 150 mm length, 120 mm wide.

Pigment samples: testing with coloring pigment content in the sample: organic pigment 0.1%, inorganic pigment 0.2%.

(2) Procedure

Colorless benchmark model: injection molding machine heat up and keep on 200℃, plastic resin injection into model.

Coloring test sample: paint with good dispersion (masterbatch or mixture processing method), according to the ratio of injection into a test sample.

All sample injection can be finished in 90℃ water 30 minutes, then cool to room temperature.

(3) Transverse shrinkage (PST)

Transverse shrinkage (PST) the following type:

$$PST = \frac{L_m - L_p}{L_p} \times 100\%$$

Where, L_m present mold size, L_p present size of injection molded parts.

(4) Evaluation

The influence of the pigment to injection products in deformability is evaluated according to the IF. The IF value 10 is seen as a boundary. When the IF values is smaller than 10, there is no deformability, and when bigger than 10, there is deformability. The bigger the IF value, the bigger the deformability resulting from pigments.

6.8.2 Evaluation of Warpage

By the way of injecting a plastic disk, set the injection port in the mid of the disk, and compare the colored and uncolored sample at the same time. The influence of the pigment to injection products in warpage is evaluated through observation according to the altitude difference resulting from the warpage of the both ends of the disk diameter. The determination in this way is instructive and meaningful to product the lid, tool boxes, barrels, furniture and so on.

(1) Materials and apparatus

Materials: low density polyethylene(LDPE).

Apparatus: injection molding machine(according to EN 12877).

Sample: round sheet with diameter of 120 mm.

Pigments: contents of pigments in simple for test: organic pigments 0.1% or inorganic pigments 0.2%.

(2) Procedure

The resin of HDPE and pigments are mixed by high speed mixer for 10min, and the mixture is molded by the injection machine. Set the the working condition temperature of the machine as following: charging barrel 200℃, injection port 250℃, mould 20℃. Put the injection molding into the water of 90℃ for 30min immediately the molding is demoulded, in order to eliminate the innerstress and prepare for evaluation.

(3) Evaluation

Put the transmutative injection disk onto the horizontal platform with the middle lower and the both ends higher. Find the direction along the diameter warpaged most seriously and press one side down to the platform. Then measure the perpendicular distance between the platform and the highest point warpaged on the other side. Calculate the value of deformation index(IF).

$$IF=(H-d)/d\times100\%$$

Where, H is real value of single sample, d is thickness of single sample.

Chapter 7
Pigment Dispersion in Plastics

7.1 The Purpose and Significance of Pigment Dispersion in Plastics

Pigment dispersion has extremely important significance for coloring plastics. The result of pigments dispersion affects not only the appearance of colored product (spots, streaks, gloss, chroma and transparency), but also the quality, for example, color strength, elongation, aging resistance and resistivity, etc. It affects plastics processing ability which includes colored master batch and application performance. Pigment particles generated during the beginning of formation are called the primary particles. Due to various reasons, these primary particles have the tendency to form agglomerates or aggregates in the subsequent manufacturing process which are several times larger than the coloration requirement for plastics. In order to reach the ideal tinting strength, transparency, brightness and other characteristics, it is necessary to study the pigment dispersion process in plastics and understand its results so that the aggregates can be better dispersed and are close to the primary particles or smaller aggregates.

Pigment dispersion in plastic refers to its ability to reduce the aggregates and agglomerates to the ideal particle size after pigment wetting. The pigment characteristics are basically realized through whether it can be dispersed ideally. Therefore the dispersion of pigment is an important indicator in pigment application.

7.1.1 Improving Pigment Coloring

For plastics coloring, the most important performance of pigment is the tinting strength. The so-called tinting strength is a measure of pigment coloring ability for giving plastics. When plastic products color with the pigments, the one using less dosage has higher tinting strength. For a specific pigment, the coloring performance mainly depends on the obtained dispersion state. When pigment is dispersed in certain medium, the coloring intensity increases with the decrease of average particle size in the submicron range. Figure 7.1 shows the relationship between the pigment particle size and tinting strength. If the pigment particle sizes is about half of visible light wavelength namely 0.2~0.4μm, the obtained tinting strength is supreme. Generally speaking, the better pigment dispersion in plastic, the higher the tinting strength. The pigment coloring and covering effect are reached through complex interactions of the surface and light. If pigment disperses well, the average particle size is small, the specific surface area is large and the effect upon the light becomes strong, the appearance of coloring product will be more uniform, brighter, shows less undispersed dots and narrowed color difference. Thus the pigment dosage can be reduced to achieve the same coloring depth and covering effect and obtains the highest of cost-effectiveness.

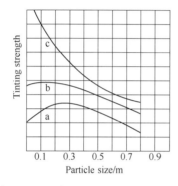

Figure 7.1 The relationship between the pigment particle size and tinting strength
a—applicable to inorganic pigments;
b—applicable to most azo pigments;
c—applicable to carbon black and phthalocyanine pigments

To mix pigments uniformly in coloring plastic is not diffcult, the key is to get full dispersion. Pigment should be fully dispersed and then evenly distributed in the plastics system. However, full dispersion is not equal to the uniform mixing, sometimes the pigment which has good dispersion, but because of uneven mixing it will also leave color line and stripe on the surface of plastic products.

7.1.2 Meeting the Requirement of The Plastic Processing For The Pigment Dispersion

The total dispersion forces of pigment to be subjected in the plastic processing is far lower than it in the ink and paint processing. The basic characteristics of plastic decide that it must be melted into molding, then cooled to plastic product. So in general machinery of plastic processing units are capable of heating. For the pigment dispersion, the most common processing machinery is single (twin) screw extruder, the shear stress experienced by pigments in the flow molten body of plastic is not enough to reach the ideal requirements of pigment dispersion. However, the processing equipment of pigment used in ink and paint are generally high speed sand mill and three roller machine. Due to the different grinding medium and means, pigment are effected not only by the high shear force but also very strong impact force and static pressure.

Thus the requirement of plastic coloring for pigment dispersion is much higher than that of paint and ink.

How big or small the particle size is appropriate after the pigment dispersion? It depends on the type of plastics as well as the kinds of processing technology applied.

General speaking, for a man who has more sensitive vision, he could easily spot the undispersed specks when the pigment particle size is greater than 20μm. The coarser particle will not only make the products surface display dots and lines, but also produces the phenomenon such as screen blocking during the extrusion and molding and thus affects the normal production. For plastic products in general, when the contained pigment particle size is more than 10μm, it can lead to poor surface gloss. When the pigment particle size is less than 5μm, it can satisfy completely the requirement of production and appearance. But for the many high performance products, the result of particle dispersion could also affect the mechanical properties, electrical properties and processing performance. For fiber (single silk diameter is 20~30μm) and super-thin film (film thickness is less than 10μm), the pigment particle size must be less than 1μm. For fiber application, the control of pigment particles is critical that is highly associated with the fineness of the final fiber and the mesh size of the filter screen. Under the premise of quality, the requirement for fiber production to reduce number of filter screen replacement as much as possible in order to prolong the continued operation and to prevent the blockage of spinning nozzle during the process of production, or to avoid the breakage of fiber filament, and to inhibit fiber strength problem because of pigment particle size. So it is important for fiber producer to ensure that the pigment particle added to the fiber processing have a certain particle fineness and particle size distribution. A rule of thumb is that the pigment particle size of less than 10% of fiber silk diameter is not desirable. Table 7.1 illustrate pigment dispersion characteristics according to different application fields of fiber products.

Table 7.1 Pigment particle size expectation for different specification of fiber

Application field	Fiber number (denier)/(g/9000m)	Fiber diameter (circular type)/μm	Expectation pigment particle size/μm
Carpet	22	58.2	6
	20	55.8	6
	18	52.9	5
Furniture	16	49.9	5
	14	46.7	5
	12	43.2	4

continued

Application field	Fiber number (denier)/(g/9000m)	Fiber diameter (circular type)/μm	Expectation pigment particle size/μm
Car	10	39.4	4
	8	35.3	3
	6	30.5	3
	4	24.9	2
	2	17.6	2
Clothes and accessories	1	12.5	1
	0.8	11.2	1.1

If pigment dispersion can't achieve the expected outcome in fiber production, it may result in frequent screen changing which will cause downtime, reduces capacity, severely affect productivity and all of which extremely impact the product quality and displays heavy waste. Therefore, the dispersibility of pigment is an important element for plastics processing and application.

7.2 Dispersion of Pigments in Plastic

7.2.1 The Types and Properties of Particles Before Pigment Dispersion

The pigment first forms crystal nucleus in the production process, which grows from single crystal into a multi crystal with mosaic structure. Of course, this particle is quite fine and the linear size is about 01~0.5μm, commonly it referred to as primary particle. The primary particles are prone to aggregation. After the aggregation they are known as secondary particle. According to the different way of aggregation, the secondary particle are divided into two categories. One is crystals between which the crystal edge or corner connects each other, the attraction between them is relatively small and loose, they are easily separated through the dispersion. This is referred as the agglomerate. Another is crystal surfaces border each other and has more attraction. The particle is separated hard and is called aggregate. The total surface area of aggregate is less than the sum of each particle surface area. So it is very difficult to disperse by common dispersion process. Pigment synthesis is shown in Figure 7.2. Structure characteristics of pigment particle in Figure 7.3.

The schematic diagram of structure characteristics of different pigment part are shown as follows.

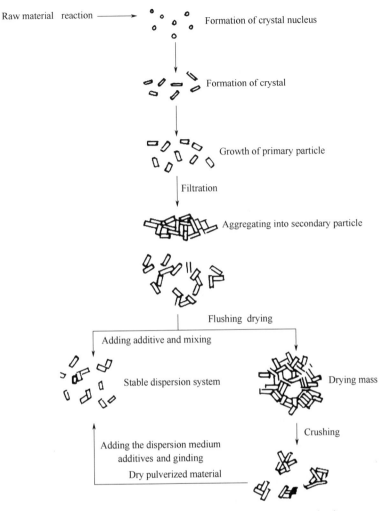

Figure 7.2 The schematic diagram of pigment synthesis

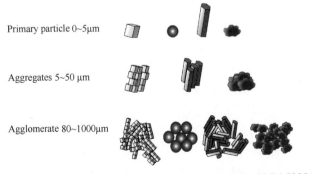

Figure 7.3 Structure characteristics of pigment particle (DIN 53206)

The type and performance before the pigment dispersion are listed in Table 7.2.

Table 7.2 The type and performance before the pigment dispersion

Name	Definition
Particle	Individual unit of pigments, it can be in any shape, with any structure
Primary particle	The particle can be identified with the appropriate method (i.e. optical or electron microscopy method)
Aggregates	The primary particles grow together and face to face arrange into aggregate The total surface area is less than the sum of each particle surface area
Agglomerate	The aggregate is formed through connection of the corners, and edges and/or aggregate of agglomerates There is not much difference between the total surface area and the sum of individual particle surface area

The commercial pigment particle are larger than agglomerate. It is usually around 250μm, therefore it can not be used directly for plastic coloration, and must go through dispersion process.

7.2.2 Dispersion of Pigment in Plastic

Dispersion of pigment in plastic usually consists of three essential stages.

Wetting: to change the interface between pigment and air or water into that of pigments and medium

Dispersion: to crush the pigment particles of agglomerates and aggregates under the action of external force.

Stablization: to stabilize pigment particles which were dispersed in a medium, and effectively preventing the coalescence again.

An important step in pigment dispersion is to wet the pigment particle, it is divided into an initial wetting of pigment and continuous wetting of dispersed particles in the process. For the plastic processing, initial wetting is a process that resin chromatophore replaces air and water on the surface of pigment, the purpose of pigment wetting is to make the cohesion between pigment particles decrease. If the surface is completely not wetted by the chromatophore that may also contains a wetting agent, then the shear stress produced by the dispersing device through the molten medium could not affect pigment agglomerate and aggregates. The difficulty and ease of pigment wetting depend upon the surface properties of pigment particles and chromatophore nature. The smaller the contact angle between pigment and chromatophore, the smaller the surface tension of chromatophore, then the easier the wetting.

The second step of pigment dispersion is to break down pigment particles. After wetting, pigment particles may complete refinement process under the action of external force and crush the pigment into the ideal particle size. At present, the mechanical dispersion force is nothing more than the pressure, impacting and shearing force which are used for crushing agglomerate and aggregate.

The third step of pigment dispersion is to stabilize pigment particles after being dispersed. The surface area increases after pigments refinement process. The pigment particles could re-aggregate and it is necessary to go through surface treatment reducing the formation of the new interface energy and preventing aggregation of refined particles during further processing.

The above described pigment dispersion process is: wetting, breaking down and stabilizing. However, they are closely related and occurred almost simultaneously in the actual operation, they are separately discussed so as to facilitate the discussion.

The initial wetting of pigment is utmost important to pigment dispersion. Usually this is achieved through some seemingly simple steps, so its importance is very easy to be over sighted, but the success of pigment particle dispersion is heavily relied on this step. If the initial wetting is not done well, the good dispersion of pigment particles is out of the question. It is because that chromatophore resin must be penetrated into the gaps of the pigment particles through the capillary effect. Due to the capillary phenomenon, pigment particles aggregation is reduced and can easily be crushed and broken down through the shear force. Therefore pigment wetting rate and capillary permeability degree play a decisive role for the dispersion rate of pigment particles and product quality.

Numerous facts have proved that the bad dispersion of pigment particles is primary derived from poor initial wetting. Once this happens, any attempt to remedy the situation is difficult. The only way to resolve the dispersion issue at this time is to properly re-wet the pigment particles. It is shown in Figure 7.4 that various degree of dispersion resulted from different wetting. When using common wetting agent in pigment dispersion, no matter how many extrusions are preformed, one or six times, the resulting properties are unsatisfactory. If suitable wetting agent is used for the same process, the result will have a qualitative leap.

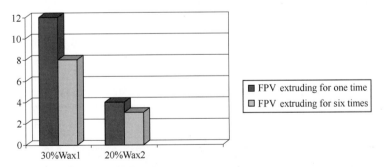

Figure 7.4 The influence of different wetting agents on dispersion
Wax 1—polyethylene wax, no polarity; Wax 2—EVA wax, 12% VA; Carrier resin: LLPPE

7.3 Wetting

7.3.1 Wetting Phenomena and its Evaluation Contact Angle and Young's Equation

When the liquid is in contact with the solid surface, newly formed solid/liquid interface gradually replaces the original solid/gas interface, this phenomenon is called wetting. When liquid contacts with a solid, it will form an angle, known as contact angle, it is measurement of wetting ability for a liquid to a solid. Figure 7.5 shows that the different beads of (a) and (b) appear on the solid's surface.

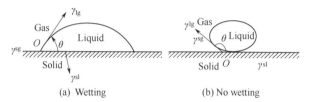

Figure 7.5 Different beads appeared on the surface of the solid

The contact angle is the angle between the interface gas-liquid and solid-liquid which is the result of balance through the three jointed tensions. The relationship between various interfacial tension can be expressed by Young's equation:

$$\gamma_{l\text{-}g} \cos\theta = \gamma_{s\text{-}g} - \gamma_{g\text{-}l}$$

In the formula: $\gamma_{l\text{-}g}$: interfacial tension between liquid and gas.

$\gamma_{s\text{-}g}$: interfacial tension between solid and gas.

$\gamma_{g\text{-}l}$: interfacial tension between solid and liquid.

Dr. A. Capelle & etc. pointed out that wetting efficiency = $\gamma_{s\text{-}g} - \gamma_{s\text{-}l}$, that is $\gamma_{l\text{-}g} \cos\theta$.

It can be obtained by the equation, the smaller the contact angle between solid and liquid, the higher the efficiency of wetting, the better the wetting effect. Figure 7.6 shows the relevance of the different wetting effect of liquid/solid interface with contact angle.

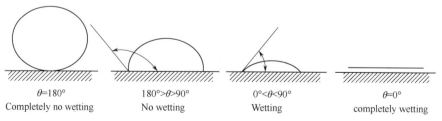

Figure 7.6 The phenomenon of different wetting on liquid and solid interface

7.3.2 Pigment Wetting

The initial wetting of pigment is to replace the surface air and water with wetting agent and pigment interface. The wetting process can be simply divided into three stages: adsorption, penetration, diffusion. When pigments contact with wetting agent, the latter is adsorbed on the pigment surrounding with small contact angle. The wetting agent is penetrated among the tiny gap of pigment particles through the capillary action, and then gradually infiltrates into the slits between pigment particles, which reduce the cohesion force between the particles, thereby decrease the energy for the crush of pigment aggregates. Figure 7.7 shows the schematic diagram of pigment wetting respectively:

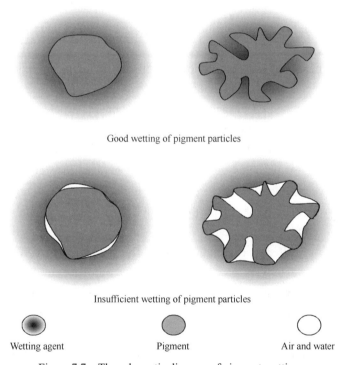

Figure 7.7 The schematic diagram of pigment wetting

Pigment wetting can be described by the famous Washbum equation. The wetting efficiency of the initial wetting is expressed by the following equation, in which the pigment particle clearance is regarded as the radius R approximately and penetration length of capillary action is L, the penetration time (wetting efficiency) is T, then:

$$T = \frac{KL^2}{R} \times \frac{2\eta}{\gamma \cos\theta}$$

In the formula: η—viscosity of wetting agent.
γ—surface tension of wetting agent
θ—contact angle between pigment and wetting agent
K—constant

$\dfrac{KL^2}{R}$ indicates the influence of geometry factors of pigment particles agglomerates on the pigment wetting rate

$\dfrac{2\theta}{\gamma \cos\theta}$ indicates the influence of wetting agent properties on the pigment wetting rate

7.3.2.1 Wetting——The Properties of Pigment Particles

By Washbum equation, the influence of pigment particles upon pigment wetting is as follows.

Surface properties of pigment particles, geometry including aggregate porosity
Pigment particles size and distribution.

(1) Pigment wetting and surface properties of pigment particles

The surface properties of pigment particles are related to the molecular packing and arrangement, different particle arrays show different surface states. When producing pigments, the primary particles generated possess high surface energy which causes strongly attractive interaction among them, the molecule increases quickly and grows into crystals. If the affinity of crystal in each direction is same, it can obtain the same cube as sodium chloride, but usually the molecular array is asymmetry, the different growth rate on the different direction leads to the formation of different crystals of lamellar, acicular, rectangles etc. Therefore, we should do our best to control the crystallization of pigment particles and lead them growing into a specific lattice according to the desired direction and getting the desired surface properties during the pigment production.

Taking azo red C.I.PR 48∶2 as an example. Figure 7.8 shows that the pigment molecular contains different polarity groups, such as —SO_3, —COO—, —Ca^{2+}, —Cl, —N=N—, etc. In the generation of lake, the molecular array usually close to a plane, they are tightly displayed together, and the primary particles generated in the formation of laking possess high surface energy which cause strong attraction and form crystal particles.

The crystals under different process conditions display the arrangement of model A, B, C (in Figure 7.9) of the cube, flake, needle, thus forming the surface characteristics of different pigment crystal.

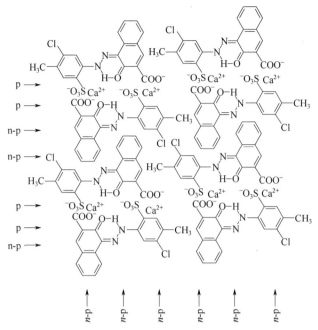

Figure 7.8　The graph of lake Pigment Red 48∶2 structure plane

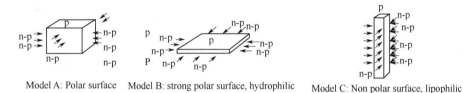

Model A: Polar surface　　Model B: strong polar surface, hydrophilic　　Model C: Non polar surface, lipophilic

Figure 7.9　Surface characteristics of different arrangement of Pigment Red 48∶2 particles

There are more polar groups on the top and bottom of the model A, it shows a strong polarity effect, it is denoted by p; whereas its side is mainly non-polar, as indicated by n-p, It only display some polarity in certain areas, so the non-polar area far exceeds its polar part, which leads to the strong lipophilicity of crystals. On the contrary, the flake crystals show larger polar area than the non-polar area and result in strong hydrophilicity. The rod crystal particles owe to that the non-polar part is greater than the polar area, so have the strong lipophilicitily. Surface characteristics of different array of Pigment Red 48∶2 particles is seen in Figure 7.9.

According to the principles of like dissolve like, if the surface of pigment particles shows high polarity, then the wetting angle between pigments and polar solvents (such as water) is small, they can be easily wetted by polar solvents. Therefore, the pigments are often applied to high polar system and present an ideal dispersion property, for example, water-based inkjet ink, waterborne paint etc. If the pigment particles surface is nonpolar, it can very well be wetted and dispersed. Figure 7.10 and Figure7.11 are imaging graph of transmission electron microscopy of Pigment Yellow 183 and Pigment Blue 15∶3. It can be seen from the graph that Pigment Yellow 183 shows needle shaped structure, the length and width (diameter) ratio can be as high as (30~50)∶1; crystalline particles of Pigment Blue 15∶3 is rod shape, its length to width (diameter) ratio is up to (8~10)∶1. It can be predicted that both products of this crystalline structure have strong lipophilicity and their dispersibility in nonpolar plastic resin should be very good, the dispersibility of Pigment Yellow 193 is better than that of Pigment Blue 15∶3.

Figure 7.10　TEM of Pigment Yellow 183　　　Figure 7.11　TEM of Pigment Blue 15∶1

(2) The wetting is related to the particle size of pigments powder

The fineness of pigment powder particles has great influence upon pigment wetting and dispersion. Usually, the gap between the fine powder particles is smaller than that of large particles, so the wetting and penetration rate of chromatophore resin to small pigment powder particles is relatively slow, which would affect the final dispersion of pigment particles.

The test results obtained from Table 7.3 validate the above conclusion. With different surface treatment, pigments of the same chemical structure (Pigment Red 122, quinacridone) produce different particle size by the same masterbatch process. The filtered value of the colored masterbatch testing through the screen of 25μm shows very different result from two groups. Experiment proves that pigments with larger particle size are relatively easy wetting and display a better dispersion.

Similarly, most inorganic pigments are composed of metal oxides and the average particle is larger than that of organic pigments. Therefore, inorganic pigments are better than organic pigments with respect to dispersion. Especially, titanium dioxide, chromium and cadmium pigments are readily dispersed in plastic.

Table 7.3 The dispersion of Pigment Red 122 with different particle size

Pigment Red 122 of fine particle size (JHR-1220K)	Filter pressure value/(bar/g)	Pigment Red 122 of large particle size (OPCO)	Filter pressure value/(bar/g)
Batches of 1	6.4	Batches of 1	1.2
Batches of 2	5.8	Batches of 2	2.6
Batches of 3	9.4	Batches of 3	1.4
Batches of 4	6	Batches of 4	1.4
Batches of 5	9.4		
Average value	7.4	Average value	1.65

(3) Wet ability and pigment particle distribution

The so-called ideal large particle not only refer to the average particle size, the particle size distribution also needs to be relatively concentrated, and minimize the finer particles. If there are more fine particle powder, owing to that the small particles will fill the gap between the larger particles and the powder accumulation becomes dense, therefore, the wetting and capillary penetration rate of chromatophore resin on the powder particles is slow. The pigment particles can't quickly be wetted. In other words, the wetting difficulty is higher, eventually because the shear stress in the dispersion process can't convey to the particle surface, it does not make the aggregates/agglomerate particles open, thereby affects the dispersion of pigment particles.

Different products of crystaline particle morphology can be obtained with different pigmentation treatment. Their particle size and aggregation morphology of the secondary particle obtained are not the same. It can be foreknown, their wetting and dispersing effect would not be the same, so the application performance will also be different.

Figure 7.12 is a photomicrograph of Pigment Yellow 95 particles revealed under transmission electron microscope. It shows the needle morphology, the length to width(or diameter) ratio can be as high as $(8\sim10):1$. The micrograph clearly shows that particle aggregates are based on edge to edge/ edge to surface /surface to surface combination. Single needle crystals have particle size mainly in the range of $0.1\sim1.0\mu m$, the particle size distribution range is wide. Figure 7.13 shows that particle of azo condensation Pigment Red 144(JHR-1440K) is agglomerate, it combines through angle-angle or angle-edge, the agglomeration is relatively loose and form needle shape on their combination. The uniform particle size is uniform and the particle size distribution is narrow. Both pigments are often applied in polypropylene fiber coloration, the red pigment (PR144) which is easily wetted and dispersed but the yellow pigment (PY95) which has wider particle size distribution is reltatively difficult to wet and disperse.

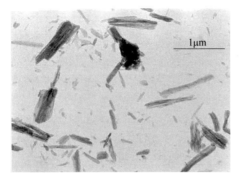
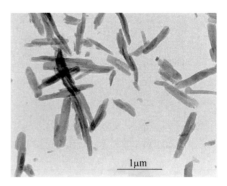

Figure 7.12　Organic Pigment Yellow PY95 (TEM picture)

Figure 7.13　Organic Pigment Red PR144 (JHR-1440K)

In addition, it can be seen from transmission electron microscopy imaging diagram of Figure 7.14 and Figure 7.15, two Pigment Yellow 139 produced by a same company: Paliotol Yellow K1841 and Paliotol Yellow K1841FP. The particle size distribution of K1841 FP is more uniform than K1841. Experiments show the dispersibility of Paliotol Yellow K 1841 FP in plastics.　FPV value (Filtration pressure value) is <1, which indicates that it is more easily wetted and shows outstanding dispersibility. It is mainly recommended for fiber industry.

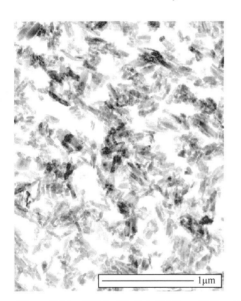
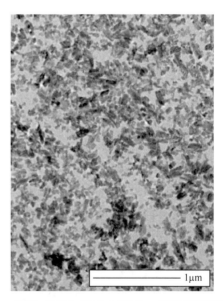

Figure 7.14　TEM of Paliotal Yellow 1841　　Figure 7.15　TEM of Paliotal Yellow 1841-FP (FPV<1)

Usually the filtration pressure value（EN 13900-5）is used to evaluate pigment dispersibility in chemical fiber industry. But in actual operation, two products with FPV values of little difference reflect the large dispersibility difference. The difference of dispersibility can be revealed through the length of time to replace the filter screen. Only when we adjust the filter screen to much finer, the difference between the product with different FPV value can be realized. Often the product with wider distribution has higher FPV value. This also reflects that the wettability and dispersibility of the narrow particle size distribution are better than that of the wider distribution.

7.3.2.2 Wetting-Wetting Agent and Its Performance

In plastic processing, it is usual to add wetting agent in order to accelerate the wetting and capillary penetration of resin to pigment surface and to achieve the good dispersion.

What is the wetting agent? The so-called wetting agent is materials which make a solid substance easier to be wetted by liquids (or melt fluid). Through the reduction of surface tension or interfacial tension, the liquid substance (or melted fluid) can be deployed on a solid surface, or through capillary permeability, the fine gaps of solid material can be wetted from the outside to inside and result in accelerated wetting of the solid material.

Wetting agent in the plastic processing industry is also known as "dispersant".

Usually, the common wetting agents used in plastic processing are resins of high melt flow rate, polyethylene of low molecular weight and super dispersant, etc.

It is found that the wetting process of pigment particles is affected by the performance of a wetting agent in the view of Washbum equation.

Effects of wetting agent on pigment particle wetting mainly coneatrates in three aspects.

Will wetting agent automatically spread to th surface of pigment particles?
Is the pigment particle surface completelly wetted?
Does the wetting process need energy?

(1) Wetting is related to the viscosity and absorption of wetting agent.

Pigment wetting is associated with the viscosity of wetting agent, this is because that the viscosity of wetting agent used in pigments is lower, the flowability is better and the wetting efficiency is higher. During the pigment wetting process, the easier the capillary permeability perform among the small gap of pigment particles, the easier the pigment will be wetted. However, shearing force in pigment dispersion at plastics is the key force, it is divided into mechanical shear and viscosity shear, and among them, the viscosity shear is the dominant factor. Therefore the overall viscosity in the processing system must not be too low which is unfavorable to the pigment dispersion and it is important to pay attention to the balance between them in the formula design.

Experimental results shown from Figure 7.16 indicates that the dispersibility of wetting agent with low viscosity agent (700cps) is better than that of high viscosity (1800cps). In addition, if the wetting agent can be micronized, it is more beneficial to the adsorption and wetting of pigment surface, the filtration pressure value is low and has better dispersion.

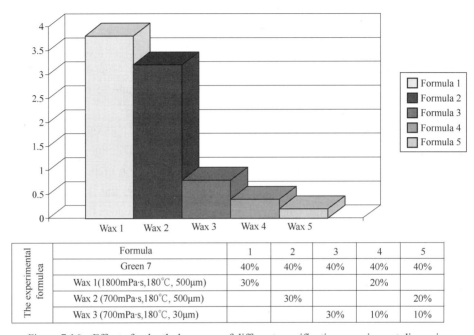

	Formula	1	2	3	4	5
The experimental formulea	Green 7	40%	40%	40%	40%	40%
	Wax 1 (1800mPa·s, 180°C, 500μm)	30%			20%	
	Wax 2 (700mPa·s, 180°C, 500μm)		30%			20%
	Wax 3 (700mPa·s, 180°C, 30μm)			30%	10%	10%

Figure 7.16 Effect of polyethylene wax of different specifications on pigment dispersion

(2) Pigment wetting needs energy (energy transfer and wetting)

In the conventional processing treatment, improving the temperature of dispersed system is in favour of the wetting and permeability of pigment particles, and reducing the temperature is more beneficial to the shear and dispersion of pigment particles. How to find the processing conditions and methods which not only facilitate wetting of pigment particles, but also can effectively disperse them, this is the problem we must seriously consider.

Pigment Violet 19 is dispersed in a milling machine, various temperature or shear stress condition are provided for grinding pigment, pigment particle size becomes smaller, the corresponding particle surface area is rapidly increasing, causing the viscosity of dispersion system to increase quickly, which can also be used indirectly to evaluate the dispersion degree of pigment particles. The experiment study the effect of temperature and shear stress with the dispersion of pigment particles through the different settings and the viscosity of the system.

By results of viscosity increase of the system in Figure 7.17 and Figure 7.18, it can be judged that the impact of temperature on pigment dispersion will do more than that of simply increased shear stress, so the temperature on the dispersion effect of pigment particle plays a decisive role.

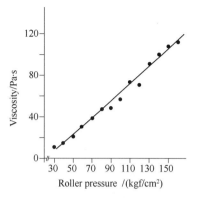 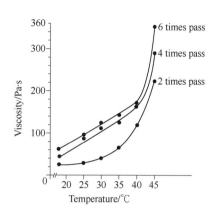

Figure 7.17　The relationship between the shear stress and viscosity of Pigment Violet 19 at constant temperature

Figure 7.18　The relationship between the shearing temperature and viscosity of Pigment Violet 19 under constant pressure

In addition, the processing machinery in the dispersion of pigment particles also plays a great role on wetting accelerating and pigment particles penetration through the motion force of melt transmission.

Therefore, wetting needs energy, the choice of appropriate wetting temperature and wetting promotion way is particularly important.

7.3.3　The Judgment Basis of Pigment Surface Properties—Wetting Angle

The surface properties of pigment particles determines whether pigment wetting is easy or difficult. As one of the important factors to influence the surface characteristics, it is determined to a large extent by the aggregated array mode of fine particles and polarity characteristics. It is important to accurately determine surface properties of pigments and properly select the wetting dispersion technology in order to achieve a multiplier effect. This can be done through measurement of pigment wetting angle as a basis for judging the surface properties of pigment.

Judgement and measurement of wetting angle need to use special instruments.

The pigment powder is pressed to form a smooth surface, it is placed horizontally on the surface of pigments. A drop of water is gently dropped, the final equilibrium shape of water drops spreaded on the surface is observed and measured, this shape depends entirely on the surface characteristics of the pigment itself. Measurement of the intersection angle of contact point of water droplets to plane can determine the wettability of water on the pigment.

Here, water is considered as the representative of polar substances, through this experiment, it is judged that the surface properties of pigment products under an environment of polar substances. If other liquid substance is substituted for water, the test and judgement can be done in the same way, the wettability of the liquid substance on the corresponding pigment products can be determined.

Figure 7.19 is EasyDrop measuring system made in Germany, it is widely used in the measurement of contact angle. EasyDrop uses the combination of computer digital processing technology. Optical system with CCD camera makes the droplet image clearly to display on the computer screen, provides instant memorized image, and through the instant analysis system (Figure 7.20), water droplets image can be accurately measured and analyzed. Its six multiply zoom lens ensures the best and full screen display, so the measurement of contact angle is fast, simple and easy.

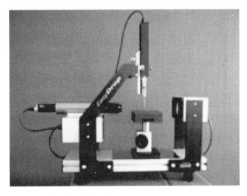
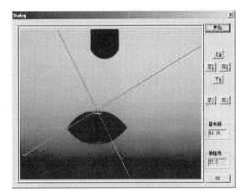

Figure 7.19 EasyDrop measurement system of Germany

Figure 7.20 Drop trigger imaging analysis system

7.4 Pigment Dispersion

Pigment dispersion is that agglomerates and aggregates of pigment particles are crushed under the action of external force, so that the performance of the pigment can be nearly perfect in the application system. The attractive force between pigment molecules must be overcome initially in pigment dispersion. The dispersed particles need to maintain stability to prevent the aggregation.

Method for dispersing pigments in plastics is mainly the melting shearing force, this is commonly used in plastic industry. The melting shearing method, as the term suggests, in which pigments are encapsulated and wetted by the melted plastic resin and dispersed by the shear stress resulting from the melt motion.

Figure 7.21 and Figure 7.22 visually display the process of dispersion and stability of pigment particles in the resin melt. First, pigment particles and resin melt pass through the shear zones through the help of dispersion equipment. When the shear stress reaches a certain level, the dispersed substrate breaks and crush, the particle size becomes smaller. When the viscosity is reduced gradually to viscous flow state through the melting and plasticizing of resins, the smaller particles penetrate into the polymer, under the action of shear stress of flow field, the particle size is further reduced, it is not uniformly dispersed in the polymer until the agglomerate of pigment particles is broken up and close to the primary particle state. Finally pigment fine particles are highly mixed and uniformly distributed under the sustained effect of flow field. The effectiveness that the melting shear stress transfers to pigment particles depends on the wetting effect of resin melt and shearing ability of equipments.

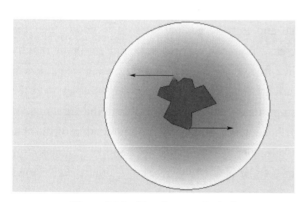

Figure 7.21 The shear method of plastic melt

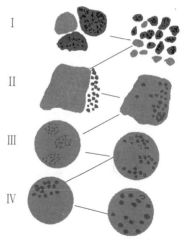

Figure 7.22 The schematic diagram of pigment dispersion in the plastic melt

In case the melt in the screw is simplified to Newtonian fluid, the viscosity decreases with the rise of temperature, therefore higher shear force is resulted from relatively lower temperature which is beneficial to the breakage of agglomerates of pigments. When employing extrusion shear, the extrusion temperature should be higher than the melting point of the resin. Once the temperature of plastic is close to the melting point, the dispersion state of melt is the best since the shear stress produced under this temperature is greatest. However, it is beneficial for pigmens to mix rather than disperse at higher temperature. As the shear stress increases with the increase of screw speed (See the formula below), higher speed will reduce the residence time of pigment particles in the shear field, so that we must choose the correct screw speed. Since the depth of spiral groove determines the size of the shear field, it will also affect the shearing.

The shear stress in the working model of screw ζ(Newtonian)

$$\zeta = \mu \gamma$$

μ—the viscosity of melt resin.

γ—the shear rate.

The melt shearing equipment commonly used in plastics are two roll mill, internal mixer, extruder (single, double screw) etc.

7.5 The Dispersion Stability of Pigments

The dispersion process of pigment particles in plastics consists of three stages: wetting, dispersion and stability. Wetting and dispersion are mainly relied on the medium and process equipment to complete. This is the premise that creates perfect dispersion, but it is enough to stabilize the fine particles of pigments in the dispersion system. Once the shear stress in the system is weakened or eliminated, the fine particles of pigment dispersed in a medium will aggregate and form larger particles. It is known that the pigment dispersion process will decrease pigment particle size and result in rapidly increase of the surface area. The surface free energy of particles is also increased, it will induce the attraction force among fine particles, which continuously increases with the dispersion process progressing. Microparticles always move toward the direction of reducing surface area for lowering surface energy, so that the re-aggregation of the fine particles is the unavoidable result. Consequently, the work to stabilize pigment fine particles in the dispersion system seems to be absolutely necessary. Figure 7.23 shows the dispersion and re-aggregation process of pigment particles.

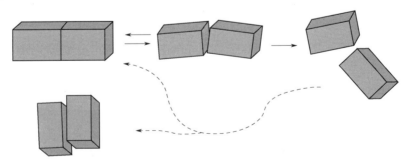

Figure 7.23   The schematic diagram of re-aggregation after pigment dispersion

Generally speaking, there are two kinds of qualitative theory of mechanism for stable dispersion of fine particles, they are the stabilization mechanisms of double electronic layer and steric hindrance, respectively.

- The stabilization mechanism of double electronic layer: it brings microparticle surface a certain amount of charges and forms a double electronic layer which makes the attractive force between particles greatly reduced by the repulsion force, so as to realize the dispersion stability of particles (Figure 7.24)
- The stabilization mechanism of steric hindrance: as shown in Figure 7.25, an uncharged high molecular compound is introduced to the system which is adsorbed on the surface of microparticles. Barrier is formed between particles and achieve the purpose of dispersion stability.

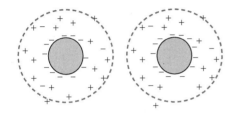

Figure 7.24 The stabilization mechanism of double electronic layer

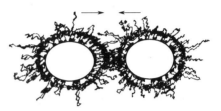

Figure 7.25 The stabilization mechanism of steric hindrance

7.5.1 The stabilization Mechanism of Double Electronic layer

Pigments dispersion in a liquid medium is a fully dispersed system for particles. Usually the particle diameter of 1~100 nm is identified as a colloid. Pigment particles at this time have a large surface area and high surface free energy, and thus become very unstable in thermodynamics. However, this instability is not necessarily immediately apparent. Depending on the particle material and its surface characteristics, they may reflect transient stability characteristics in a limited period of time.

There is repulsive potential energy between micelles, and also attractive potential energy. The sources of two potential energies can be obtained and interpreted from the analysis of micelle structure.

In Figure 7.26, any point outside two positively charged micelles is unaffected by the positive charge. Any point in the diffusion layer, owing to that the positive charge effect is not completely offset, still shows positive. But when the two micelle diffusion layers do not overlap, both have no repulsion.

When two micelle diffusion layers overlap, the negative ion concentration in the region of overlap as shown in Figure 7.27 increases. The symmetry of diffusion layer of two micelles have been damaged, thus making the excess negative ion in the overlap region spread to unoverlapped region. Due to the damage of the electrostatic equilibrium of double electronic layer, thus resulting in osmotic repulsion. Both repulsive potentials can raise with the increase of the overlap area.

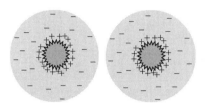

Figure 7.26 The diffusion layer is not overlapping, repulsion is not produced between the two micelle

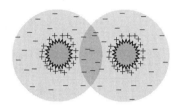

Figure 7.27 Overlap of diffusion layers, damage of equilibrium, emergence of osmotic repulsion and electrostatic repulsion

The attractive potential energy between the particles in the dispersed phase of the micelles has the property of Vander Waals forces, but it is a Vander Waals force of long distance, the attraction increases with increasing the particle diameter. The action range is thousand times larger than that of general molecular, it is inversely proportional to one square or two square distance between particles (The Vander Waals force between general molecules or atoms is inversely proportional to six square distance between particles). The attaction makes particle condensation and precipitation. Vander Waals force of long distance decreases with increasing particle separation distance. The so-called electrostatic repulsion, here mainly refers to the repulsion between charges, it is the key factor to maintain the micelle stability. The electrostatic repulsion increases with increasing the particle size, double electronic layer thickness of particle surface, but decreases with increasing the distance between particles. It has yet a close relationship with parameters that affect the electrical performance characteristics of the system, such as pH value, salt concentration and ionic valence number etc.

According to DLVO theory, it considers that total potential energy E between particles is equivalent to the sum of double electrostatic repulsion Ea formed around particles and Vander Waals force, when the repulsion between changes is great, the sum of repulsion and gravitation is larger than 15KT (K: Baltzmann constant, T: absolute temperature), it will produce an energy barrier in order to guard against particle condensation and keep the dispersion stable. Repulsion potential, attractive potential energy and total potential energy are respectively function of distance between particles.

Making a drawing of repulsion, attractive potential energy and total potential energy between particles on their distance and get the potential energy curve (see Figure 7.28)

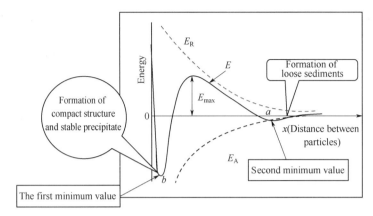

Figure 7.28　The schematic diagram of theory of DLVO double electronic layer

As shown is Figure 7.28, when the distance x between particles is reduced, there appears the smallest value a, then reach to second minimum value b, its value is a few KT. Particles will form a loose and unstable coagulated sediment here. When the environment changes, the coagulated sediment can then separate to obtain micelle. Between a and b, the repulsive force plays a major role, when x continues to shrink, the maximum value E_{max} (potential barrier) appears on the potential energy curves. Generally, particle kinetic energy can not overcome it, so that the micelle is in a relatively stable state. When the kinetic energy which is accumulated in the thermal motion of two colloidal particles is more than $15KT$, it may exceed the energy value, passover E, result in the first minimum value b on the potential energy curves, particles fallen into this trap produce an irreversible coagulation and form a compact and stable polymer.

In practical application, the dispersion stability of pigment particles of liquid colorant in adhesive glue and vinylon depends mainly on the energy relationship between particle-carried charge repulsion and Vander Waals attractive force. Therefore, the charged amount of pigment dispersed particles is an important factor of stabilizing dispersion system. If the pigment particles carry positive charges on the surface, the surface will adsorb negative charges and constitute double electrical layers. The number of the adsorbed negative charges is less than that of positive charges carried by the particle surface, its thickness is about radius of one ionic, the charge adsorbed layer is called the fixed layer. The rest negative charges diffuse toward the direction of the main body of the dispersion medium, the number of negative charge gradually reduces along the outward of the radius. The energy barrier generated by the whole double electronic layer structure prevents mutual condensation between particles, so as to maintain the stability of the dispersion system.

7.5.2 The stabilization Mechanism of Steric Hindrance

After dispersion of pigment particles, with the particle size becoming smaller the pigment surface area rapidly increases, resulting in surface energy increasing. Thus reducing the stability of pigment particles, agglomeration between particles is produced once more, which prompts the surface energy to reduce until it reaches the steady state. Figure 7.29 illustrates this process.

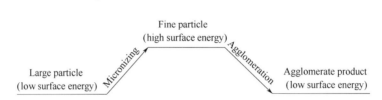

Figure 7.29　The process of energy variation during micronizing pigment particles

When the pigment particles are dispersed to form new surface, a layer of coating is attached to the surface at the same time, so that the surface energy of the pigment particles can be reduced. When pigment particles with the coating layer collide again, the coated particles effectively reduce surface energy and the steric function of surface coating layer, pigment particles will no longer produce agglomerate so as to achieve the pigment dispersion and stabilization. Figure 7.30 gives a good interpretation to the mechanism.

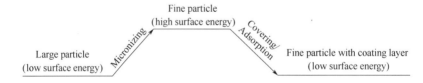

Figure 7.30　The variation process of energy at covering/micronizing of pigment particles

The coloration of thermoplastic plastics is largely based on the principle of stabilization mechanism of steric hindrance. The most comprehensive and effective embodiment of this theory is widely used in the color master batch manufacturing and application technology of plastic processing.

After adsorption of polymer compounds on the surface of pigment particles, on the periphery of the particles it forms a barrier layer which effectively cut off the collision and agglomeration among particles, thus playing a stabilizing role. The structure can seen in Figure 7.31.

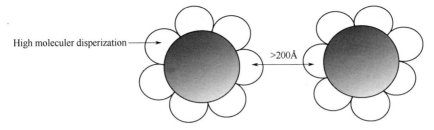

Figure 7.31 The schematic diagram of the adsorption of pigment for polymers

The thickness of the adsorbed layer determines whether the distance between particles is large enough to overcome the Vander Waals force between molecules. For the majority of particles, the adsorption layer thickness of greater than 100 Å is considered to be the ideal. Figure 7.32 shows the relationship of the amount of polyethylene wax which covers pigment with the tinting strength and pressure.

The factors affecting the adsorbed quantity on pigments are as follows.

(1) Coverage of the adsorbed layer The relatively high coverage of the adsorbed layer is beneficial to the stabilization of particles. In general, the molecular weight of adsorbed substance is larger, the amount of adsorption is larger.

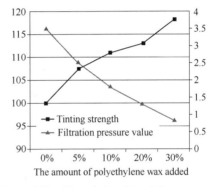

Figure 7.32 The relationship of the amount of polyethylene wax which covers pigment with the tinting strength and pressure

(2) The surface properties of pigment particles The surface properties of solid pigment is mainly physical properties (shape, porousness, pore size and depth, etc.) and chemical properties (polarity, non-polarity, surface tension).

As everyone knows, the atomic force field of pigment surface is not saturated, and it has half the remaining valence force. Usually the solid surface is not a true smooth plane, it has a lot of concave and convex and is somewhat rough. So the degree of saturation in different parts of the surface of the atomic valence force is different. Generally in the area of edges, corners, sides, concave and convex, the residual valence force is strong. It has larger adsorption capacity. For example, carbon black, though has a porousness nature and easily penetration theoretically, the adsorption of polymer is relatively difficult, this is because the polymer is in a curly state, is not easily embed into pores and adsorbed, so the bigger its molecular weight, the less the adsorbed amount. Therefore the dispersion stability of ordinary carbon black is relatively poor.

(3) The influence of coverage of pigment co-adsorbed layer When two or more than two kinds of pigments are in a formula, it is possible that different pigment particles may possess opposite charges, and make the pigment particles attracted each other and agglomerated. Therefore, it is necessary to choose appropriate adsorption material to neutralize charges, at the same time, to promote the absorbed layer thickness and increase adsorbed layer coverage for the sake of effective barrier.

(4) The influence of pigment concentration Since the pigment concentration increases, the probability of pigment particle collision raises greatly. In addition the covering of adsorbed layer requires energy and time to be implemented, those particles which did not get completion of absorption will agglomerate quickly in order to reduce surface energy. Simultaneously, the stability of pigment particles is in need of sufficient adsorbed materials to ensure the adsorbed layer thickness and coverage. That is why pigment concentration in a colour masterbatch should not be too high. Usually the weight ratio of organic pigment in a color masterbatch in the actual production does not exceed 40%.

There are two ways for the polymer fixed on the surface of pigment particles. One is an adsorption, its mechanism has been described above. The form of adsorption can be displayed as:

Line up shape: lying flat on the surface.

Tail shape: spreading in a medium.

Circle shape: returning to the surface after unfolding.

Bridge shape: connect two particles and so on.

The second fixation method is anchoring, which is connected with molecular on the particle surface by chemical function. Anchoring type may produce a mushroom, pancake and comb shape. Various forms of the above two fixed modes are shown in Figure 7.33.

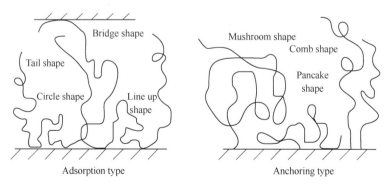

Figure 7.33 The formation of polymer compounds on a particle surface

In recent years some dispersion stabilizers known as hyper-dispersant are launched internationally, which are resemble partially to the traditional polymer layer adsorbed on pigment surface, but have obviously different mechanism, thus the produced dispersion and stabilization effect are not the same.

Hyper-dispersant is a highly efficient polymer additive for pigment dispersion and stabilization.

The molecular structure mainly consists of two parts: one part is anchor groups, there are $-NR^{3+}$, $-COOH$, $-COO-$, $-SO_3H$, $-SO^{3-}$, $-PO4^{2-}$, polyamine, polyalcohol and polyether etc. Hyper-dispersant takes the respective anchor group as basic point, is adsorbed firmly on the surface of pigment particles through the mutual attractive action of ionic bond, covalent bond, hydrogen bond and Vander Waals force etc. The another part is a solventized polymeric chain. There are polyester, polyether, polyolefin and polyacrylate etc. They can be classified into three categories according to the polarity: low-polarity polyolefin chain; middle-polarity polyester chain or polyacrylate chain etc; strong-polarity polyether chain. In the dispersion medium matched with the polarity, the chain has good compatibility with the host dispersion medium, it can merge into one (Figure 7.34).

The anchor group of hyper-dispersant is firmly absorbed on the surface of pigment particles, the solventized polymeric chain easily extends in the dispersion medium and forms a protective layer of enough thickness(approx 5~15nm). When solid particles adsorbed with two or more than two hyper-dispersant molecules draw near and collide against each other, due to the steric hindrance of the extended polymeric chain, it makes solid particles sprung off and prevents agglomerate and keeps the stable dispersion state.

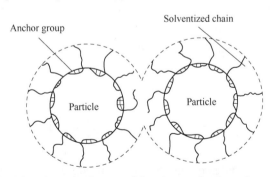

Figure 7.34 The schematic diagram of function mechanism of super dispersant

Different hyper-dispersant has a different structure: some has an anchor group of strong adsorption force, which is absorbed on a solid particle, and some have a plurality of anchor groups which the adsorption force is not strong and absorbed on the particle at the same time. This can be seen in Figure 7.35 and Figure 7.36.

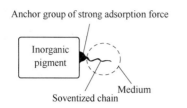 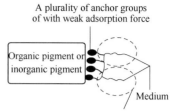

Figure 7.35 Hyper dispersant of single anchor group

Figure 7.36 Hyper dispersant of a plurality of anchor group

The mechanism of hyper-dispersion is not only to change the electrical properties of solid particle surface, increase the electrostatic repulsion but also to improves the steric hindrance by increasing the thickness of polymer adsorbed layer. When comparing with the electrostatic repulsion, the advantages of steric hindrance is its effectiveness in polar and nonpolar medium and the steric hindrance stability is thermodynamically stable, and the electrostatic repulsion is thermodynamically metastable. Therefore, compared to traditional dispersant, the effect of dispersion stability is greatly improved.

7.6 Color Masterbatch and Pigment Preparation

7.6.1 The Definition of Masterbatch

Mankind use pigments(dyes) in plastic coloring has been a very long history, but the application of color masterbatch of pigments in plastics and its product coloration was only started in the 50s of the last century. The industry of the first application of color masterbatch is cable and synthetic fiber industry, this is because even at that time in accordance with the quality requirements of products and technologies, the direct use of pigment powder to process and mold, the shear force of equipments is not up to the requirements.

The so-called color masterbatch is colored particles of high concentration which are prepared by well dispersing in resins, this is used for plastics coloring. The weight ratio of pigments (or dyes) contained in the color masterbatch is generally 20%～70%. According to the principle of cost optimization and processing feasibility, the ratio of the color masterbatch and the main resin is 2 : 100.

In addition to meeting customer requirements for the final color, color masterbatch mainly is to solve the problem of pigment dispersion in plastic resin, and to meet customer requirements for particle fineness in products. Therefore, the production of color masterbatch is not a simply process of mixing, extrusion and pelleting of pigment and resin, it is a highly integrated system engineering of pigment chemistry, resin melt processing and surface chemistry etc.

With the rapid development of petrochemical industry, the plastic production in China increases greatly year by year, the quality is continuously improving, new varieties are emerging. Therefore, the demand potential of color masterbatch closely related to plastic processing industry is self-evident. In 1994~2012, the average annual growth rate of Chinese color masterbatch reached 15%. According to reports from china chemical industry newspaper in 2007~2010, the Chinese consumption of color masterbatch in different kinds of resins and consumption demand of various markets can be seen in Table 7.4 and Figure 7.37.

Table 7.4 The composition of Chinese consumption of color masterbatch in different kinds of resins

Resin	Percentage	Resin	Percentage
LDPE	25	Nylon	6
HDPE	22	PET	4
PP	19	PVC	3
ABS	13	EVA	2
PS	6		

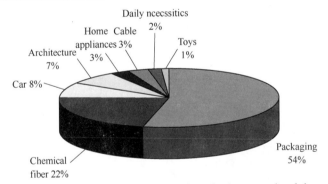

Figure 7.37 The composition of Chinese consumption of color masterbatch in various markets

7.6.2 Pigment Dispersion in the Color Masterbatch of Polyolefin

Polyolefin is the largest species in the production and usage of the entire plastics market, occupy more than 60% of the entire plastics market amount. Polyolefin resin has been widely used for plastic film, pipe, plate, hollow parts, textile and other industries. More than 80% in the polyolefin plastic products coloration are colored with color masterbatch process. So this section takes polyolefin color masterbatch as an example to discuss the dispersion and application problems of pigments in plastics.

Currently many manufacturers of color masterbatch at home and abroad adopt the process of high-speed mixing and twin screw extrusion technology, of which the flow can be seen in Figure 7.38. It was first pioneered by joint development of the former West Germany Hoechst company and equipment manufacturers.

This process is that the raw materials (carrier resin, polyethylene wax, pigment and additives) according to the formula amount are added in a high speed mixing machine and mixed. The coarse particles of pigments are pre crushed through the interaction of high-speed mixing and flow blocking plate, then pre-wetted and fused with polyethylene wax of relatively low melting point, at the same time, it uniformly pastes on the surface of the plasticizing resin particles because of high speed stirring and frictional heating. After finishing the above process and through the cooling of low temperature mixing machine, thus removing the stickiness of particles. The mixture is added to a twin screw extruder of same direction and parallel, melted and plasticized (this step includes further wetting on the pigment particles), blended, grinded, followed by extrusion through die and cutt into color masterbatch. Experience proves that the pre-wetting of pigments during the high-speed mixing process determines to a great extent whether the eventual dispersibility of color masterbatch product is good or not.

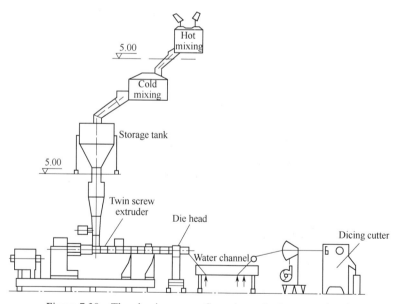

Figure 7.38 The classic process flow chart of color masterbatch

As everyone knows, the primary problem to solve is the pigment dispersion in the color masterbatch process. It is impossible to solve the problem that powder pigment is used in direct processing during the production process of plastic products. It is very difficult to disperse the particle to an ideal stage with only one method. Many factors are involved, the comprehensive factors must be considered according to each particular formula. It must review all situation and solve their respective problems. These problems include the pigment itself, processing, equipment, product application, etc.

7.6.2.1 Selection of Well Dispersed Pigment Varieties

(1) Choosing the pigment which is easily dispersed by resins EN13900-5 filter pressure test can be used to confirm that the pigment surface properties showed by the wetting theory described in Washbum equation have a decisive significance for pigment dispersion in a plastic resin.

Under the same test conditions, if the pigment can be wetted by plastic resins and is quickly dispersed at a relatively low shear conditions, it is getting through a specific screen package in which the filtration pressure value (FPV) is less than 1 bar/g, which indicates that the dispersibility of this pigment is excellent. The color masterbatch made in accordance with the appropriate process is suitable for coloring products which has very high requirements of dispersion. On the other hand, with the same test processing conditions the measured filtration pressure value is greater than 1 bar/g, which shows that the performance of wetting and dispersion of pigment is poor, the processed color masterbatch may not be able to comply with the follow-up application requirements.

The filtration pressure test is based on the use of settings of different fineness screen package. The test process and the ascending value of pressure from front-end screen package reflect higher ratio of the larger particles size of which the dispersed fineness do not meet the application requirements in the pigments dispersion. The size and number of large pigment particles filtered out after passing through the filter pressure test are shown in Figure 7.39.

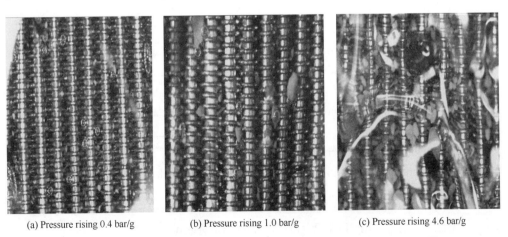

(a) Pressure rising 0.4 bar/g (b) Pressure rising 1.0 bar/g (c) Pressure rising 4.6 bar/g

Figure 7.39 The state of pressure filtration value and residual pigment particle on the filter screen

The filtration pressure test can directly reflect dispersion properties of pigment products in practical application. The reproducibility and actual processing of the test results are very high, it has been more and more widely accepted by domestic industry colleagues. This is the reason why most of the international trader and customers request that the suppliers should provide filtration pressure testing data of related products.

In addition to the filtration pressure rising test, of course, there are some other methods for the evaluation of pigment wetting and dispersion performance, for example, in the third (6:3) section of the sixth chapter, the tinting strength evaluation of double roller milling and the color specks detection of blowing film, etc., all the detection method methods are selected depend on the different practical application. However, these methods used in the practical application is not very common in their use, the reason is that all the test must be conducted by professional equipments. But some of the manufacturers do not furnish such instruments, it requires professional knowledge to perform reasonable test methods and evaluate the relevant quality requirements of their own products.

In addition to the above test methods, there is another simple method to judge and test, that is the selection of pigment varieties with relatively low oil absorption for coloring plastics. Because most of the plastic resins are low polar or non-polar, therefore select oil with the same polarity (low or non-polar) and perform oil absorption test on pigment, based on the "like mix like" principle, the lower the oil absorption of pigment the more the lipophilicity, which also means that the polarity of pigment surface is lower, and can predict that the wetting and dispersion of pigments are relatively easy in the coloring and processing plastics.

As mentioned above, as a general principle of plastic coloration, it is applicable to the universal processing and application. However, the principle has the particularity in the processing of plastic products. In the actual application, the resins with high polarity may be also used and this kind of application will be more and more with the advanced technology and special product requirement. Under such condition, the pigment of adequate polarity should be chosen according to the actual situation to facilitate the implementation of wetting and dispersion, ensure product quality.

(2) In the same kind of pigment, powders with relatively large particles are beneficial to wetting and dispersion.

(3) Generally, selecting the product with lower bulk density among the same varieties is in favor of wetting and permeability and then to help disperse.

7.6.2.2 Choosing Low Viscosity Wetting Agent

According to the wetting theory described by the famous Washbum equations, it is known that the viscosity of wetting agent and functional group are the vital link to pigment wetting and dispersion. Here is to illustrate the example of taking polyethylene wax as wetting and dispersion agent for the production of polyolefin color masterbatch.

(1) The viscosity of wetting agent　The general rule is that, the selection of wetting agent with relatively low viscosity is in favour of the wetting of pigment. Taking polyethylene wax as an example, the simple discrimination method is to choose the polyethylene wax of density of $0.91 \sim 0.93 g/cm^3$, its relative viscosity is low, the resulting dispersion is somewhat ideal. When the density is $0.94 g/cm^3$ or more, the viscosity significantly become higher so that the wetting of pigments appears poor. In addition, when using polyethylene wax as wetting and dispersing agent, it serves also a typical external lubricant. The lubrication of system can greatly strengthen when the density is too high and is unfavourable to the dispersion of pigment.

(2) The molecular weight distribution of wetting agent　This is a very important factor and is often ignored. In the case of polyethylene wax, the polyethylene wax with the relative narrow distribution of molecular weight must be chosen. If the molecular weight distribution of polyethylene wax is wide, it must contain more chains whose molecular weight is either too large or too small. If the wax viscosity of large molecular weight is too high, it will inevitably affect the wetting effect. However, if the molecular weight is too small, the viscosity and viscosity shear will be low which lead to affect the dispersion. In addition, low molecular weight may also increase the amount of volatile matters to pollute production environment or bring the peculiar smell. So, it is very necessary to select the wetting agent of appropriate molecular weight distribution. Generally speaking, the polyethylene wax should be manufactured with the polymerization process, the degree of polymerization can be effectively controlled, this process will produce the product of relatively narrow molecular weight distribution. But the wax that is produced with pyrolysis production, due to the uncontrollability of the fracture site, thus leading to the characteristics of wide molecular weight distribution of manufactured product.

(3) Focusing on "the low temperature viscosity" and "high temperature viscosity of polyethylene wax　Here the mentioned "low temperature" and "high temperature" refer to the concept of relative temperature. When an ideal wetting and dispersing agent treats pigment particles in the wetting stage, the viscosity (relatively low temperature) is relatively lower in favour of wetting of pigment particles. While it is in shear dispersion stage (relatively high temperature), we hope for higher viscosity, which would be helpful to shear dispersion.

(4) The powder fineness of polyethylene wax products　Finer wax particles can be mixed more uniformly with pigment powder, which would facilitate the wetting more

rapidly and effectively. Thus, the mircowax powder is getting more attention in the industry.

The above mentioned content only discusses the effect of wetting agent on the dispersion in the actual production, but there are some other factors that must be also paid close attention to. For example: environmental factor and product safety. It is known that the wax product of too low molecular weight will increase volatility and cause the pollution of the environment and release bad smell. If such wax is used in sensitive applications such as plastic food packaging area, children's toys, etc., it will cause very serious consequences. Therefore, the selection of main auxiliary material must be cautious.

7.6.2.3 The Energy Required by Dispersion Begins at Wetting

We all agree that pigment dispersion requires sufficient energy. But many people may think that the required energy is for pigment grinding only. It is important to point out that this understanding is not correct. Because in the previous sections of this chapter, we have already mentioned that the pigment wetting must need energy to achieve good wettability of pigment particles: Appropriate temperature can melt the carrier resin and wetting agent to ensure the melt viscosity required by fast wetting (or flow performance) is realized. So properly increasing the melt temperature is beneficial to wetting. Such simple temperature effect already illustrates the importance of energy in wetting, not to mention the mechanical agitation would further accelerate the progress of wetting.

There is no doubt that the viscosity shear and mechanic shear applied for the dispersion of pigment particles are provided by the processing equipment.

In Figure 7.40, it illustrates the effect of processing temperature on pigment particles in the wetting stage and in grinding shear stage.

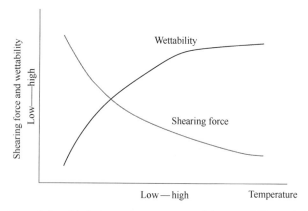

Figure 7.40 The relationship between temperature and the wettability and shearing force

7.6.2.4 The Choice of Carrier Resin

(1) The selection of powdered and granular resins

Taking the polyolefin color masterbatch process described above as an example, the statistics of experimental data and analysis show that the final result of the dispersion got with powdered resin production is better than that of using the granular resin. The obvious difference is that the powdered resin with small size can quickly complete the melting process in equipment, that is to say the pigment particles can be wetted faster and moved into the process of dispersion from wetting sooner. At the same time the powders can be mixed more uniformly and effectively to improve the wetting surface In granular resin, because of its large particle size, the melting process from the outside to the inside requires some time, thus pigment wetting speed is far less than that of powder resin. This problem seems to be footy, but it has direct impact on the dispersion results, it should pay more attention. However, not all the resin products supplied in the market are in the form of powder, therefore, a lot of color masterbatch manufacturers buy milling equipments and grind all their resin into powder form for color masterbatch production to promote and ensure the quality of products.

(2) Selection of the resin with high melt viscosity

The higher is the molten viscosity of resin, the larger is the shear viscosity of system. The shear force transferred to the pigment particles by the molten resin is also larger. This is beneficial to the dispersion of pigment particles. So for polyolefin color masterbatch process, which resin is more beneficial to the pigment dispersion? Generally we would use linear low density polyethylene(LLDPE).

The so-called linear low density polyethylene (LLDPE), is a copolymer manufactured by polymerizing ethylene monomer with a small amount of α-olefins (such as butene-1, haxanene-1, octene-1, tetramethyl pentene-1 etc.) through high or low pressure under the action of catalyst. Its density is in the range of $0.915 \sim 0.935 g/cm^3$. LLDPE is a non-toxic, tasteless, odourless material and the molecular structure is characterized by its linear backbone. LLDPE has no long branched chains, which leads to its high crystallinity. In the case of the same molecular weight, the melt viscosity of the linear structure LLDPE is higher than that of the non-linear structure LDPE. However, in the case of the same melt index, the melt viscosity of LDPE is lower than that of the LLDPE. Therefore, the choice of the linear low density polyethylene as the carrier of polyolefin color masterbatch is not surprising.

7.6.2.5 Extrusion Dispersing Equipment and Process Control

(1) The design of screw combination of high shearing force

Presently, the key tool of color masterbatch production process is twin screw extruder (Figure 7.41) as the main dispersion equipment, it is important to understand and reasoning allocation the process performance of twin screw since it is related to the product quality and production capacity.

Color masterbatch manufacturers generally select building block type co-rotating twin screw extruder. Unlike traditional extruders, the full set of screw extruder consists of various types of screw thread and grinding components like building blocks aligning together. This provides great flexibility for different product processing, and help to achieve the best balance of product quality and production capacity.

(a) Parallel twin screw thread block element (building block)

(b) Example of twin screw configuration after assembly

Figure 7.41 Example of parallel twin screw element and assembly

(2) Screw speed and shear fore

For the screw of specific configuration, the shear force produced is decided by the rotation of screw: the higher the speed, the greater the shear generated, the bigger the dispersibility of solid phase, and the finer the obtained dispersed phase size. But if the speed is too high. Due to the temperature rising generated by the severe friction, this easily leads to the thermal degradation of polymers. High speed also makes the residence time of materials in the equipment become short, if there is no intervention, it may produce uneven mixing of materials. Conversely, the lower the speed, the smaller the shear, this may lead to the insufficiency of dispersion and uneven mixing. Due to the long retention time of material in the mixing tube, this will intensifies the degradation of polymer. Screw assembly and speed are closely related with the shear dispersion, so we must put the two as a whole to consider.

(3) The reasonable temperature control

The setting of extruder operating temperature is related to the performance of resins processing, the index of characteristic properties (fluidity, viscosity etc.), the wetting dispersion of solid particles, and the insurance of the minimum degradation, etc. Therefore, the principles of temperature setting must take into account of all the factors mentioned above.

Generally speaking, the temperature of main area of extruder should be higher than that of the resin's melting point, so as to ensure that the resin is in its best of molten state and provide adequate strength of shear stress.

Of course, the extruder temperature should also be set up according to the different demand of characteristics of processing materials and process stages, As to the different process of melting, wetting, grinding, and shearing, we must set up and adjust flexibly in order to achieve the best results.

In addition to a reasonable setting of extruder, the manufacturing of high quality color masterbatch should pay special attention to the control of production process. This is not only referring to the control of process operation, but also the formulation reasonability, especially some easily overlooked additive composition. When comparing with the main resin, the vast majority of additives are compounds of low molecular weight, it can lead to not only waste, but also poor processing or the change of product properties once added in excess.

7.6.3 The Jewel on the Crown of the Color Masterbatch Technology—Chemical Fiber Dyeing

Except color matching, the most key function of color masterbatch is to solve the problem of pigment dispersion. According to different product requirements, pigment dispersion requirements in products are not the same. Among them, the challenge of pigment dispersion for chemical fiber spinning industry is the most harsh one. With the sustained development of chemical fiber industry, spinning speed is getting faster, spinning fiber size becomes finer and finer, and new technology emerge in an endless stream, requirements of products thus derived for pigment dispersion are increasingly high. It always meet new requirements or new problems which must be solved. Therefore, the special discussion conducted in this chapter becomes really necessary.

There are two different routes in the chemical fiber dyeing. One is to add the dispersed pigment in the resin polymerization stage and be colored (such as vinylon and viscose), this process is called polymerization coloration. The other is to disperse a pigment (dye) in the polymer carrier which can mix with chemical fiber polymer and manufacture color masterbatch which is then added to melted fiber. This process is polymer coloring route (polypropylene and polyester), Common chemical fiber dyeing process and colorant dosage form are shown in Table 7.5.

Table 7.5 Colorant dosage form of chemical fiber dyeing

Chemical fiber and spinning process		Colorant morphology	
Variety name	Spinning form	Liquid	Granular
Viscose	Wet spinning	Aqueous dispersion	
Vinylon	Wet or dry spinning	Aqueous dispersion	
Polypropylene fiber	Melt spinning		Color masterbatch
Polyacrylonitrile (Nitrilon)	Wet or dry spinning	Solvent dispersion	
Polyamide fiber (Nylon)	Melt spinning	Aqueous dispersion	Color masterbatch
Polyester (Terylene)	Melt spinning	Glycol dispersion	Color masterbatch

The spinning process after dyeing exists three kinds of process route, it is respectively: wet spinning, dry spinning and melt spinning(Figure 7.42).

Wet spinning   The so-called wet spinning is to dissolve the polymer resin in the solvent, pressure through a spinneret and spin into a coagulation bath with a stream of filaments. The synthetic file such as polyacrylonitrile (nitrilon), polyvinyl alcohol (vinylon) and viscose or other artificial fiber products are produced by this process.

Dry Spinning   Different from the wet spinning, the dry spinning is to dissolve the resin in a solvent, pressure through a spinneret and spin into an air bath with a stream of filaments. Consolidation of filament by hot air flow. Dry process is more suitable for spinning of long filament products.

Melt spinning   Polymer resin is heated and molten into the melt, which is suitable for spinning viscosity, filtered and extruded through the spinneret plate, cooled by the air flow and drawn into high fine filaments. Melting process is commonly used in polyester(Terylene), polypropylene fiber, polyamide fiber(Chinlon, commonly called Nylon), etc.

Melt spinning originally uses the process of autoclave heating and pressurized into filament, While the method of extruding into filament with an extruder is used today, it has the advantages of simple, low energy consumption and consistent quality.

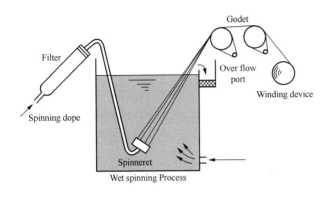
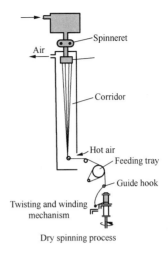
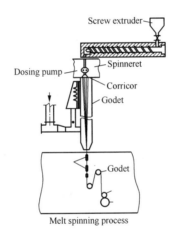

Figure 7.42   Three kinds of spinning process

The above processing method either wet or dry process, both make use of a large number of solvent at the beginning for producing resin solution, and coloration is also conducted in this process, in which colorant is held up for a considerable time, therefore, the solvent resistance and deflocculation performance of pigment must be considered in the selection of pigment. In addition, pigment is added as paste or pre-dispersed form due to lack of dispersible grinding process and related equipment for pigment particles in the process. Melt spinning has no need to consider the problem of solvent resistance during processing due to no solvent participation, but in the process, the resin is treated in the heat molten state, especially in spinning stage the temperature is generally at 300℃ or so. So we must seriously judge whether or not the selected pigments can withstand such high temperature.

Regardless of dry/wet or melt spinning, all of them can get the fiber length in the kilometer, we call it the filament. Filaments contain a monofilament, multifilament and cord wire etc, a linear density of conventional filament is 1.4~7dtex; and the fine monofilament is about 0.55~1.3dtex, the very fine one below 0.11dtex, it mainly is used in artificial leather, precision filtering material and other special applications. The requirements for pigment dispersion in making these products are extremely harsh. It will being great trouble and waste to the production operation once the spinneret is blocked by the larger particles of pigment.

The tow after spinning is cut into several cm to 10-odd cm, it called staple. Staple fiber can be divided into cotton, wool carpet and long staple according to cutting strength. Staple fiber can be used in pure spinning and non woven cloth, can also blended with different proportion of natural fiber or other fiber to make yarn, fabric and felt.

The traditional fiber weaving process must undergo the process of scotching, carding, drawing roving, spinning, winding, warping, sizing, reeding, weaving, etc, then it can be made into cloth. In recent years, rapidly developed chemical fiber non-woven technology abandons the complex process completely. It simplifies into spinning, the process of net laying and weaving and is completed in one step.

In case of spun-bonded nonwoven fabrics, the resin melt is extruded, injected into a spinning box through measurement and ejected into filament. Through the silk separated and paved by airflow after the drafting and sizing, it is shaped into cloth with hot pressure (The process is shown in Figure 7.43). In the base materials of spun-bonded non-woven fabrics, polypropylene is regarded as the majority, followed by polyester non-woven fabrics. Spun-bonded non-woven fabrics are widely used in engineering, agricultural, and packing materials, decorative materials and clothing lining materials, etc.

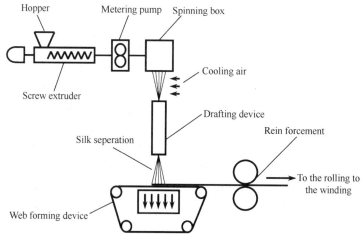

Figure 7.43   The schematic diagram of flow process of spun-bonded non-woven fabrics

In recent years, melt blowing non-woven fabric which is fast developed proceeds with the change of raw material. The resin with ultra high melt index (MFI is from hundreds to over one thousand) is melted and ejected, it becomes a micronized fiber of non-continuity through the stretching at high-temperature and high-speed airflow which is sprayed from the spinneret side, after spreading into webs, it is pressed into cloth. The resin currently applicable in the melt blowing process is the polypropylene and polyester, the resulting non-woven material is composed of micronized fibers at 0.2~0.4μm and has excellent filtering characteristics. It is made into all kinds of filtering materials, such as filter paper, filter cloth, filter cartridge, filter cotton and so on. They are widely used in the industrial filtration, water treatment, gas filter, health care and insulation materials, etc. The melt blowing method (Figure 7.44) has the advantages of short process flow, quick production speed and excellent performance characteristics.

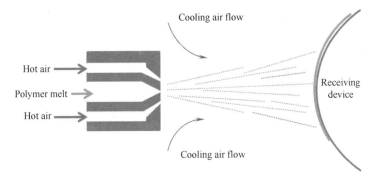

Figure 7.44 The schematic diagram of melt blowing process

In addition, as viewed from application, the combination with non-woven fabrics produced in different process or with other films can obtained superposition of properties and expand the new application characteristics. For example, the composite non-woven fabric (S/H/S) made of spun-bond layer and melt blowing layer(M). It takes the spun-bond layer(S) as a whole framework to ensure the application strength of products, but also reflects the excellent filtering performance given by the melt blowing layer(M). Therefore, it is widely used in medical and health field like masks, surgical clothes, etc. Academically the non-woven fabrics are collectively referred as the non-woven clothes. The S/M/S process flow diagram is shown in Figure 7.45.

7.6.3.1 Coloration of Viscose and Vinylon Before Spinning

The liquid colorant of viscose and vinylon fiber usually is the form of prefabricated agent which takes water as a medium. It is different from the conventional preparation of plastic pre-dispersion. The color paste of homogeneous body formed through ultrafine

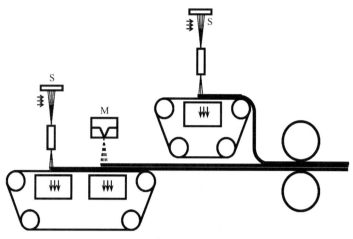

Figure 7.45 S/M/S process flow diagram

dispersion after mixing pigments and water with proper amount of dispersant. As a colorant for chemical fibers, the most important point is how to disperse pigments with the aim of meeting the requirements of operating process and product quality. Typically, the pigment size in the dyeing is controlled around 0.5~2μm. For special products, the demand of the fineness is more harsh. If the prefabricated agent or coloring system contains larger particles, or the poor stability of fine particles of pigments in the system induces agglomerate, the colorant could fail to mix with the polymer, and cause the spinning nozzle blockage which will affect the spinnability or produces all kinds of problem such as broken wire, fluff, stiff blocks during the stretching treatment. Because the colorant takes water as a medium and water is a polar solvent, the pigment of high hydrophilicity should be chosen for dyeing of viscose and vinylon fiber which is beneficial to the dispersion and stability of pigments in the aqueous phase.

After the ultrafine dispersion of liquid colorant for viscose and vinylon fibers, the dispersed particle of pigments in essence is in an unstable state, there is always a trend of reducing surface area and surface energy, producing agglomerates between particles which will cause the separation of color paste (sedimentation or flooding), flocculation and aggregation, directly affect the normal operation and product quality of spinneret.

The pigment particles in the aqueous phase go through superfine dispersion to achieve the required fineness, timely filtration is very important to maximize the efficiency of dispersion and grinding and shorten the grinding time, moreover also can help to separate the possible impurities.

In addition, for pigments of coloring vinylon and viscose fiber, we need to pay attention to choose the one with good acid and alkali resistance. Whether in wet or dry spinning, all of them should be able to withstand the treatment process of 20% sulfuric acid at 20℃ or 5% sulfuric acid at 100℃ and 5% NaOH treatment. Therefore, the azo lake pigment such as ultramarine blue, iron oxide yellow, iron oxide black and chrome yellow pigment are not suitable for use in this field. In view of the process requirements, when doing dyeing of vinylon, hot water resistance of pigments must also be considered. Of course, in view of application of final product, the colorant for viscose fiber and colored vinylon must also have good fastness properties, such as light (weather) resistance, heat resistance, abrasion resistance, migration resistance, solvent resistance and other chemical properties, etc. As the further consideration, the impact of colorant on the physical characteristics of products should also be minimal, tensile strength, tensile elongation and so on.

7.6.3.2 Terylene Coloration Before Spinning

Terylene scientific name is polyester fiber (polyethylene terephthalate fiber), referred to PET. Its performance is excellent and the raw material is easy to get with wide range usage. After the development of more than half a century, it has become the largest production of synthetic fiber, various varieties and has broader range of application.

Polyester color masterbatch is prepared by fully dispersing the excess pigment or dye and added uniformly to the polymer carrier which possesses good compatibility with polymer. Since addition of the color masterbatch is prior to the spinning, we must carefully study and judge the influence of the colorant on spinning properties and fiber performance. Therefore, to choose appropriate colorants is a most important step.

The glass transition temperature of polyester is as high as 81℃, so you must choose a proper pigment or dye for polyester color masterbatch. Because of the addition of the masterbatch to the polymer melt is prior to the spinning, it will be subjected to the high temperature of 285~300℃ of spinning melt, the colorant which consists of either pigments or dyes must be tested under the temperature of 285~300℃ for 10~30min without discoloration. Obviously, they must have excellent heat resistance. For the dye, it also requires good sublimation resistance and permeability resistance. In view of this, we can choose the stable pigments such as phthalocyamine blue and green, perylene, anthraquinone, dioxazine carbazole and some high performance inorganic pigments, etc. In addition, high performance organic pigments such as the products of DPP, quinacridones can be considered according to their properties. Beside the heat resistance of pigments, we also need to consider the performance of light(weather) resistance, chemicals resistance, and the requirement of environmental protection and safety when using the final products.

7.6.4 Pre-Dispersed Pigment Preparations for Plastics

With the sustained development of industry, today's colored plastic processing and forming are developing toward the direction of a large-scale equipment, highly automated production, high speed running, more refinement and standardization products. Many ultra-fine, ultra-thin and ultra-micro products, emerge as the times require. The requirements and standards for pigment dispersion have been improving. In addition, the voice for the production meeting the high efficiency, environmental protection, energy saving and reducing the cast becomes more and more popular. Because the conventional plastic forming and processing equipment (such as injection molding machine, spinning machine or single screw extruder, etc.) can not provide the shear force required for the pigment dispersion, therefore, the work of pigment dispersion is often done by professional production firms-pigment suppliers or masterbatch manufacturers.

Pigment prefabricated preparation (often referred to SPC — single pigment concentration) is a predisposing of single pigment with high concentration. Based on the characteristics of different pigments, generally pigment prefabricated preparation contains 40%~60% pigments. It is made with the special process by the specific processing equipment. Effective dispersion methods and strict quality control make the pigments dispersed in the finest particle morphology and reflect to the best color performance. The resulting product is fine particles of approximate 0.2~0.3mm and which can be also made into an ordinary color masterbatch. Because the pigment prefabricated preparation has such obvious characteristics, it has been more and more applied in the making of color masterbatch.

The pigment prefabricated preparation has the following advantages.

- Because of the complete dispersion of pigments, it has a relatively high tinting strength. In comparison with the use of powdery pigments, generally it can increase 5%~15% of tinting strength.
- It needs only minimal shear force to achieve the homogeneous condition. A simple equipment (for example: single screw) can be used to make high quality color masterbatch products.
- It is suitable for various kinds of extrusion equipment, produces stable quality and offers flexible production scheduling.
- Reflecting the color properties of perfection: color brightness, transparency, gloss etc.
- Preventing the dust in the production process. Improving the working environment. Reducing the pollution.

- No equipment contamination, simplifying the equipment cleaning in the color conversion process.
- Pigment particles are fine and homogeneous, prolonging the using time of the filter screen, reducing the replacement frequency of filter screen, improving the production efficiency.
- Uniform appearance of the preparation. No mutual adhesion. Suitable for all kinds of feeding machine. Neither bridge nor blockage during transportation.
- Without the need for pigment dispersion. This can greatly improve the production capacity of color masterbatch.
- It can be used in conjunction with other colorants, strong applicability.
- Various preparation formulation. It can be suitable for different forms of the carrier resin, good compatibility.

Although in comparison with the direct use of powdery pigments, the cost of used prefabricated preparation will be somewhat higher, but in comparison with finished product characteristics, such investment is worthwhile especially for those demanding superfine and ultra-thin products of high quality, there is no way to produce them using common dispersing methods. In respect to the increasingly function of market and industry today, the emergence and development of such specialized processing industry is an unstoppable trend.

Chapter 8
Practical Technology and Quality Control of Color Matching for Plastics Coloring

Only colored plastics can become a commodity and serve the society. The color matching for plastics coloring is to prepare satisfactory color based on three basic colors: red, yellow and blue, which must comply with the requirements such as heat resistance, light fastness, weather resistance, migration resistance and deformation resistance in the processing and using. What's more, coloring of plastics may also give plastics with a variety of functions, such as, aging resistance, electrical conductivity and antistatic properties.

With the improvement of people's life level, the requirements of color have become increasingly demanding. Modern society likes an ocean of information which make people feel confused and may momentarily pour a large amount of information. In this case, what allows users to receive information instantly and make accurate reaction is color, graphics and text. Color orientation aims to highlight the aesthetic feeling of goods, and makes people see the characteristics of goods from appearance and color.

At present the books for coloring of plastics are rare and the personnel who are excellent in color matching are very scarce. In order to prevent the color staff run away, many factories often to do a lot of secret documents for their color matching technology which not conducive to the progress of the industry. Therefore, there have a brief introduction to the basic technology of coloring of plastics in this chapter and also want to provide a guide for the newcomers.

8.1 Basic Theory and Principles of Color Matching

8.1.1 Basic Theory

Coloring of plastics means the subtractive color mixture for sunlight by using dyes, pigments etc. Different colors can be obtained by changing the light absorption and reflection. For example, if all of the light is absorbed, the color is black. If only part of the light (i.e. only absorb certain wavelengths of light) is absorbed, and the number of scattered light is very small, then the plastic becomes colored and transparent. If all of the light is reflected, the color is white. If all of the light, which is not absorbed, scattered, then the plastic becomes colored and opaque.

By mixing the three primary colors of red, yellow and blue, all sorts of colors can be get. All sorts of color can be regard as mixing the three primary colors of red, yellow, blue with a certain percentage amount. The three primary colors blend is shown in Figure 8.1.

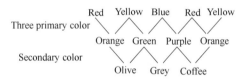

Figure 8.1 Color matching chart of three primary colors

Usually, defined red, yellow and blue with better saturation and large brightness as primary colors (basic color). By mixing any two kinds of basic colors, a variety of other different colors can be get which called secondary color (inter-color). For example, bright red, orange and apricot can be obtained by the mixture of red and yellow. Green, lake blue, light green, grass green and emerald green can be obtained by the mixture of yellow and blue. Purple, pale purple and dark pink can be obtained by the mixture of red and blue. Three-color (or re-between colors) can be obtained by the mixture of a primary color (or secondary color) and secondary color. For example, olive is mixed by orange and green, gray is blended by green and purple, and brown is mixed by orange and purple. Triangle color matching method of the visualization of above principles is shown in Figure 8.2.

In addition, on the basis of the primary or secondary colors, different shades of colors such as light red, pink, light blue and lake blue can be configured by white reduction. Different lightness of colors like brown, dark brown and dark green can be configured by adding various amounts of black. The required special attention is to avoid forming black in the central area of the triangle by mixing red, yellow, and blue at the same time.

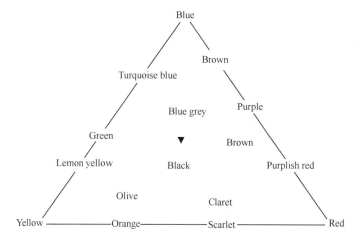

Figure 8.2 Chart of triangle color matching

At present, the trial and error method of visual color is mostly used in coloring of plastics. Though it is simple, but unscientific, time-consuming, and requiring the operator having a wealth of experience, which is difficult to operate. Practicing painting is the best way to improve the speed and accuracy of matching colors. The founder of the famous Munsell color scale, Munsell himself, is an artist. But the result of color matching is affected by many factors, such as climate, season, quality, experience, emotions and health conditions of staff. Computer color matching is still an inevitable direction in improving accuracy.

8.1.2 Basic Principles of Color Matching

The process of color matching for plastic coloring is very complicated, and to get the desired effect, being fast, accurate and economical, the following principles should be followed.

"Small amount" principle The varieties of colorants should be used as less as possible for plastic color matching. In short, it is better to use two mixed colors than three mixed colors, not only because too many varieties of colorants will cause trouble, but also bring complementary color to lead to dimmed. In addition, the more varieties of colorants, the more system errors are, because of the factors of dispersion and tinting strength and so on.

"Similar" principle The similar heat resistance of species should be chosen in coloring of plastics. Otherwise, the change of temperature during the processing can cause color change, because the heat stability are too different.

The similar dispersion of species should be chosen in color matching of plastics.

Otherwise, owing to the large difference of dispersion ability, the change of shear force in the process of production may causes color to change, such as color matching by organic pigments with bad dispersion and inorganic pigments with good dispersion.

The pigments with similar performance, similar light and weather resistance should be chosen in color matching of plastics. Otherwise, the color of plastic products would be unrecognizable after the outdoor exposure because of the large difference of performance.

"Complementary colors" principle The complementary colors should be avoided in color matching, otherwise, the shade of original color would become dim and gray, so the brightness of color is affected. Complementary colors are also known as complementary hue and color excess. Usually, we call them complementary colors which apart from the primary colors used to compound intermediate color. For example, orange is made of red and yellow, and blue is complementary color. Green is made of yellow and blue, and red is complementary color. Violet is made of red and blue, and yellow is complementary color. In the process of preparing violet with red and blue, yellow should be added as complementary color, and the color of violet will be dim, because adding more complementary color will turn gray.

The different colored lights should be noticed in color matching. For example, carbon black and titanium dioxide, used in coloring of plastics, are presented as blue and yellow shade because of different particle size, and make larger difference of pigment color light. It is advised that the complementary pigments should be avoided in color matching, otherwise it would reduce saturation and make the color dark.

8.1.3 Color Matching for Plastics Coloring is a Complex Process

It's well known that adding a single species of colorants for color matching is impossible. Although there are a wide variety of plastic colorants, but the supply is limited in the market. The only way to solve the problem that limited varieties against unlimited requirements of colorants is color matching. Namely, the desired color is matched by mixing several colorants reasonably. On the other hand, considering the costs and reduce inventory, it also need to use the limited colorants to satisfy the requirements of customers.

It is an extremely complex issue that adjusting each variables (species and amount) of coloring agent to achieve a known color vision characteristic, which is reasonable in price and meet the requirements in molding and using. Therefore, it's wrong to consider coloring of plastics is a simple dyeing granulation. Coloring of plastics is a systematic project, and only make a detailed understanding and elaborate design of every factor in this system, can achieve low cost and high quality objectives in the fierce market competition, which is shown in Figure 2.5. This requires a multi-industry skills, knacks and professional knowledge. For the same color, if the pigment has been properly selected, the quality of coloring is good and the cost is low. In contrast the quality of coloring is poor and the costs are high. To achieve the purpose of the best coloring, it should ensure the quality of coloring and select the suitable pigment of comprehensive considering the application performance, the application object, the application formulation and the application process. Therefore, color matching of plastics is an engineering system, combined integrating pigmenting chemistry, polymer chemistry and physical chemistry.

8.2 Basic Knowledge That Color Matching Staff Should Have

Color matching of plastics is a specialized technique which involves a very wide range of knowledge. Therefore, the professional engaged in color matching not only need good color resolution, but also need a wealth of professional knowledge and a lot of experience. In addition to the professional should also have a very serious and patient work attitude and a very flexible mind. An excellent professional personnel must fully grasp the following knowledge.

8.2.1 Comprehensive Understanding the Performance of Colorant

Colorant is the most important raw material in color matching of plastics which need coloring professionals have a full understanding and flexible utilizing.

Color matching staff must well know not only the various properties of colorants, such as tinting strength, colored light, dispersion, heat resistance and other indicators, but also be the performance in plastics application, such as light fastness, weather resistance, migration resistance, and security. An excellent color matching staff also should be quite understand about the performance of colorant in different concentration and resins, because the performance of colorant in this area is very large. Color matching staff also need to know some colorant can't be used in certain resins. Such as quinacridone pigment (phthalocyanine red) which have good heat resistance can't be used in ABS. Cadmium pigments can't be used in PVC because it can react with lead and produce black lead sulfide.

Because the economy of color matching is very important, the color matching staff should know the price of colorant, client always requires products which is cheap and good quality.

8.2.2 Comprehensive Understanding the Performance of Plastics

Plastic resin is the object of color matching, so color matching staff must be well aware of the various performance.

The main ingredient of plastic is a synthetic polymer or natural polymer. Plastics have many varieties, such as polyvinyl chloride, polypropylene, polystyrene, polyvinyl chloride, polycarbonates and so on. Plastics of different varieties have different molding conditions and requirements for colorants.

The composition, basic features and color of plastic is different, so it is necessary to have a comprehensive understanding of the color luster, transparency, heat resistance, aging resistance, strength and other indicators of plastics. Even for the same kind of plastics, if the grade is different, the performance is different. In recent years, with the rise of plastic modified market, the color matching is more complicated.

8.2.3 Fully Grasp the Process Conditions of Plastic Molding Process

The purpose of the plastic molding process is to use all possible molding conditions based on the original performance of plastics to make them possess an application value of products. There are many plastic molding methods, different molding methods have a large different in processing temperature and dispersion. Thus, being well aware of colorant is necessary to keep the quality.

8.2.4 Fully Understand the Type of Plastic Processing Aids and Master the Content

Plastic processing aids as raw materials are necessary for plastic processing industry. Its characteristic is that a small adding amount show a large effect. It is very appropriate that compared it to the MSG of plastics industry. The variety of plastic processing aids are various, such as plasticizers, antioxidants, ultraviolet absorbers, light stabilizers, flame retardants, antistatic agents, antibacterial agents, opening agents etc.. The effect of combined processing aids and colorants for plastics should be fully proficient. For instance, some resins can cause yellowing of titanium dioxide and pearl powder when adding antioxidant BHT, and then it causes the quality problems of colored products.

8.2.5 Comprehensive Understanding of the Safety Regulations of Plastic Products at Home and Abroad

Plastic is widely used material in the world, particularly in the area of consumer goods in which a variety of different types, colors and properties plastic products play an important role. In order to meet the requirements of product safety, environmental protection and health, plastic products must meet the regulatory requirements of each country and regions in the world, especially for chemicals. Due to the state, regional and product types are different, the requirements for colorants are different, some for plastic colorants, some for plastic materials and the other for the product's versatility. It involves a very wide range of consumer products, mainly include toys, electrical and electronic products, food containers and food contact materials or products and automotive materials. Therefore, it is necessary to understand the domestic and foreign regulations, and the selected colorants and additives should meet regulatory requirements.

8.3 Basic Steps of Color Matching

8.3.1 Preparatory Work of Color Matching

The first preparation of color matching for plastics is to establish the color systems and databases. Firstly, the system of enterprises should be set up. In other words, what kinds of colorants as the assortment of color matching series should be chosen to meet the ever-changing needs of the market must be clear. Of course, color matching staff could choose the top products of international well-known multinational companies which have high coloring power, high color saturation, the excellent performance and have a quality advantage. However, the prices of these products is too high to have much market competitiveness.

Color matching staff can also choose some products in the market without any investigation to match color, which may cause quality complaints, and even the claim.

Establishing a set of effective color matching system will make company occupy the commanding heights of the coloring of plastics in the fierce market competition.

Currently, many companies treat their color matching system as confidential, and some companies even take common used colorants represented by code to prevent leakage.

How to establish the color matching system? The following are several suggestions for reference.

① Perusal and master the world famous "Colour Index" and other pigment and dyes manual which have the technology data about application of colorants in plastics.

② Fully understanding and grasping the product samples of multinational companies, and carefully reading the various types of application data.

③ Grasping the science and technology books of applications of pigments at home and abroad.

Theory must be combined with practice. The color matching system of enterprise should be continuously improved in practice. After a color matching system is established, the enterprises must establish their own chromatography library. The first step to establish a chromatogram library is to make every breed to swatch based on the nature color and one-third depth in the confirmed color matching system. The second step is to mix colors among these species, then establish a model.

Companies should have a set of chromatographic formulations which are adapting to coloring of different resins, a set of weather-resistant for outdoor use, one for indoor and low cost and a set of colorants chromatographic formula satisfy the requirement of regulations domestic and international.

In addition to the injection of swatches, enterprises should build film chromatogram library for film monochrome and multi-chrome according to the customer's requirements, so does fiber. The more varieties and more complete of chromatography library, the more convenient while seeking contrast in color matching.

The establishment of chromatography library will not happen overnight, but companies accumulate over a long period in the process of color matching.

8.3.2 Specific Steps of Color Matching

When professionals of color matching facing a sample provided by the clients, they should use the following steps, as shown in Figure 8.3.

8.3.2.1 Fully Understand the Request of Sample

The first step of color matching is to make a fully understand of the samples provided by customers. When you get the sample, three conditions have been identified.

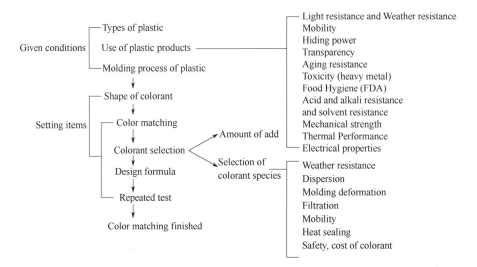

Figure 8.3 Process diagram of coloring of plastics

Plastic types of samples Polyvinyl chloride (PVC), polyolefin, ABS, or engineering plastic, etc.

Application of plastics Indoor and outdoor? Food packaging, automotive plastics, household appliances, toys, etc.

Molding process of samples Blown film molding, blow molding, injection molding and fiber spinning and so on.

According to the three conditions, the requirements of colorants for plastic coloring can be clear, such as heat resistance, light resistance, weather resistance, migration resistance and security. In addition, the state of colorant depend on the requirement of customers should be confirmed. For example, pigment powder, master batch, sand-like state, modified materials and so on, and the proportion(adding volume) of colorant should also be confirmed.

8.3.2.2 Designed the Preliminary Formulations According to The Customer's Requirements

First professionals should have a comprehensive understanding and careful thinking about customer's samples, then search the sample as a reference similar to the standard sample in the chromatographic library established by their own companies based on the requirement of overall design of plastic products. The reference selected appropriate or not directly related to the coloring effect. In order to search for better coloring reference, the chromatographic library should accumulate more and prepare multi colored plastic swatches or other related films, fibers for reference. They should also write their own experiences and lessons of color selection into the corresponding color matching formula for reference at the same time.

(1) Selection of pigment

Colorants should be selected according to transparent, opaque or translucent of samples. If the sample is transparent, the dye of hard rubber plastic need to meet requirements of heat resistance, light resistance and migration resistance, while soft plastics need to choose the organic pigment which have excellent transparency. If the sample is opaque, it is necessary to determine whether the currently selected color pigment meet the color requirements of the sample. If the sample is translucent, we need to know whether the color is the same in the reflected and transmitted light, and how to irradiate. If the color provided by customers is a piece of cloth, a color number of US PATTON color cards, a painting or a photograph, it's necessary to ask the customer whether the sample can be observed at any wavelength of the light immediately, and consult customers how much of the color difference should be, since the plastic products comparing with the sample provided by customers will be a little bit different in color. According to the sample to confirm whether there is a fluorescent pigment or metallic pigment.

There are several considerations for special requirements of sample as follows.

If the sample is used outside, it's necessary to judge whether the selected pigment meet the requirements of weather resistance.

If the sample using the lamination process, it's necessary to judge whether the selected pigment meet the requirements of heat resistance.

If the sample is soft PVC, it's necessary to judge whether the selected pigment meet the requirements of migration resistance.

If the material of sample is PP cap, it's necessary to judge whether the selected pigment will cause warping.

If the sample is fiber, it's necessary to judge whether the selected pigment meet the requirements of dispersion.

If the material of sample used in food packaging, toys, household appliances, it's necessary to judge whether the selected pigment meet the requirements of security.

In conclusion, the color matching system of corporation has detailed information about these, and the color matching staff need to master expertly.

(2) Comprehensive consideration of coloring costs

Cost is also an important factor. The price of the pigment is as important as the performance. A pigment can fully suitable for the performance of plastic processing, and meet international safety regulations, while usually the better the performance, the higher the price. Under the condition of meeting the requirements, choosing pigment at a good price-performance ratio is a reflection in the level and value of color matching staff.

The unit price of different colorants are different. It is not enough to just focus on its kilo price. In addition, the tinting strength and application are important. Pigment dispersity should ensure that it could fully disperse in the applied medium, which is the prerequisite of playing full value of the pigment and using the pigment economically.

Selecting pigment is a highly technical work, the effective using value of pigment to the user depends on color performances, fastness properties and processing properties of the pigment. For a high color value of pigments, if the processing performance is poor, such as dispersion, rheology, etc. it can't exert color value under the condition of application processing and the using value is low. Similarly, for a pigment which has good color performance and processing properties used in the applications that its fastness properties are inappropriate, its use value of the application is low.

Choosing a suitable pigment is definitely not an easy task, which requires industry expertise, knack and expert knowledge. For the same application, if the pigment has been chosen properly, the quality of coloring is good and cost is low. If pigment is inappropriate, the quality of coloring is poor, while the cost is high. The basic principle of selecting a suitable pigment are as follows. To ensure basis of coloring quality, optimally considering pigment performance, application formula, application equipment and technology, application conditions and comprehensive changes interact in the environment to achieve the goal of optimal coloring costs.

(3) Work out a preliminary formula

Through repeated studies, repeated comparisons with standard color sample and reference difference from the terms of hue, brightness, gradation, etc. the colorant formulations of reference can be corrected based on them and work out a preliminary formula.

It should be carefully to observe and analyze the shade, hue and brightness, etc. of plastics (sample) color without reference to determine the color attributes. Then according to the color of Munsell color system to draw up the preliminary formula.

8.3.2.3 Repeatedly Prepare Sample to Approach the Client's Sample

At present, the way of color matching using in enterprises is actually trial and error method, which have no great scientific, but so far it is still very recognized in the color matching of plastic industry and widely used for the vast majority of businesses.

Color matching staff carry physical coloring tests according to the proposed preliminary formula, and compare coloration sample with standard color swatches and reference to further adjust the color matching formulations. Then prepare samples to compare according to the adjusted formulation.

An experienced color matching staff should know that there must be a large number of trials before finding the correct colorant formulations. When the color matching trial is approaching to the color and luster of customer samples, it need to fine-tuning, which is often more complex than the approximate color matching at the beginning. The dosage is difficult to control at the time, which test patience and endurance of the color matching staff who must repeat trail until get the desirable color. Sometimes, the color also should be palette with other attached colors. These additional colors can change other features required in formulation. When laboratory ingredients in color and all other features have meet the requirements, we should make the hue of sample the same as standard color sample or reach the extent of closest to the standard color sample, and trial production samples to provide to the customer.

If customer found the sample have bias via testing, we need to further revise until the customer satisfied according to customer feedback comments and sample.

When each sample color matching completes, the information of color sample, formulas, dates and user can be deposited in a computer, which can easily search, find and as a reference when modifying, and also convenient, efficient, improve work efficiency and ease of confidentiality.

8.3.2.4 Comparison of Colors

It is very important to compare experimental sample with the sample provided by customers especially the color. According to customer's sample production process to make corresponding sample comparisons, for example thin film products is often compared by blown film method, injection products is compared by injection molding method, fiber products adopt spinning method comparison, also can use two rolled sheet method comparison. It's necessary to pay special attention to the following points when compared samples color.

(1) Effect of light source

The natural light is the best when visual inspection, also can use standard light source box. Otherwise, observation of two colors is the same in some light, but there is a big difference in natural light, because of the "metamerism" phenomenon.

The chromatic aberration of sample and customer sample are appreciably different, which is because the color matching is very difficult to do the same, generally are metamerism. So even in a certain light looks the same, but it may have minor differences in other sources, so a qualified color matching should be the same in a number of typical light sources, such as sunlight, tungsten, fluorescent, ultraviolet light. However, it is generally to choose the light in practical application based on the use of products, and can be consistent in such light.

(2) Effect of thickness

The feeling of different thickness of the products to human eye is different. Smaller thickness easier pervious to light and the color become shallow, so the thickness of the sample and the customers sample should be as far as possible consistent when color matching, or used the extrapolation. Such as when the color of 0.4mm film samples and 0.2mm film samples is consistent, the dosage of various kinds of pigment in sample formulation can be doubling, and get roughly formula.

(3) Effect of the surface state

The light reflection on the different surface are different, such as on the same sample, the color of high-gloss surface and matt surface clearly gives a different feeling; coarse surface is relatively deep, dark compared with smooth surface. So it should compare the similar surface state parts as far as possible when colorimetric observed. Such as when PVC tape color matching, because of the different surface state (gloss, embossed, etc.), the trial sample should be side by side with the sample, and post transparent tape or immersed in water to compare, which are taken the surface state into a consistent.

The material of sample and plastic of color matching is different. For example, when using the "International color card" as the standard color sample to coloring of plastics, due to the "International color card" is as a reference to the color of ink in printing industry, and the material is different with plastic, so it bring some difficulty to contrast observation.

(4) Effect of resin

The color of same pigment have different color when used in different resin, which is due to the different ecru color and transparency of resin. It should be noted that using the resin provided by customers to color matching as far as possible to ensure the accuracy of color. Especially color matching of modified material should pay more attention.

(5) Misjudgment of Colorimetric

The most prone to problem is misjudgment of colorimetric in the process of checking. Misjudgment of colorimetric is mainly caused by lack of experience in general, usually only by constant practice and enrich the experience to solve, appropriately apply some instruments, such as colorimetric, also can help to get correct judgment.

8.3.2.5 Metamerism

A pair of samples are prepared by different colorants. When under specific light source and viewed by a particular viewer, the color is matched. If the lighting changes, the object will no longer match. We define the object which have different spectral reflectance curves but have a same color coordinates under certain conditions as a metamerism object, or conditions to match color, as shown in Figure 8.4. There are four examples which have the same tristimulus values, but have four different spectral curves shown in Figure 8.5, Figure 8.6.

Figure 8.4 Metamerism

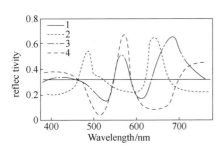

Figure 8.5 Four different spectral curves

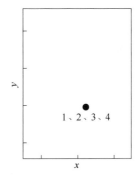

Figure 8.6 Same tristimulus values

The phenomenon of metamerism is very common, and sometimes unpleasant, but metamerism can make many different pigment color matching possible. Poisonous colorants can be replaced with non-toxic colorants. Expensive colorants can be replaced with cheap colorants to reduce costs. Due to the existence of metamerism phenomenon, using different colorants to give a re-formulation is possible. Color matching staff can impart a coordinates to a product in a color space, and make their color matching, although they are obtained by different colorants. The color matched by metamerism essentially poor. A group of "match" color which obtained by highly metamerism can make customer complaints. In the long term, the unique way to eliminate the phenomenon of metamerism is to use same colorants.

8.4 Quality Control

When customers confirm the color matching samples need to mass production, the batch of colorants should be same with the small test formula, which is difficult to achieve. There may be errors on tinting strength and shade because of the different batches of colorants. Due to the different equipment, processing conditions are hard to consistent, which may cause color difference even if the same batch of colorants. Therefore, plastic coloring products should be quality control.

8.4.1 Chromatic Aberration

Chromatic aberration literally refers to the color difference which is observed. Economic benefits are always taken into account in color matching, so the range of allowable deviation of color is prescribed. In order to meet the requirements of customer, the color deviation and tolerance must be very small.

Evaluating the color difference by human eyes is quick, cheap and very simple, but not reliable. Human visual can be divided into three types. The vision of rod cells is sensitive at night (scotopic vision). The vision of cone cells is sensitive during the day (photopic vision), intermediate vision is the conversion between photopic vision and scotopic vision, which relys on human eyes to judge the chromatic aberration. However, the color difference observed is also related to the psychological and physiological factors of the observer. In addition to the subjective consciousness and living habits of observer, chromatic aberration also has a great relationship with product specifications. If there is no regularity to follow and just to warn people of "good" color matching, the range of chromatic aberration could be explained too small or too large.

8.4.1.1 Definition of Chromatic Aberration

It is a great advance in science to use data to describe the differences in color. To avoid misjudgment caused by human and environmental factors, colorimeter can be used for the colorimetric examination. The $CLE^*L^*A^*B$ color system implemented by International Commission on illumination in 1976 is currently the general CLELAB color system in plastic industry, which shows the difference between two colors in color vision by specific data.

Chromatic aberration can be represented by DE of which D is a greek alphabet indicating differences in certain aspects and E is a german word indicating sense, or ΔE.

Chromatic aberration ΔE represents the distance between two colors in the color space.

$$\Delta E_{CIE}1976 = [(\Delta L^*)^2 + (\Delta a^*)^2 + (\Delta b^*)^2]^{1/2}$$

Lightness Difference: $\Delta L^* = L^*_{Sample} - L^*_{Standard}$

Red-Green Difference: $\Delta a^* = a^*_{Sample} - a^*_{Standard}$

Yellow-Blue Difference: $\Delta b^* = b^*_{Sample} - b^*_{Standard}$

ΔE can obtained by ΔL, Δa, Δb, the greater the value, the greater the chromatic aberration, and the smaller conversely (Figure 8.7).

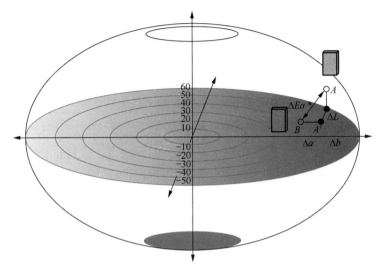

Figure 8.7 Definition of chromatic aberration

8.4.1.2 Evaluation of Chromatic Aberration

In 1939, American National Standards Institute (ANSI) adopted the Judd's advice and implemented chromatic aberration formula by which the color difference could be calculated. The absolute value 1 is used as a unit in this formula and as a unit of chromatic aberration. The NBS (National of Standards) unit is about 5 times of the visual color difference recognition threshold. If compared with the chromatic aberration values of adjacent levels in Munsell system, 1 NBS unit is approximately equal to 0.1 Munsell lightness value, 0.15 Munsell chroma value and 2.5 Munsell shade values (chroma is 1). The chromatic aberration values of adjacent levels in Munsell system is about 10 NBS. The relationship between NBS and visual is shown in Table 8.1.

Table 8.1 Relationship between chromatic aberration values and munsell system

Chromatic aberration value of NBS	Feeling degree
0.0~0.5	Trave
0.5~1.5	Slight
1.5~3	Noticeable
3~6	Appreciable
Above 6	Much

Table 8.2 shows chromatic aberration acceptance of different products in coloring of plastics.

Table 8.2 Chromatic aberration evaluation in coloring of plastics

Chromatic aberration	Feeling
Below 0.1	Indiscernible
0.1~0.2	Experts distinguishable
0.2~0.4	General distinguishable
0.4~0.8	Chromatic aberration range of strictly controlled on parts
0.8~1.5	Chromatic aberration range of commonly controlled
1.5~3.0	Seems to be one color separated
Above 3.0	Obvious chromatic aberration
Above 12	Different colors

Residul difference is also known as mimetic isobathic chromatic aberration, which refers to color difference of consistent depth calculated in accordance with a certain relationship by two samples of difference depth. There are different degrees of change in the brightness and saturation of color when the tinting strength changes, therefore it is necessary to make isobathic-processing before calculating chromatic aberration.

8.4.1.3 Application of Chromatic Aberration

(1) Quality control of product

It should set corresponding limits respectively for $\Delta L^*, \Delta a^*$ and Δb^* around standard samples and form a rectangular "box" (Figure 8.8). The samples in the space are considered qualified, others are unqualified.

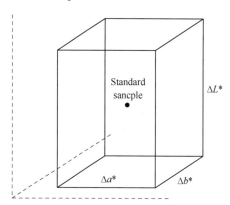

Figure 8.8 Chromatic aberration control chart

(2) Intermediate control of product

The values of component chromatic aberration can be used to control the quality in production. The depth and brilliance of samples can be understood by component chromatic aberration, and the proportion of shade in the total chromatic aberration would also be known.

- Adopting the combination of ΔC and ΔH or the combination of Δa and Δb to evaluate the color characteristics such as brilliance and shade.
- Adopting the combination of ΔC and ΔH to evaluate product with high saturation, and the shade angle h can be used as characteristic value.
- Adopting the combination of Δa and Δb to evaluate color with low saturation, such as black, grey, brown, and so forth. However, these colors are difficult to regard the shade angle h as characteristic value.

The sensitivity of human eyes to three component chromatic aberration is different, the most sensitive is shade, followed by saturation, and the last is lightness. According to the variation of human eyes can feel, the limit values of the three parameters are usually controlled: ΔL=0.8, ΔC=0.4, ΔH=0.2.

Manufacturers provide the necessary information for determining whether the deviation between different batches within the allowable range, which may contribute to analyze whether the production process is in normal condition.

(3) Metamerism color matching

A color can be reached by a variety of formulas, and it can be achieved by the chromatic aberration control in order to meet customer requirements and pursue the best interests.

(4) Evaluation of application fastness

The contrastive results between gray card or gray card stained with color and visual observation are usually used to evaluate the weather resistance and migration fastness, which may cause some errors. The fastness levels of samples can be determined by comparative table composed of chromatic aberration value and staining fastness. The smaller chromatic aberration value, the higher staining fastness. The comparison of chromatic aberration with staining fastness is shown in Table 8.3.

Table 8.3 Comparison of chromatic aberration and staining fastness

Fastness grade	Chromatic aberration	Deviation
5	0	0.2
4~5	2.2	±0.3
4	4.3	±0.3
3~4	6	±0.4
3	8.5	±0.5
2~3	12	±0.7
2	16.9	±1.0
1~2	24	±1.5
1	34.1	±2.0

8.4.2 Quality Control of Formula Design

(1) Minimize pigment varieties

More pigment varieties selected in plastic color matching may not only cause troubles in matching, but also lead to gray by introducing complementary color, thus the varieties of colorants selected should be minimized. In addition, due to the influence of factors such as the dispersion and tinting strength, more pigment varieties could cause greater systematic errors in color matching and production.

(2) Select pigments of different tinting strength

It should select high tinting strength pigments when preparing dark products, because adding a small amount can achieve the same depth. For example the tinting strength of organic pigments is higher than inorganic pigments, therefore organic pigments are generally selected.

It should select low tinting strength pigment when preparing light products. On the one hand adding pigment as much as possible can reduce error in color matching. On the other hand some inorganic pigments with low tinting strength also show excellent heat resistance and light fastness in the case of adding a small amount.

(3) Select pigments of similar dispersion

It should select pigments of similar dispersion in plastic color matching. If selecting an organic pigment of hard dispersed and an inorganic pigment of easy dispersed to match, it could cause chromatic aberration since the change of shear stress in the process which brings about shade changes.

(4) Notice the thermochromatic effect of some pigments

It should be aware that the shade of some pigments would change with the temperature changes in plastic color matching, for example, when preparing pink with lake red pigment and titanium dioxide, the shade may turn into deep orange under the processing temperature, and this orange would turn into pink only after 24 hours at room temperature. This phenomenon is caused by the slow speed of some pigments with different crystal structures to achieve the color balance at different temperatures.

In fact, most pigments have this thermochromics effect but not obvious. It should avoid using the pigment with serious thermochromic effect, especially not mix with other pigment.

In addition, it should note that to compare the color of pigments with thermochromic effect, it is necessary to cooling a period of time before observed.

8.4.3 Quality Control of Production

Supplier must provide products with consistent color of samples after customer confirm the samples provided by laboratory and formally place an order. Production will be repeated depending on the sales. However the actual situation differs somewhat from ideal, especially in color, when the color stability is expected in each production. Color deviation is a comprehensive term that would lead to a color change in one or all of the production process. To achieve good quality, various factors must be strictly controlled and the interaction should be understood. The changes of color in production can be controlled from the following aspects.

(1) Strictly control the raw materials

In order to ensure color consistency of product, the raw materials used in production must be consistent with the small test samples in principle. When using the new procurement of raw materials, to ensure the stability of production the main projects such as tinting strength, color light, dispersion, moisture content and other indicators must be tested by standard, and whether raw materials to be used would be decided.

It should be noted that the different dispersion of each batch of pigment could lead to the difference of shade and saturation, especially the organic pigments show more obvious change than inorganic pigments.

(2) Strictly control the mixing process

Mixing process is a decisive factor to determine the quality of product, and the final dispersion of pigment is determined by the degree of wettability. Powder carrier is conducive to the dispersion of pigment in mixing formulas. If selecting granular carrier resin and high mass fraction of pigment, the poor wetting effect of pigment may cause poor dispersion, besides the stratification of granular carrier resin and the unevenness in extrusion feeding or blanking could lead to chromatic aberration.

(3) Strictly control the extrusion process

In order to avoid chromatic aberration, it is necessary to control feeding speed, extrusion temperature, and maintain the stability of shear force and process. Some difficult dispersed pigments may cause color deviation because of the shear force in plastic processing, the best method to solve the problem is pre-dispersion or constant process conditions. The results of pearlescent or metal effect pigments would be affected by shear force. In order to avoid bearing excessive shear force the process conditions must be improved.

(4) To prevent some pigments from being affected by high temperature, the head-temperature of extruder should be strictly controlled and the residence time should also be reduced under high temperature. For some light-colored varieties reducing the process temperature or adding antioxidants can avoid chromatic aberration caused by high temperature.

(5) Prevent human errors by establishing strict management system and

strengthening intermediate control

In mass production, each procedure should be strictly controlled. Human errors include incorrect weighing and feeding, wrong processing, dirty machine, polluted products, calculation errors, and so forth.

(6) Establish good production environment

Because most colorant are powders which would affect the quality of product and cause environmental pollution floating in the air, so it is necessary to isolate and clean production line.

8.4.4 Establish Quality Control System

(1) Formulate tolerance of chromatic aberration

The standard of tolerance of chromatic aberration is a problem never can be evaded in plastic color matching, and it refers to the accuracy of color between different batches. Chromatic aberration is not necessarily caused by processing problems. There will be some changes in the performance of the product between different batches, so a certain chromatic aberration is allowable.

Supplier and customer should reach an agreement on the tolerances. On the one hand, the tolerance limit should have economic feasibility and on the other hand a constant color for each batch t of product. These contrary requirements may cause a conflict of interests, and experience shows that direct comprehensive dialogue between both sides can avoid such a conflict of interests.

In addition, suppliers and consumers have different color measuring devices. In this case, the common way is to measure the same colored sample on both color measuring devices. By comparing the data, a tolerance can be determined separately for each device. The tolerance is for each color measuring device of the same colored sample, and both parties must agree to the tolerance of the other color measuring device. This procedure should cause no problems, because it is the same sample only measured on two different devices. Even if both sides have the same color measuring device, the procedure is recommended. Color measuring devices are industrial product with certain tolerances, although all efforts of the manufacturer are made to minimize the tolerance as far as possible.

The tolerance should be smaller than $\Delta E= 0.8 \sim 1.0$. Such a difference can be observed from the shade, especially the entire difference lies on one coordinate. This indicates that it is usually not enough to determine only a ΔE, but determine a tolerance limit for each coordinate. According to the shade the sensitivity of eyes for special color varies. Consistent with different sensitivities, the tolerance may be different on each coordinate, so it is not necessary to ask for the same tolerance on each coordinate.

Experience shows that only ΔL, Δa, Δb are respectively less than 0.7 can make the ΔE less than 1.

(2) Storage and replacement of raw materials and standard samples of finished product

The test of raw materials and finished product is an important part of the quality control system, so the storage and replacement is vital. Standard samples should be stored carefully in a lightproof and good sealed container. The good sealed excludes the influence of gaseous pollutants and dust in the industrial environment. These conditions are very important to ensure the pure of standard samples during storage. The packing material should not contain any plasticizer, optical brightener, or other additives that may migrate.

Due to polymer age, the storage time of standard sample is limited even under these strict conditions of storage. Therefore, the recommendation is to replace the standard samples approximately every two years.

(3) Establish standard test method

Standard test method is not only the eye to control the quality of product for the enterprise, but also the language to communicate the quality of product with foreign.

To improve technical and be in line with international norms, the national standards should be adopted.

8.4.5 Problems and Solutions

(1) Color specks

Color specks in the final product are caused by pigments of incomplete dispersion and sometimes by dyes of incomplete dissolution. Pigments are not dispersed by the system, for example, it is not enough shear in the single-screw extruder and insufficient dispersing agent. Another reason for color specks is bad pigment wetting, which may cause difficult dispersion in polymer melt.

(2) Color streaks

Color streaks in the final product are caused mainly by an incomplete mixture of the molten polymer with the masterbatch melt. There are many reasons, for example, the too short mixing section of screw, too low processing temperature, too short residence time, too large difference in the viscosity of melts, and a too small coloring concentration of masterbatch. In addition color streaks in the amorphous polymers could be caused by dyes that are incompletely dissolved in the masterbatch.

Color streaks in the final product are more likely to occur in two types of color masterbatch which contain a high concentration of organic and are difficult to disperse pigments. For example, the blue and green phthalocyanine pigments. Generally the content of organic pigment is 40% and inorganic pigment is 50%~70%.

(3) Thermal damage

Thermal damage in plastic processing can lead to brownish color streaks or black spots. Brownish color streaks are the sign of thermal damage of the molten polymer. Heaters and their structures (dead spots, too small size etc.) are the source of these problems, but not exclusively. If the thermal damage of the polymer melt always exists in every shot, the thermal damage continues and black spots are the final result. This process takes more time, so the appearance of black spots indicates thermal damage after a long processing time. Black spots appeared at the beginning of the processing are caused by impurities in the polymer or other components.

Thermal damage is also associated with incorrect processing parameters, for example, too long residence time, too fast injection, faulty structure regarding the position and dimension of the gate.

A long residence time sometimes can't be avoided, which occurs in some tiny quality products. The structure of each injection machine requires a minimum size and screw volume, but this may be too large for the small products. Another reason for extending the residence time is that some polymers tend to stick to the wall of barrel, and this thin layer can't be replaced with every shot, and a slow damage of this layer is inevitable. But the addition of lubricant or the use of other alloys as lining for the barrel can reduce these problems.

(4) Manifold source of impurities

Impurities can be caused by abrasion. The impaired barrel and screw can cause not only impurities or a gray tinting of the color, but also problems in the processing, the production of masterbatch or the coloring of polymer on the injection molding machine. Impurities can be detected quickly in the raw materials, however, it is difficult to detect the sources by a later contamination. Impurities also can be caused by bad operation habits and not enough care, for example, during the cleaning of auxiliary devices with a color changes. In addition, contamination with other colors may also cause color streaks. Table 8.4 contains the problems, reasons and methods of elimination in the coloring of plastics.

Table 8.4 Problems, causes and methods of elimination in the coloring of plastics

Problems	Reasons	Methods of elimination
Color specks in the masterbatch and final product	Incomplete dispersion of pigments	Improve the wettability of pigments
		Increase the quantity of dispersing agent
		Use a mixing head and increase the shear
	Worn out plasticizing screw	Replace the screw
	Incomplete dissolving of dyes	Changing the processing parameter of the masterbatch

continued

Problems	Reasons	Methods of elimination
Color streaks in the final product	Incompletely mixing of polymer melt and masterbatch melt	Use a mixing head or similar devices
	a. Not enough mixing of extrude	Check the function of each heating section of extruder
	b. Too low mixing viscosity caused by high extrusive temperature	Check/adjust the temperature settings of heating section
	c. Too large difference in the melt viscosity of both components	Using a batch with a better adjusted melt viscosity
	d. Too low masterbatch	Reduce the concentration of colorant and increase amount
Black spots	Thermally damaged colorants	Change/reduce the processing temperature
		Reduce the residence time
		Avoid dead spots in nozzle or hot runner
		Avoid the production of small parts in a too large extruder (too long residence time)
	Thermally damaged polymer	Reduce the melt viscosity of the barrel, screw, etc.(add lubricant or use other alloy)
	Contaminated colorants or polymer (dirt, impurities, etc.)	Use clean pigment
		Check the source for the contamination
	Contaminated recycled material	Strict management
Bubbles/streaks in flow direction (colorless)	Humidity in the raw material	Check moisture content of components
		Check drying (increase the drying time and temperature)
		Check the function of dryer
Bubbles/streaks in flow direction (brownish)	Too high processing temperature or too long residence time	Reduce the processing temperature
		Reduce the residence time
		Check the temperature of the hot runner
	Too high shear (frictional heat because of too small feeding system)	Check the hot runner diameter, if necessary enlarge it
		Reduce the injection rate

8.5 Practical Technology of Color Matching

8.5.1 How to Formulate Special White Colorant for Plastic Coloring

To get a beautiful white product, firstly, the titanium dioxide which have blue shade should be selected, such as the titanium dioxide produced by chlorination process. The particle size is small, the distribution of molecular weight is uniform. Use titanium dioxide witch has blue shade to get white pigment will make people feel freshness, on the contrary, if the titanium dioxide have a large size and with yellow shade, the beautiful pure white cannot be got.

Because the white degree of titanium dioxide is not enough, some additives often be added to increase the whiteness, such as a small amount of blue and purple pigment or fluorescent whitening agent, the common and easiest way is to use ultramarine diaobaikuai, the whitening effect known as magnetic white in the market. The whitening effect of fluorescent whitening agent is best, but the cost is the most expensive.

8.5.2 How to Formulate Special Black Colorant for Plastic Coloring

It should be attention to the problem of black color when use black carbon to formulate a special black products. Under the incident light, usually small particle size of carbon black is bluer than large particle size carbon black, but under the transmitted light (colored transparent) and fight gray the color with brown tone. To get a black shiny plastic products, small particle size and low structure carbon black should be chosen. Because the blackness is based on the light absorption. So the smaller the particle size, the higher the degree of light absorption, the weaker the reflectance of light, the stronger of blackness. Using the carbon black pigment mentioned above to obtain satisfactory coloring effect should pay special attention to the dispersion of carbon black. Only solved the dispersion of carbon black the highest tinting strength of carbon black can be achieved.

After the oxidation of carbon black, some polar groups such as —OH and —COOH were introduced, which will improve the dispersion of carbon black, and there is a volatile matter index in the performance indicators of carbon black, the higher the value, the higher the degree of oxidation.

There always has some yellow shade in carbon black, that can use blue pigment to palette, with a dosage of 8~10%, so the black blackening will greatly increase. The pigment blue 15:3 are commonly used. Adding some purple or red colorant on the basis of the blue colorant, the black blackening may be blacker.

8.5.3 How to Formulate Gray for Plastic Coloring

When preparation gray tone with carbon black pigment, shade issues of carbon black need to be noticed. When adding titanium dioxide, if the particle size of carbon black is small, the color would present yellowish gray. If the particle size of carbon black is big, the color may present bluish gray. Titanium dioxide shade depend on the size and distribution of the particles, large particle size exhibits yellow shade, small particle size exhibit blue shade. Therefore, the relationship between the particle size and hue of carbon black and titanium dioxide must be clear when preparation gray, otherwise making a wrong hue, and adjusting with pigment, may lead to much more complex.

When prepare gray, in addition to carbon black and titanium dioxide, sometimes need additional colors to palette in order to achieve the user's requirements. Due to the small amount, it should choose some low tinting strength varieties, such as iron oxide (Pigment Yellow 119, Pigment Red 101), metal oxides (Pigment Yellow 53, Pigment Brown 24), that can make color matching system has a better color stability.

8.5.4 How to Formulate Colorant for Outdoor Plastic Products

(1) Deep color

Plastic products such as chairs of the stadium, the cable sheath of bridge, construction material, advertising boxes, pass box, plastic parts of auto, etc., it must require good light fastness and weather resistance due to long-term use outdoors. Inorganic pigments generally have good light stability, and can be used for outdoor plastic products coloring, but it is limited by the low tinting strength.

Firstly, the pigment that with good light fastness and weather resistance should be chosen, secondly, as little as possible with or without titanium dioxide, to make the concentration of pigment as high as possible. In addition, a sufficient amount of ultraviolet absorber and antioxidant in the coloring formulation to ensure that the color of resin does not change. The organic pigments used for outdoor plastic products are shown in Table 8.5.

Table 8.5 Main varieties of organic pigment for outdoor

	Color and lustre	Pigments index number	Chemical structure	Pigment 0.1%		
				Heat resistance /°C	Light fastness grade	Weather resistance grade
1	Yellow	Pigment Yellow 110	Isoindoline ketone	300	8	4~5
2	Yellow	Pigment Yellow 181	Benzimidazolone	300	8	4
3	Orange	Pigment Orange 61	Isoindoline ketone	300	7~8	4~5
4	Red	Pigment Red 254	Diketone-pyrrole-pyrrole	300	8	5
5	Red	Pigment Red 264	Diketone-pyrrole-pyrrole	300	8	5
6	Blue	Pigment Blue 15:1	Phthalocyanine	300	8	5
7	Blue	Pigment Blue 15:3	Phthalocyanine	300	8	5
8	Green	Pigment Green 7	Phthalocyanine	300	8	5
9	Green	Pigment Green 36	Phthalocyanine	300	8	5
10	Brown	Pigment Brown 41	Disazo condensation	300	8	5

(2) Light color

Light color was prepared by adding large amount of titanium dioxide and small amount of pigment, since the weather resistance and heat resistance of most organic pigment will have different degrees of decline after joining titanium dioxide. Due to less addition of pigment in light colorant varieties, the influence is bigger, since many varieties with bad weather resistance, causing plastic fade, customer complaints, and have very bad influence. Due to high tinting strength of organic pigment, the amount in a light- colored varieties is less, due to the presence of error propagation can cause color deviation in production and application.

Inorganic pigments should be used to modulate light colorant, mainly using the good heat resistance, and low tinting strength. Optional inorganic pigment varieties are shown in Table 8.6.

Table 8.6 Main varieties of inorganic pigment for preparation light color

	Color and lustre	Pigments index number	Chemical structure	Heat resistance /°C	Pigment 0.5% Light fastness grade	Weather resistance grade
1	Yellow	Pigment Yellow 119	Iron oxide	300	8	5
2	Yellow	Pigment Yellow 53	Metallic oxide	320	8	5
3	Yellow	Pigment Brown 24	Metallic oxide	320	8	5
4	Yellow	Pigment Brown 184	Pucherite	280	8	5
5	Red	Pigment Red 101	Metallic oxide	400	7~8	5
6	Blue	Pigment Blue 29	Ultramarine	400	7~8	
7	Blue	Pigment Green 28	Cobalt blue	300	8	5
8	Green	Pigment Brown 50	Cobalt green	300	8	5

If there is no heavy metal safety requirements, coated chrome yellow and chrome red can be chosen, the varieties are shown in Table 8.7.

Table 8.7 Main varieties of coated chrome inorganic pigment for outdoor

	Color and lustre	Pigments index number	Chemical structure	Heat resistance /°C	Pigment 0.5% Light fastness grade	Weather resistance grade
1	Yellow	Pigment Yellow 34	Chrome yellow	260~280	8	4-5
2	Orange	Pigment red 104	Molybdate bed	260~280	8	4

8.5.5 How to Formulate Colorant for Transparent Products

The first thing for preparation of light-colored transparent article is the resin itself is transparent and colorless, for example, polypropylene and transparent styrene. It need to select a colorant with good transparency and consistent shade. Generally, inorganic pigments cannot be applied for the poor transparency, organic pigments pigment varieties with good transparency are shown in Table 8.8.

Table 8.8 Main varieties of organic pigment for transparent product

	Color and lustre	Pigments index number	Chemical structure	Pigment 0.1%			Recommentded varieties
				Heat resistance /°C	Light fastness grade	Weather resistance grade	
1	Yellow	Pigment Yellow 139	Isoindoline ketone	240	8	4~5	Paliotal Yellow 1841
2	Yellow	Pigment Yellow 199	Anthraquinone	300	7-8	3~4	Cromophtal Yellow GT-AD
3	Red	Pigment Red 149	Perylene	280	8	4	Paliogen Red K3580
4	Red	Pigment Red 254	Diketone-pyrrole-pyrrole	300	8	5	Cromophtal DPP Red BOC
5	Blue	Pigment Blue 15:1	Phthalocyanine	300	8	5	Heliogen Blue K6911
6	Green	Pigment Green 7	Phthalocyanine	300	8	5	Heliogen Green 8730
7	Green	Pigment Green 36	Phthalocyanine	300	8	5	Heliogen Green 9360
8	Brown	Pigment Brown 41	Disazo condensation	300	8	5	PV Fast Brown Hfr Clariant

Additionally, if plastic itself is covered with color, the color to be formulated and the color of plastic already must have not been mutually complementary color, or the color will darkening.

8.5.6 How to Formulate Pearlescent Colorant for Plastic Coloring

The addition ratio of the pearlescent pigment in a plastic injection molding product typically is 1%, 4~8% in the extruded film products, that depend on the thickness of the plastic film. In co-extrusion lamination, the content of pearlescent increases, the adding amount is 5% to 10% based on the thickness of the plastic layer of pearlescent.

(1) Pay attention to the choice of pearlescent pigments and organic pigments

Using pearlescent pigments in plastics processing some suggestions need to be noticed.

The transparency of plastic resin must be good.

As far as possible to choose the pigment with good transparency, such as organic pigments, and solvent dyes.

Not compatible with titanium dioxide, in order to achieve a certain degree of opacity, small particle size of pearlescent pigment can be used simultaneously, and try to avoid the use of high opacity pigments.

It need to consider the weather resistance of pearlescent pigments when used in outdoor plastic products.

(2) Pay attention to the processing and molding process of pearl pigment

Increasing back pressure during injection to enhance the mixing of screw, and to improve the dispersion of pearl pigment. The processing temperature of injection general election in the upper limit of recommended temperature range, which will ensure the dispersion of pearl powder. In the molding process, the melt flow driven automatically directed of pearlescent pigments lamellae, and achieve good pearl effect.

The surface smooth of the mold is very important. The better smooth of mold, the higher the possible of having a uniform direction arrangement and smooth pearl color.

The design of the mold gate is also very important. Selecting a single gate is better than multiple gate to reduce die flow line. The gate should at the thick place where are away from the flow obstacle. The distance between the gate terminal and runner system should be as small as possible, in order to reduce the dispersion unevenness caused by the differences of fluid resistance.

Since the shape of pearlescent pigment is sheet, in the process of plastic processing, due to the influence of shear force, the particle size of pigment becomes smaller, and pearlescent effect may be reduced. Using large aspect ratio and appropriate fineness of the filter to increase the head pressure, which will minimize shear damage of pearl pigments during processing.

When pearl pigment used in PMMA, PC, PA system, they must do some drying process in advance. When gold and bronze pearlescent pigment products used in PVC plastic, because it contains free iron ions which will accelerate the decomposition of PVC resin, must pay attention when using. Silver pearlescent pigment in some plastic products will be yellow, so anti-yellowing products should be applicative.

8.5.7 How to Formulate Gold, Silver Colorant for Plastic Coloring

Silver powder is, in fact, aluminum powder. Because of the aluminum surface can strongly reflect the entire visible spectrum, including blue light, so the aluminum pigments can produce very bright blue-white mirror reflected light.

Aluminum powder is used in plastic with different species. The average particle diameter is 5 μ m, its tinting strength and hiding power is excellent. The average particle diameter is 20~30 μm, which can be used with color pigments. The average particle diameter is 330 μm with a special coarse sparkle effect.

The melting point of aluminum is 660℃, but when contact with air at high temperature directly, its surface can oxidized to an off-white, therefore when aluminum coated with a silicon oxide protective film, the heat resistance, corrosion resistance and acid resistance can be increased greatly.

Gold powder is actually copper powder, which can easily be oxidized at high temperature. In addition the weather resistance is not good, and will become dark when exposed in the outdoor for a long-term. As using of treatment technology makes copper surface coated with silicon oxide protective film, which can greatly improve its heat resistance, weather resistance and acid resistance. The RESIST grades of gold powder provided by the company of Germany ECKART have been treated by silicon oxide, using raw materials trial make masterbatch which does not have dark phenomenon.

In order to formulate a colorant with special metallic effect, some organic pigment with better transparency and similar shade can be appropriate to add a small amount, but need to be aware that if too much titanium pigment and pearl powder added, can make products darken.

Because silver powder and gold powder showed a sheet-like structure, therefore the use of metallic pigments and pearlescent pigments are easier to produce die outflow line in injection molding, and then affecting the appearance of plastic products. The following methods can be used to reduce the above: adopting large diameter metal pigments, increasing the amount of metallic pigments, selection of high-viscosity resin, increasing the injection aperture and speed.

8.5.8 How to Formulate Fluorescent Colorant for Plastic Coloring

Fluorescent pigments used in the plastic coloring is prepared by dispersing fluorescent dye in a particular polymer resin, which can improve the migration resistance. Since 1970 using complex system polyamide resin, the thermal stability of fluorescent pigment reach 305℃, recently use polyester as the carrier of fluorescent pigment is popular, and its thermal stability also can reach more than 285℃.

Fluorescent pigment is sensitive to UV light, so its light fastness was poor. Method to improve light fastness of fluorescent products is to add same tone non-fluorescent colorant.

Reference

[1] 周春隆. 有机颜料技术. 北京: 中国染料工业协会有机颜料专业委员会, 2010.
[2] 周春隆. 有机颜料化学及进展. 北京: 全国有机颜料协作组, 1991.
[3] 周春隆. 有机颜料——结构、特性及应用. 北京: 化学工业出版社, 2001.
[4] 周春隆. 有机颜料百题百答. 台湾福记管理顾问有限公司, 2008.
[5] 沈永嘉. 有机颜料——品种与应用. 北京: 化学工业出版社, 2007.
[6] 莫述诚. 有机颜料. 北京: 化学工业出版社, 1991.
[7] 冈特·布克斯鲍姆. 工业无机颜料朱传棨等译. 北京: 化学工业出版社, 2007.
[8] 朱骥良. 颜料工艺学. 北京: 化学工业出版社, 2001.
[9] 阿尔布雷希特. 塑料着色. 乔辉等译. 北京: 化学工业出版社, 2004.
[10] 吴立峰等. 塑料着色和色母粒. 北京: 化学工业出版社, 1998.
[11] 吴立峰等. 色母粒应用技术问答. 北京: 化学工业出版社, 2000.
[12] 吴立峰等. 塑料着色配方设计. 北京: 化学工业出版社, 2002.
[13] 宋波. 荧光增白剂及其应用. 广州: 华东理工大学出版社, 1995.
[14] 刘瑞霞. 塑料挤出成型. 北京: 化学工业出版社, 2005.
[15] 张京珍. 泡沫塑料成型加工. 北京: 化学工业出版社, 2005.
[16] 胡浚. 塑料压制成型. 北京: 化学工业出版社, 2005.
[17] 赵俊会. 塑料压延成型. 北京: 化学工业出版社, 2005.
[18] 汉斯·茨魏费尔. 塑料添加剂手册. 欧育湘等译. 北京: 化学工业出版社, 2005.
[19] Roys Berns. 颜色技术原理. 李小梅等译. 北京: 化学工业出版社, 2002.
[20] 周春隆 穆振义. 有机颜料索引卡. 北京: 中国石化出版社, 2004.
[21] 周春隆. 塑料着色剂(有机颜料与溶剂染料)的特性与进展. 上海染料, 2002(6).
[22] 周春隆. 有机颜料工业技术进展. 精细与专用化学品, 2007(7).
[23] 周春隆. 有机颜料制备物技术及其应用. 上海染料, 2013(3).
[24] 张合杰. 有机颜料的晶型特性——塑胶应用. 上海染料, 2012(4).
[25] 宋秀山. 苯并咪唑酮颜料回顾. 上海染料, 2012(4).
[26] 宋秀山. 高档有机颜料的研究. 上海染料, 2011(4, 5).
[27] 章杰. 塑料着色用新型有机颜料. 上海化工, 1994(5).
[28] 章杰. 高性能颜料的技术现状和创新动向. 上海染料, 2012(5).
[29] 杨新纬等. 国内外溶剂染料的进展. 上海染料, 2001(1).
[30] 高本春等. 蒽醌型溶剂染料. 染料工业, 2012(4).
[31] 张慧等. 铜酞菁型溶剂染料的合成及性能测试. 青岛大学学报, 1998(4).
[32] 陈荣圻. 有机颜料的助剂应用评述. 上海染料, 2011(4).
[33] 乔辉等. 中国色母粒行业调查与分析. 塑料, 2012(2).
[34] 孙贵生等. 粘胶纤维原液着色超细紫色色浆分散性及纤维性能. 人造纤维, 2010(5).
[35] 黄海. 尼龙用着色剂. 染料与染色, 2009(5).
[36] 刘晓梅等. 汽车内饰涤纶织物着色剂. 染料与染色, 2010(5).
[37] 章杰. 化学纤维原液着色用新型着色剂. 湘潭化工, 1995(1).
[38] 杨蕴敏. 聚酯纤维纺前着色技术的进展. 合成纤维工业, 2008(6).
[39] Vaman G Kullkarni. 化纤用色母粒和功能母粒最新进展. 合成纤维工业, 2006(5).
[40] 张恒等. 户外测试检验加速测试. 装备环境工程, 2010(4).
[41] 张正潮. 浅谈耐日晒色牢度的测试标准. 印染, 2005(3).
[42] 章杰. 有机颜料安全性探讨. 上海染料, 2011(5).